HOW THE
MOUNTAINS
GREW

Garden of the Gods

Great sandstone fins jut out more than 300 feet into the air west of Colorado Springs, Colorado, the remnants of two mountains ranges: the Ancestral Rockies, which formed more than two hundred million years ago, and the modern Rocky Mountains, which began to form eighty million years ago. *Photo used with permission: Tonda at iStock.*

HOW THE MOUNTAINS GREW

*A New Geological History
of North America*

JOHN DVORAK

PEGASUS BOOKS
NEW YORK LONDON

Pegasus Books, Ltd.
148 West 37th Street, 13th Floor
New York, NY 10018

First Pegasus Books paperback edition July 2022
First Pegasus Books cloth edition August 2021

Interior design by Maria Fernandez

Library of Congress Cataloging-in-Publication Data is available.

ISBN: 978-1-63936-215-8

10 9 8 7 6 5 4 3 2

Printed in the United States of America
Distributed by Simon & Schuster
www.pegasusbooks.com

To Joyce and Sarah,
who continue to inspire me

CONTENTS

NOTE TO THE READER

Charles Lyell, working in the nineteenth century, remarked that there were three things one must do to understand the Earth: travel, travel, travel. I agree. To this end, throughout this book, I have described numerous places where one can go and see the rocks that record the major events that have shaped our planet. If, during the course of reading this book, the reader ventures to one or more of these places—and, once there, surveys the landscape and contemplates how dramatically the surface of the planet has changed—and, from that, realizes that the Earth has a deep history that can be read and understood—then the writing of this book will have been worthwhile.

There was nothing but land; not a country at all, but the material out of which countries are made.

—Willa Cather, *My Ántonia*, 1918

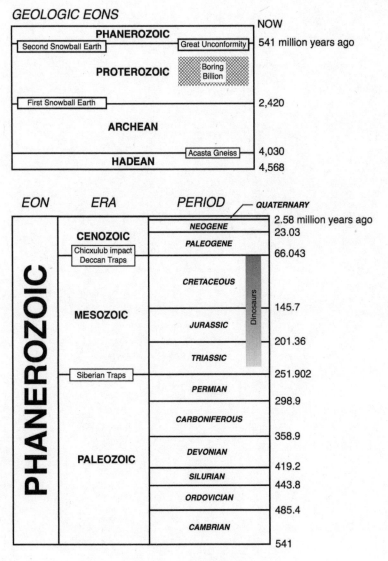

Geologic Time Scale

Mount Rushmore, South Dakota

I n the far western reaches of South Dakota, among the many towers and great blocks of hard rock that are the Black Hills, the faces of four former presidents stare out from high on a rocky cliff face. These four faces—George Washington, Thomas Jefferson, Theodore Roosevelt, and Abraham Lincoln—colossal by any standard, were literally blasted out of the hard rock with dynamite. They were then chiseled into their final forms with jackhammers manned by a small army of men who suspended themselves down the steep cliff face at the ends of long steel cables and who sat on broad leather straps that resembled bosun's chairs to do their work.

In all, it took fourteen years to complete the stone carvings at Mount Rushmore. The artist who conceived and directed the work, Gutzon Borglum, would tell politicians and newspapermen who could promote his work that it was his intention to make "something more than the 'biggest' in the world." It was his goal to produce a monument that would last through the ages, one that would rival the great stone statuaries of ancient Egypt and of ancient Rome. In that, he succeeded. But there is something more here, something decidedly different if one takes the time to study the four great stone carvings and to explore the surrounding countryside.

Look at George Washington. He and the other three are carved out of a light gray rock. Immediately below Washington is a band of a much darker rock with a distinct pattern of loose lines that sweep up and to the left. The light gray rock is a granite, once a bulbous mass of molten rock that rose up out of the Earth's crust, where it cooled and solidified. The dark rock was originally composed of ocean sediments that are much older than the granite. Long after those sediments were laid down, they were buried, then heated when the granite was molten and rose up and moved into the crust. The effects of the burial and of the heat changed the mineral content so that a new suite of minerals formed. That transformed the original ocean sediments into a different type of rock, a metamorphic rock that, from the degree of transformation, is known as a schist.

Now hike the trails and drive the roads that run through the Black Hills. One soon discovers that the dark schist forms a ring around the granite. More exploration reveals the ring of schist to be surrounded by three more rings. The first, moving outward, is a yellow limestone, the next a red shale, and finally, a white sandstone. From a high, overhead view, the entire assemblage looks like a giant blister that has risen out of the Earth and had the outer layers of skin scraped away. That, in fact, is not far from the truth.

At the center of the blister is the granite. It and the schist and the other three rings of rock have been pushed up so that a broad dome has formed, the Black Hills. The amount of upward movement has been considerable. The highest point in the Black Hills is Black Elk Peak. It stands several thousand feet above the broad surrounding plain, making it the highest point of land between the nearby Rockies and the distant Pyrenees.

That is the puzzle. Why do the Black Hills, one of the youngest mountain ranges in North America, stand so high, and why do they lie so isolated in the middle of a continent? Most mountain ranges, at least most that are rising today, lie along the edge of a continent where other geologic activity is high, meaning where earthquakes are frequent and where volcanoes are erupting. For North America, that is along the West Coast where the Cascades, the Sierra Nevada, and the Traverse Ranges of southern California are rising. So why did the Black Hills form where they did? Why so recently? Why is there a mass of rocky cliffs and high peaks standing in the middle of North America?

For a long time, geologists could say little more than the obvious about why mountains grew: There were forces acting inside the Earth that pushed the surface upward, and that formed mountains. Then came the theory of plate tectonics.

This theory, developed in the 1960s, proposed that the surface of the Earth is divided into a dozen or so rigid plates—tectonic plates—that are in constant motion. As the plates move, the boundaries between the plates push or pull or slide against each other. It is where the pushing or the pulling or the sliding is taking place that earthquakes are frequent and where volcanoes are now erupting.

It is where two plates are pushing against each other that mountains form. The Himalayas are rising because the tectonic plate that includes the subcontinent of India is pushing against the massive continent of Eurasia, another tectonic plate. The Andes are a product of a tectonic plate, one that underlies part of the Pacific Ocean, slamming into South America.

The theory of plate tectonics revolutionized our understanding of the Earth. It united many disparate observations. A ring of volcanoes have formed around the Pacific because the tectonic plates that lie beneath the Pacific are slowly sliding into the Earth along the edge of that ocean, producing magma that is supplied to those volcanoes. Earthquakes are frequent in California because the boundary between two tectonic plates—the Pacific and the North American plates—runs through this state. Rocks in Labrador are similar to those in Scotland and rocks in Florida are similar to those in West Africa because those two pairs of places were once adjacent; that is, it was once possible to walk from the East Coast of the United States to Europe and Africa. That they are now far apart is due to a slow spreading open of the Earth's surface that has led to the formation of the Atlantic Ocean.

And yet, though the theory of plate tectonics has been remarkably successful at explaining how ocean basins formed and why most of the earthquakes and volcanoes occur where they do, even as the theory was being developed, it was already clear that there was much about the Earth that place tectonics could not explain. The existence of the Black Hills, for example. And if the Black Hills seem too localized and limited, consider this: The Black Hills are part of the Rocky Mountains, the topographic backbone of North America. And those multiple ranges of mountains also formed far from a plate boundary.

The theory of plate tectonics, as it was originally envisioned, could also not explain why thick layers of chalk that had formed at the bottom of an ancient sea now stand as high walls in western Kansas. Or why the western end of Lake Superior follows an arc, and why the western part of Lake Superior is the deepest part of the Great Lakes. Why is there a diamond mine in Arkansas? How was it that a giant sea wave once rolled across Nebraska? Why are Oregon and Washington rotating slowly clockwise? Why are the Sierra Nevada Mountains in California or the Adirondacks in New York rising? What is the reason that the world's largest reserves of oil and of coal are in North America? And why do major earthquakes occur far from the West Coast beneath Missouri and South Carolina, and why are earthquakes scattered beneath New England?

Plate tectonics has these limitations because it is an incomplete theory. It is primarily a theory about the origin and development of ocean basins and says much less about the origin and evolution of the continents.

Moreover, the continents are much older than the ocean basins—most continental rocks are older than two billion years, while the rocky floors of the ocean basins are less than a tenth of that age. So, if one wants to understand the Earth and its long history, one needs a theory about the continents and how they formed and evolved.

And that has eluded us—until now.

◆

Two recent advancements have changed our perception of continental evolution. It is now possible, with microscopic precision, to probe individual mineral grains within rocks and to determine, with a remarkable degree of accuracy, the age of individual grains and, just as remarkably, to determine the environmental conditions on the Earth when a mineral grain formed. The former relies on nature's most reliable timekeeper, the decay of radioactive atoms. The latter, the determination of past environmental conditions, requires measuring the slightly different concentrations of isotopes—atoms with the same number of protons and different number of neutrons—within a mineral grain and, from that, determining the temperature and the chemistry of ancient oceans and of the ancient atmosphere.

The Great Oxygenation Event, Snowball Earth, the Carboniferous Rainforest Collapse, the Carnian Pluvial Event, the Paleocene-Eocene Thermal Maximum, and the Younger Dryas Event—little, if anything, was known about these events until recently. Each one was a major shift in climate, a turning point in the evolution of the continents—and of life. The ability to understand these events, to bring them into tight focus, has only come recently—from the determination of accurate ages and of past environmental conditions.

The other advancement came from looking deep into the Earth. The history of the planet is not only written on the surface, but also in the interior. Now, at last, it is possible, with dense networks of instruments that record every slight shaking of the Earth's surface, to produce detailed images of the Earth's interior. It is analogous to how detailed images of the human body are made using CAT scans. In the case of the Earth, seismic waves are used to probe the planet.

No longer is the Earth's interior seen as a series of concentric shells separated by featureless expanses. Instead, there are great slabs of ancient rock that are torn and warped, and that lie deep beneath our feet. There are giant plumes of hot material rising from deep inside the Earth. And there are great cold chunks on the bottom of the continents that are slowly settling down into the Earth.

Of particular interest is the remnant of an ancient tectonic plate—the Farallon Plate—that now lies deep beneath North America. That ancient plate—now regarded as a slab—is slowly sliding west to east beneath the continent. It is highly contorted—and it is those contortions and the movement of this slab that have made it one of the most important factors in the history of North America, giving rise to a variety of features, including the Rocky Mountains—and the Black Hills.

There is one more important point. Geology and biology are no longer considered separate sciences. The geological evolution of the Earth—creation of ocean basins, growth of mountain ranges, and so forth—has had an undeniable influence on biological evolution. And biological evolution—the development of photosynthesis, the first forest, the expansion of grasslands, and so forth—has been a major factor in determining how the rocky parts of the Earth have evolved.

This intertwining of geology and its various components—glaciology, volcanology, seismology, geochemistry, geochronology, and more—and of biology and its components—genetics, biochemistry, population dynamics, paleontology, and more—is known, rather dryly, as Earth System Science.

But it is not dry in content. For example, mass extinctions were once considered incidental in planetary evolution, merely blips in the long evolution of life. They are now known to be important parts of the geological evolution of the planet. So, too, are the activities of human beings, which, when viewed from the perspective of the long history of the planet, have clearly become—and continue to be—a major geologic force.

◆

Such claims require evidence. That is the centerpiece of this book. What follows is a narrative of a geologic field trip, one that will follow a chronological history of the Earth. So that the narrative can be confined between the covers of a single book, it is necessary to focus on one large region. North America has been chosen because, within its many and highly diverse records of rocks, it contains evidence of many of the planet's major events.

It is that evidence, where it can be found, where specific rocky outcrops can be seen and touched, and where the landscape can be surveyed and walked upon, that is of prime importance because it is physical evidence that tells the story. It is the best way to gain a sense of distant times and of distant events, such as, when seas once swept over much of the land, when the entire planet was encased in ice, when conditions shifted wildly from flourishing life to barrenness and back again, and when, through the inexorable whims of nature, the mountains grew.

1

The Relics of Hell

Hadean Eon: 4.568 to 4.030 Billion Years Ago

I t is a two-hour drive from Minneapolis to the small town of Morton, Minnesota. Along the way one passes through farmland rich in deep soil. Here people boast that, given the abundance and the beauty of the place, they must be living in God's country. That certainly seems to be true. And the people of Morton have an additional boast to make. At the western end of their town is a low rocky knoll. When one stands atop that knoll, one is standing on the oldest rock in the United States.

The Morton Gneiss[1] gained its initial notoriety from its colorful swirls and attractiveness. The swirls are lustrous and are a blend of pinks and blacks that course through the rock, reminding one of the patterns produced when heating and stirring marbleized candy. But these swirls have been frozen in place for billions of years. And it is the swirls—and the fact that this stone is remarkably resistant to weathering—that has made the Morton Gneiss a favorite of architects who are constructing large buildings.

The Morton Gneiss can be found as the facing stone that lines the base of the AIG Building along Pine Street in New York and the base of the Exchange Building on Second Avenue in Seattle. It covers the oddly angular walls of art-deco-styled Adler Planetarium in Chicago. The swirls grace the walls of large buildings in Boston, Houston, Detroit, and Cincinnati, and the walls of modest ones in Lincoln, Nebraska; Portland, Oregon; and Norfolk, Virginia.[2]

Near Morton, it has been a common choice for headstones in local cemeteries. It is found lining the edges of major streets as curbstones. It is the rough-cut stone that covers the exterior of the local Zion Lutheran Church and appears as a dozen or so highly polished panels on the front of a local tavern on Main Street. In nearby Redwood Falls, this oldest-of-all rocks in the United States serves as countertops in the main room and in the restrooms of a local McDonald's.

Given its wide use, it may come as a surprise that natural outcrops are rare. That they exist at all is consequence of a recent geologic catastrophe.

It was the end of the latest ice age. A giant sheet of ice extended over the northern regions of the Earth. As the planet warmed, the sheet gradually thawed and retreated northward. The melting ice formed a large glacial lake over what is now Manitoba, Ontario, and much of northern Minnesota. The ice sheet continued to retreat and more water was added to the lake. Eventually, the lake broke and a massive flood poured across the landscape.

The fast-flowing water scoured the land. It dug through the soil and deep into the sediments. Eventually, the flow of water concentrated into a single long spillway. And it continued to scour and dig and erode until hard rock was reached.

When the water finally receded, what was left behind was a long new valley with a flat floor and steep sides—the Minnesota River Valley. It is along the floor of this valley where the fast-moving water scoured down to hard rock that patches of the Morton Gneiss are found today.

But how do we know that this rock is so ancient? There is nothing to indicate from its location that it has a record-breaking old age. The few exposed patches of the Morton Gneiss do not lie at the base of a great pile of rock, indicating, by being at the base of pile, that they must have a particularly old age. Instead, the age of this rock, that is, how much time has passed since it solidified from a molten state, is known quite precisely because of the decay of radioactive atoms in its minerals.

The radiogenic dating of rocks works this way: Half of the original number of a particular type of atom in a mineral, say, uranium-235,[3] will decay after a specific length of time, known as the half-life. Because of this regularity, by measuring the relative amounts of different atoms, it is possible to determine how much time has passed since those atoms came together to form each of those minerals.

But great care is required when doing such measurements. When a rock is heated or compressed, atoms may move around and reset the radioactive "clock." What is needed is a highly durable mineral—a tight container—that holds atoms in place and does not allow them to move around. Fortunately, nature has provided such a container.

Not to be confused with *zirconia*, a commercially produced, low-cost substitute for diamond that is often sold through magazine advertisements and television commercials, the naturally occurring mineral that provides a tight container for atoms is *zirconium silicate*, or, as it is more commonly called, *zircon*.

Zircons are incredibly durable. They are insoluble in the strongest acid. Their atoms resist movement even under extreme physical and chemical weathering. And they endure high temperatures, holding their atoms in place, even when temperatures are high enough to melt most of the surrounding rock.

The durability comes from the arrangement of the atoms. Zircons consist mostly of oxygen and zirconium atoms. The oxygen atoms are much larger than the zirconium ones. Four oxygen atoms surround a single zirconium atom in such a way as to make a geometric shape known as a *tetrahedron*, a four-sided pyramid, which is the strongest geometric shape possible. Furthermore, the tetrahedrons are linked in a pattern reminiscent of the way steel trusses are arranged to strengthen bridges. The combination of the tetrahedron shapes and the pattern of linkages essentially locks the small zirconium atoms in place. The fact that an atom of radioactive uranium, which is about the same size as a zirconium atom, may occasionally substitute for a zirconium atom when the mineral forms means that there are radioactive atoms within a zircon to determine a radiogenic age.

It is also important to note that zircons are often extremely small, smaller than the period at the end of this sentence. To study and analyze such small mineral grains, a new scientific instrument had to be developed: the SHRIMP (the sensitive high-resolution ion microprobe).

Fewer than a dozen SHRIMP instruments have been built—an indication of their complexity and their expense. The operation of this highly precise instrument requires the focusing of a tight beam of high-energy ions onto tiny mineral grains. The impact of the ions on the surface of the grain vaporizes some of the atoms, leaving a hole less than a thousandth of an

inch in diameter and about a ten-thousandth of an inch deep. The vaporized atoms are then passed through a mass spectrometer that can sort and
count the atoms, and from that the age of the mineral can be determined.
The ages of tens of thousands of minerals taken from thousands of different rocks have been determined using this technique. That includes the
Morton Gneiss. That analysis, done in 2006, gave an age of 3.524 billion
years. For comparison, the age of the oldest rocks exposed at the bottom
of the Grand Canyon, which were determined in the same way, is 1.750
billion years, less than half the age of the Morton Gneiss.

Much effort has been made to find the oldest rocks on the planet. After
such effort, only a dozen or so sites have been found where the rocks are
older than the Morton Gneiss.

One such site is in South Africa near the Tugela River, hundreds of
miles east of Johannesburg. Others have been found in remote regions
of Siberia and Australia. A small outcrop of rock in northeastern China
has yielded a radiogenic age of 3.850 billion years. A significantly older
age has been determined for samples collected from a small rocky peninsula that juts out from Antarctica into the Indian Ocean. Those rocks
have an age of 3.96 billion years. But where has the oldest rock on the
planet been found?

The search for the oldest rock is highly competitive. Every few years, a
new contender is announced and effort is made to either confirm or deny
it. The current titleholder was announced in 1988. It was the result of
years of work that required much travel and the unraveling of a complex
history of the local geology.

The sample was obtained from the barrens of the Northwest Territories of Canada. The site is hundreds of miles from the nearest
settlement—the capital city of the Northwest Territories, Yellowknife.
To add to the intrigue, no roads lead to the place where the oldest rock
on the planet has been found. The site can only be reached by floatplane
and canoe.

The rock is a *gneiss*, meaning that it, like the Morton Gneiss, has undergone considerable heating and compression since its formation. But it lacks
the colorful swirls that have made the Morton Gneiss so popular. It is a
drab rock with thin bands of white and gray that show some contortion.
The outcrop where it was found is hundreds of miles north of Yellowknife,

near the Acasta River and at the edge of Great Bear Lake. For that reason, it is known as the Acasta Gneiss. The zircons that it contains have been analyzed by several different investigators. They have decided that it has an age of 4.030 billion years.[4]

Rocks older than the Acasta Gneiss might exist, but, in the opinion of those who search for such early relics, if such rocks do exist, they are probably not much older than the Acasta Gneiss. The reasoning is this: Four billion years ago, the Earth's interior was much hotter than today, and so the Earth's surface was much less stable. Any rock that did form and solidify was probably pulled back into the planet, where it was remelted and later solidified. Hence, few, if any, outcrops with rocks much older than four billion years probably exist today.

The opinion is so strong that in 2015, members of the International Commission on Stratigraphy, the scientific body that makes such judgements, decided that the 4.030-billion-year age of the Acasta Gneiss is a major milestone in the Earth's history, that it would mark the beginning of a major geologic time period, the Archean Eon, the time in the history of the Earth when the earliest part of the rock record was preserved.

In a continuation of their work, members of the commission also decided that the time period before the Archean Eon would be the Hadean Eon. Both "Archean" and "Hadean" are derived from Greek words. Archean means "the beginning." "Hadean," while it is an obvious reference to Hades, means "a time that is unseen" because no evidence of this earliest-of-all geologic eons survives today in the rock record.[5]

This would seem to limit an ability to know anything for certain about the Earth before 4.030 billion years ago. How could it be otherwise if no rocks from an earlier time still exist? Where else might evidence of the Hadean Eon be found? Would it ever be possible to determine what that distant time and place was like? Was it, as is often portrayed in mythical stories, a hellish place? And when did the Hadean Eon begin?

A few details are known about this earliest-of-all geologic eons. But most of what is known did not come from a study of the Earth. Instead, one must look skyward and step back in time and understand how the Earth formed and how the material that it is made of originated.

◆

In the beginning there was only energy and no matter. There were no atoms. And there were none of the protons or neutrons or electrons that would later comprise atoms. That was 13.799 billion years ago, a time that has been precisely determined by a host of different measurements. It was the time when the universe began to expand.

The expansion of the universe was discovered in the 1920s by American astronomer Edwin Hubble. He was then using the largest telescope in the world, the 100-inch Hooker Telescope at Mount Wilson in California. Since then, the expansion has been confirmed many times and with a wide range of instruments, some at Earth-based observatories and others on space-based satellites.

Soon after the universe began to expand—the initial expansion is usually referred to as the Big Bang—protons, neutrons, and electrons began to form from the energy. And from them, the first and simplest atoms began to form. Lots of hydrogen, some helium, and a very small amount of lithium were then the total lot of all of the matter in the universe.

Even at the beginning, the universe was not perfectly smooth and homogenous. There were small variations in the amounts of energy (in the form of radiation) and in the amounts of matter, or mass. It was from those small variations that gravity acted and pulled matter together to form galaxies. As those galaxies condensed under the pull of gravity, as matter became more and more concentrated, clouds formed. And as the clouds formed, as they became denser and denser, individual stars formed. It was within those stars that hydrogen and helium fused together through nuclear reactions—which is the reason stars shine—and it was from that fusion that many of the chemical elements that are common in our lives formed.

Oxygen, carbon, and nitrogen were created in this manner—three chemical elements that are common in us and in all living things. So did silicon, magnesium, and iron, which, together with oxygen, comprise the bulk of the Earth's mass. But many chemical elements that are familiar to us, such as gold and silver, cannot be made inside stars. Even the huge amount of energy released during the explosion of a supernova is not sufficient to produce these elements.

The reason is that the production of oxygen and carbon and silicon and iron *releases* energy, while the production of gold and silver *consumes* energy. And so gold and silver and many other metals can only be produced by a process that yields incredible amounts of excess energy. And nature has such a process: the collision of neutron stars.

When a star more massive than the Sun has depleted its supply of hydrogen and helium so that the fusion of these elements is no longer possible, the continued contraction of the star under gravity makes the star unstable and it explodes as a supernova, spewing material out into space and leaving behind a dense ball of neutrons—a neutron star. When two stars, each more massive than the Sun, are in orbit around each and after each has exploded as a supernova, what is left behind is pair of orbiting neutron stars, something that happens more often than might be expected because a galaxy is filled with hundreds of millions of stars.

Then, as the two neutron stars continue to orbit around each other, some of their orbital energy is lost by the emission of gravitational waves. That lowers the orbital energy of the whirling pair. And that causes them to spiral closer and closer to each other. Eventually they merge. And when they do—that is, when two neutron stars collide—a tremendous explosion happens. And that releases enough energy in a sufficiently short period of time to produce gold and silver and other precious metals.

Until recently, the idea that such collisions occurred was purely theoretical. The only sense anyone had that such collisions might be real was the endless scribblings and calculations done by astronomers who wondered how such collisions might be observed, what type of energy they might send out. It was generally agreed that some of the energy would be released in the form of gravitational waves. And so large, highly sensitive detectors were designed to record the passing of such waves. After the detectors were built more than a decade passed before the first gravitational waves were detected on August 17, 2017. They came from the direction of the constellation Hydra. Since then, several more sets of waves have been detected, confirming that the collision of neutron stars does occur and such collisions are scattered across the galaxy.

This seemingly unrelated astronomical discovery is relevant to the question of the origin and the age of the Earth. First, it reminds us that we are star stuff. The oxygen, carbon, and nitrogen in our bodies and in all living things were produced inside stars. So was the iron, silicon, and magnesium that make up much of the Earth. These chemical elements were produced in abundance. But the rare ones, such as gold and silver, as well as platinum, iridium, and uranium, are rare because the process that produced them—the collision of neutron stars—does not occur uniformly across the galaxy.[6] The relative abundance of these rare elements has left a chemical fingerprint that helps determine how the solar system formed.

Look around the solar system. Different objects have different bulk compositions. The Sun, Jupiter, and Saturn are mostly hydrogen and helium. The planet Mercury has a substantial amount of iron, more than the other planets. Comets are composed mostly of water and carbon dioxide, both in the form of ice. And the Earth is mostly iron and oxygen and a significant amount of silicon and magnesium. But these diverse objects do have something in common. They all have the same chemical fingerprint, that is, they all have the same relative abundance of rare chemical elements, such as gold and silver, platinum and iridium. That means these many and highly diverse objects all formed from the same galactic cloud. And so, if it is possible to find the oldest objects in the solar system—regardless of what they are—and determine their age, then the age of the solar system will have been established. But where might these oldest-of-all objects be lurking?

There are regions of our galaxy, the Milky Way, where galactic clouds are contracting and where stars are forming today. And, when studied in detail, these young stars have a common feature: There are hot disks of gas and dust orbiting around them. This is what the solar system looked like when it began to form. Which means that the first objects to condense from the clouds—the oldest objects in the solar system—would have been small mineral grains that formed in such hot disks of gas and dust. And such grains have been found—inside meteorites.

They appear as small white beads. The mineral content of these beads includes unusual minerals, such as melilite and hibonite, as well as more common ones, such as perovskite and pyroxene. What these minerals have in common is that they all formed at high temperatures. They also contain large amounts of calcium and aluminum, and so these tiny white beads—the first parts of the solar system to condense and form—are known as calcium-aluminum inclusions, or CAIs. And when their age is determined from the amount of radioactive elements that they contain, they yield an age of 4.568 billion years.

That is the age of the solar system. That is also the age that the International Commission on Stratigraphy has assigned to be the beginning of the Hadean Eon. It is the time when the first solid objects started to form, when hot gas and dust started to condense around a young Sun. And from that disk of gas and dust, all of the planets, the moons, the comets, and everything else in the solar system eventually formed. The Earth is obviously younger than that age; some length of time had to pass between the formation of the first mineral grains, the CAIs, and the accumulation of enough material to form a planet. But how much time?

It was once assumed—in part, because there seemed to be no evidence to the contrary—that the Earth grew to its final form by a slow accumulation of material, by continuing to sweep up debris that was orbiting the Sun. But a different story is now told. The age of the Earth can, in fact, be pinpointed in time because our planet reached its final form—that is, its current size, mass, and composition—soon after a catastrophic collision.

◆

It was once thought to be so simple. The Earth and the other planets formed in the nearly circular orbits that they follow today around the Sun. The hot disk that encircled the Sun long ago condensed into individual rings. And each ring became a planet.

The building of a planet was thought to be a slow, orderly process. It began with the condensation of solid grains from a gas-and-dust-rich cloud. Small grains grew into larger ones.[7] And as the grains grew, the larger ones had the stronger gravity, and so they grew faster until each of the planets formed. This idea has now been shattered. And the shattering came after other planetary systems were discovered.

The first discovery of a planet outside our solar system—an exoplanet—was made in 1992. Since then, thousands of exoplanets have been discovered. And they show something surprising: Most planetary systems do not resemble our solar system.

The eight major planets of our solar system are arranged in a simple pattern. Four of the planets—Mercury, Venus, Earth, and Mars—are rocky in the sense that they have hard surfaces and are almost devoid of hydrogen and helium. They also orbit much closer to the Sun than the other four. Beyond the four rocky planets are two gas giants—Jupiter and Saturn—and beyond them are two smaller, though also gas-rich planets—Uranus and Neptune. Most planetary systems differ from this simple pattern in two significant ways.

First, most exoplanets orbit close to their central star, much closer than the innermost planet in our solar system, Mercury, orbits the Sun. Second, many of these close-orbiting planets are giant planets the size of Jupiter. It was the discovery of these "hot Jupiters" that caused astronomers to reconsider how planetary systems form.

A gas-rich Jupiter-sized planet could not have formed close to a young star because the hot winds produced during that star's formation would have swept the innermost region around the star clear of gas. Instead, such a planet must have formed at a great distance, comparable to how far Jupiter is now from the Sun. But how did a giant gas planet move close to a central star and become a hot Jupiter?

The inward migration occurred because, as the Jupiter-sized planet was forming, it was also disturbing the disk of gas and dust around it through its gravity. And it did so in such a way that much of the gas and dust was thrown outward, away from the central star, causing—because of the conservation of angular momentum, a basic law of physics—the Jupiter-sized planet to spiral closer and closer to the central star. And that would be the end of it: Either the giant planet would spiral downward and be consumed by the star, or it would end up in a very close orbit, which is the reason that many planetary systems have hot Jupiters. But something else happened in the early evolution of the solar system, something that pulled Jupiter back out and away from the Sun. And that something was Saturn.

Saturn formed farther from the Sun than Jupiter and it formed later and slower. As it grew, it, too, had a gravitational effect on the disk of gas and dust, and so Saturn also began a downward spiral. But, as the two planets moved closer and closer to the Sun, moving faster and faster in their orbits, a point was reached when their orbital motions started to resonate; that is, the two giant planets started to exchange angular momentum. That had the effect of reversing their motions from spiraling downward to spiraling outward. This inward-to-outward transition has been dubbed the Grand Tack, a reference to the motion a sailboat takes when it changes course by tacking around a buoy. In the case of Jupiter and Saturn, the reversal of the spiraling motions continued until they reached the distances from the Sun that they have today.[8]

It was during the outward migration of Jupiter and Saturn that the Earth and the other three rocky planets formed. They did so from material—mostly dust—that had survived being scattered by Jupiter's immense gravity. Exactly how dust particles came together to form planet-sized bodies is still debated. It probably involved making thousands of intermediate-sized objects known as *planetesimals*. As the planetesimals grew, sweeping up whatever was left of the dust, they occasionally slammed into each other. After some of the collisions, the two planetesimals would stick together and larger objects would form. Other, more severe collisions would cause planetesimals to splinter away, adding to the reservoir of dust, which was then swept up by other planetesimals, which would continue to grow in size.

Eventually, rocky planets were built, a process that took a few tens of millions of years. One of those early rocky planets was nearly the size that the Earth is today, a proto-Earth, but its growth was not yet completed. One more major event would happen before it reached the current mass and size that the Earth has today. The evidence for this last major event in the Earth's formation is not found here on our planet. Instead, evidence for it has been found on the Moon.

The origin of the Moon has puzzled philosophers and scientists for centuries. The problem is the large size of the Moon compared to the Earth. Other planets have moons—Mars has two, Jupiter and Saturn each have several dozen—but those moons are considerably smaller than the host planet. The Moon, however, has a diameter that is a quarter the diameter of the Earth, making the Earth-Moon pair a bona fide double planet, two objects of planetary size that orbit each other. How did these two objects of similar size end up in such close proximity?

For a long time, there were three possible explanations. One was based on the idea that the Moon had once been part of a rapidly rotating early Earth. The rapid rotation had caused the Moon to be flung off away from the Earth. But for that to have happened, the early Earth must have been rotating so fast that the length of a day would have been less than an hour, much faster than any other planetary body. Another was that the Moon had formed away from the Earth and had been captured by the Earth's greater gravity when the two objects happened to pass close to each other. But that would have required the Moon to have lost almost all of the orbital momentum it had when orbiting the Sun, something that is virtually impossible. The third was that the Moon and the Earth happened to form close to each other and had always remained close, but how this could have happened in the orbital chaos—the Grand Tack—of the early solar system remains unexplained. Then, after the first lunar rocks were returned in the 1970s, a fourth idea was proposed.

An analysis of those first lunar rocks showed that, early in its history, the Moon was covered by a deep ocean of molten rock, a magma ocean. None of the original three explanations of the Moon's origin could explain how enough heat was produced to suddenly melt the entire outer shell of the Moon. But that would have been a natural outcome if the Earth had been hit by a huge object and the Moon had formed from the debris of that collision. That idea is known as the giant impact hypothesis, now the favored idea for the origin of the Moon.

When first proposed, the idea seemed contrived because such a colossal collision between two large objects seemed unlikely. But now there is evidence that such huge collisions happened when the planets were forming. The planet Uranus lies on its side—its north and south poles are almost perpendicular to the planet's orbital axis—which is probably the result of the planet being hit by a huge object. And there are giant scars on almost every planetary surface and on the surfaces of many moons—the Caloris Basin on Mercury, the Valhalla crater on Callisto, a major moon of Jupiter—further evidence that huge collisions were common in the early solar system. And it has been recently reported that the star BD+20 307, a star that is beyond the early hot-dust-ring stage, has a cloud of dust around it that probably formed by the impact of two rocky planet-sized objects.[9]

The giant impact hypothesis is also supported by the chemical composition of lunar rocks. Much of the Moon has a chemical composition that is similar to that of the Earth's mantle, suggesting that some material from the proto-Earth went into forming the Moon. The Moon is also almost devoid of water, almost all of it has boiled off, which is consistent with the Moon forming after a heat-generating impact.[10]

The giant impact hypothesis is now so widely accepted that a name has been given to the object that collided with the proto-Earth. It is called Theia, the name of a mythical Greek goddess who gave birth to Selene, the Moon.

The energy of the giant impact would have produced a dense cloud of hot debris that would have encircled the Earth and out of which the Moon would form in a few million years. As mentioned, there is evidence in the lunar rocks that a high amount of heat was required to form the Moon, specifically in rocks of the lunar highlands, the bright white areas of the Moon that are visible from Earth. These are the oldest rocks on the Moon, rocks that formed from a lunar magma ocean.[11] And we know how long ago the lunar magma ocean solidified: that is, we can date the collision with Theia: The age is recorded in zircons that have been found in rocks from the lunar highlands.

Hundreds of pounds of lunar rocks were returned by the six Apollo missions that landed on the Moon. It is from samples returned by the third successful Apollo landing that zircons have been found that date the lunar magma ocean. Those few tiny grains have been subjected to a host of close study and scrutiny. And the conclusion is clear: The oldest surface of the Moon solidified from a molten state 4.51 billion years ago. That is the time of the Theia impact.

The impact increased the mass of the proto-Earth by about 10 percent. After the impact, the Earth was in its final form. Its mass and its composition were set and have changed little since then. And so the time of the collision is the age of the Earth: 4.51 billion years.

◆

Five hundred million years passed between the impact with Theia and the formation of the Earth's oldest rock, the Acasta Gneiss, a considerable length of time, similar to the amount of time that has passed between the appearance of the first animals on the planet and now. Much must have happened. But we are seriously hampered in understanding exactly what happened because of a lack of a terrestrial rock record. In short, physical evidence is lacking. But that could change in the near future.

The pockmarked surface of the Moon shows that the lunar surface continued to be hit by impacts long after it formed. Each impact flung material high above the Moon. In the case of the larger impacts, some material was actually thrown off the Moon and went into orbit and was then swept up by the Earth and fell to its surface. Some of this lunar material has been found and resides today as some of the meteorites in museums and private collections.

The reverse also happened. The Earth has also been subjected to a barrage of meteor impacts, similar to what happened to the Moon. Some of these impacts would have occurred early in the Earth's history, sending early pieces of the Earth out into space, where some of it was swept up by the Moon. Because regions of the lunar surface are much older than the Acasta Gneiss—the oldest lunar rock returned by the Apollo missions has an age of 4.46 billion years—it is assured that there are boulder fields sitting on the lunar surface today strewn with pieces of the Earth's early crust. The question is: When will these rocks be searched for and collected and brought back to Earth for study?[12]

◆

That is for the future. For now, we must accept the limitation imposed by the lack of rocks from the earliest period of the Earth's history. But we are not without some ancient treasures, microscopic in size, that date from this very early period. It should come as no surprise that these are zircons. And they exist in fragments, an indication of the long length of time they have endured. And the oldest among these microscopic fragments are found, quite fittingly, in one of the remotest places on the planet—in the Jack Hills region of Western Australia.

Remembered by those who have visited the region as a place of extreme heat, ample dust, and endless barrenness, the Jack Hills of Western Australia are situated among a complex of hills whose rocks formed about 3.6 billion years ago, about the same time when molten material was solidifying into the Morton Gneiss. But the rocks of the Jack Hills are not igneous. They did not solidify from molten rock. They were originally sedimentary rocks that accumulated by the slow addition of sand-sized and smaller particles. Every rock, so it is said, has a story to tell. And the story told here is a special one. After much crushing and sorting of rocks taken from the Jack Hills, after tedious hours of sieving through the tens of thousands of tiny particles of rock, a hundred or so individual mineral grains of particular interest were found. These were zircons that did form from molten rock hundreds of millions of years ago and were eventually freed by weathering and erosion, then blown around by the wind before settling down and contributing to the formation of the sedimentary rocks at the Jack Hills. And when a radiogenic age was determined for these precious mineral grains, it was found that each one was significantly older than four billion years.

One zircon from the Jack Hills has received special attention. It is deep purple in color. Its designation is W74/2-36. It measures less than a hundredth of an inch in any direction. Microscopic examination shows that it is a fragment of a slightly larger mineral grain, and that it has been marred by being transported a great distance, probably by wind. Its age is 4.404 billion years, which makes W74/2-36 the oldest known fragment of the Earth.

W74/2-36 and the hundred or so other ancient zircons that have been found in the Jack Hills have been subjected to intense study, not only to confirm their great ages, but also to determine what can be learned about environmental conditions on the planet when they formed. That any environmental conditions can be determined from such a small fragment of a single mineral grain—or from any rocky material—might come as a surprise, but such studies are routine and are based on a simple fact: Atoms of the same chemical element will behave differently from other atoms of that element if they are slightly more massive; that is, if they are slightly heavier. This is a key point in modern geology. And so a brief summary is needed to explain how it works.

All oxygen atoms contain eight protons in the nucleus, but some oxygen atoms have eight neutrons and others have ten neutrons. Both are still oxygen, but they are different *isotopes* of oxygen. Those with eight neutrons are known as oxygen-16 (8 protons and 8 neutrons) and those with ten neutrons are oxygen-18 (8 protons and 10 neutrons). Because an atom of oxygen-18 has slightly more mass—the two additional neutrons make it slightly heavier—than an atom of oxygen-16, it takes slightly more energy to move around an atom of oxygen-18 than one of oxygen-16.

Consider a tub of water. The oxygen atoms of some water molecules, H_2O, consist of an atom of oxygen-16 and others of oxygen-18. Those with oxygen-18 are slightly heavier, and so require slightly more energy to evaporate than those with oxygen-16. Now heat the water. More energy is available for evaporation. It is now easier for a molecule with oxygen-18 to evaporate than when the water was colder. The water vapor above the tub of heated water has more molecules with oxygen-18 than before the water was heated. It is a *very* slight increase, but enough to be measured by the SHRIMP instrument, which can measure the amounts of different isotopes very precisely.

This also works for isotopes of other atoms, notably carbon-12 and carbon-14, strontium-86 and strontium-87, and many others. In short, the measurement of isotopic ratios of different types of atoms serves as a proxy for the measurement of past environmental conditions, such as temperature. Such measurements are at the heart of much of our knowledge of past climates. And it is such measurements made on tiny fragments of zircons recovered from the Jack Hills that tell us about conditions on the Earth soon after the planet formed.

The results are stunning. First, the relative amounts of oxygen-18 and oxygen-16 measured in W74/2-36 and other ancient zircons show that within a hundred million years of the Theia impact, the surface of the Earth was *not* the hot and hellish place of rampant volcanic activity and bubbling molten rock that many scientists had assumed. Instead, the temperature on the surface was about 200° Fahrenheit (about 90° Celsius), which is still hot, but more like a very hot sauna than a fireball, and much less than molten rock. Moreover, these same measurements indicate something else: Liquid water was present on the surface, not as large puddles or isolated lakes, but as a global ocean.

This has profoundly changed our perception of the Earth's earliest history. The existence of a vast ocean on the planet from almost the beginning means there were rainstorms, and that landmasses were being eroded. There were ocean tides. (And those tides would have been much larger than the ones today because the Moon was much closer.) There would have been ocean currents that in turn would have equalized the temperature across the planet. And there was ample water on the planet where life might have originated or, at least, where life might have harbored itself from what, if there was land, was a barren and hostile place.

And there is one other important point to make about an early global ocean. It means that the oldest feature that we can see on the planet today—much older than any mountain or any record of ancient mountains—is the ocean itself.

That is not to say that the ocean has remained unchanged. The coastlines have changed as the continents shifted. The chemical composition, including the salt content, has changed based on rates of erosion and evaporation. And great volumes of water have always cycled from the ocean to the atmosphere and have precipitated out as rain or snow, then run down streams and rivers back to the ocean. But water-filled oceans have always been there.[13]

Given the longevity of the oceans—and that they play a key part in the evolution of the planet—it is worthwhile to pause and consider where the water came from.

◆

Contrary to what one might think when looking at a globe of the Earth and seeing three-quarters of the planet covered by water, the Earth is not a water-rich planet. It is a rocky planet composed mostly of iron and oxygen, silicon and magnesium. It also has significant amounts of calcium, aluminum, nickel, and sodium and lesser amounts of potassium and sodium. That composition is consistent with the idea that the Earth formed from material that was near the Sun. During the Sun's early evolution, it spewed out hot electrified winds that energized and swept away the lighter, more volatile chemical elements such as hydrogen, helium, carbon, and sulfur form the inner solar system. And so water, which is a volatile compound, must have come to the planet after it had formed.

Furthermore, the delivery of water must have happened soon after the Earth formed, within the first hundred million years, in order for a large ocean to have existed by the time W74/2-36 and other zircons of similar age had formed. So how might sufficient water be delivered quickly and early in the Earth's history?

For a long time it was thought that comets might be the source. Comets are mostly water ice—they are often described as "dirty snowballs"—and they are abundant in the outer reaches of the solar system, numbering in the trillions. But there is a problem with comets being the main source of water for the ocean.

For most water molecules, the hydrogen atom contains a single proton. But some water molecules have an isotope of hydrogen—deuterium—that has a proton and a neutron. And it is easy to determine the relative amounts of single-proton hydrogen and deuterium in a sample of water—or of ice.

Four comets observed in recent years have been bright enough to determine how much single-proton hydrogen and deuterium they have in their water ice: Halley viewed in 1986, Hyakutake in 1996, Hale-Bopp in 1997, and Churyumov-Gerasimenko in 2004. In each case, the amount of deuterium was twice what is found in the ocean. And so comets seem an unlikely source for ocean water. A different source is needed. Where else in the solar system might an abundance of water have existed that could have been delivered to the Earth early in its history?

It would have happened during the Grand Tack when Jupiter and Saturn were slowly spiraling away from the Sun and their gravitational fields were scattering objects to different regions of the solar system. That would have sent some ice-rich objects that had formed far from the Sun down into the inner region of the solar system, where a few of these ice-rich objects are orbiting the Sun today.

Ceres, the largest object in the asteroid belt, is one of those objects. Today it follows an orbit around the Sun that brings it closer to the four inner rocky planets than to Jupiter. In 2015 the *Dawn* spacecraft began a close-in study of Ceres and revealed that this tiny world—Ceres is now considered a dwarf planet—was composed mostly of water ice. During the early history of the solar system, there were probably many other such objects. Most fell into the Sun. Some, such as Ceres, continue to orbit the Sun. And a few would have collided with one of the four rocky planets or with the Moon.

Given its size—Ceres is about three hundred miles in diameter—it would have taken only a half-dozen Ceres-sized, ice-rich objects to supply the necessary amount of water to fill the oceans. And so this may be the source of water for the Earth early in its history. But we cannot get ahead of ourselves. The amount of deuterium in the water ice of Ceres and similar objects has not yet been determined. Nor is the history of water on the other three rocky planets known in sufficient detail to decide whether they had once been hit by several Ceres-like objects (though it is known that Mars did have a shallow ocean early in its history).

All of this comes with a reminder: Much of the pursuit of the Earth's earliest history is tied to a study of the other planets.

◆

Because the beginning of the Hadean Eon, as it is currently defined, corresponds to the formation of the solar system—the formation of CAIs and not to the formation of the Earth—it has been suggested that the beginning of the Hadean Eon should be changed.

A new eon, the Chaotian Eon, would be the time between the formation of CAIs and the collision with Theia. It would be the time when Jupiter and Saturn were scattering objects around the solar system at a frantic pace. It would be a time period that was dominated by the formation of the four rocky planets, when external events were controlling Earth's evolution.

The eon after the Chaotian would retain the name Hadean and be the time when the Earth began to evolve mostly in isolation of other objects in the solar system, though the occasional meteor impact would still occur. It would mark the time since the Earth had acquired the mass and composition that it has today.

The International Commission on Stratigraphy has not yet taken up the question of a new eon, but it will almost certainly do so in the future. More information about the Earth's very early history is needed before a step as radical as creating and formalizing a new geologic eon can be taken.

Regardless of what action the commission may take, the end of the Hadean Eon is secure. It is marked as the point in time when the rock record begins on the Earth, when the earliest known crustal rock, the 4.030-billion-year-old Acasta Gneiss, formed.

And that is the story that will be told in the remainder of this book, a story that will increase in detail and in certainty as it moves closer to the present because more and more of the rock record has been preserved and has been studied.

2

Bombardment
and Bottleneck

Archean Eon: 4.030 to 2.420 Billion Years Ago[1]

The Earth is very old. And the evidence is all around.

One of the most striking pieces of visual evidence is the Grand Canyon itself. Thick layers of colorful rocks dominate the scene. Beneath them and continuing to the bottom of the canyon are harder rocks that form the steep walls of a narrow and inner canyon. The total relief is almost a mile. A vast amount of time was required for erosion to cut downward through those many thick layers to form the main expanse of the upper and outer canyon, and then additional time to cut into the harder rocks and create the deep inner canyon.

To that must be added an immense length of time that was required for the individual layers of rock to accumulate. These are sandstones and siltstones and mudstones. Each one has a myriad of tiny features. There is cross-bedding and graded bedding. There are ripple marks and small dunes and antidunes. These are all signs that wind and water worked a long time to form these specific features.

These features—the bedding, the ripple marks, and the dunes—are familiar to anyone who has walked along a shoreline or a river bank and looked into the water. They form slowly and incrementally. And so, the production of the colorful bands, to build the myriad of individual small features in the sandstones, the siltstones, and the mudstones, requires the passage of a long period of time.

If anything, the scene at the Grand Canyon is at too large a scale—it is too majestic—to truly comprehend the vast amount of time that has elapsed for this layer-cake assemblage of colorful bands to accumulate and erode; for the harder and deeper rocks to be cut through and for the complete canyon to form. What is needed is a more intimate setting, one where it is possible to place a hand on a rocky outcrop and be assured that this section of the rock record represents a long period of time. A place is needed where one can see the long series of repetitive ticks of a geologic clock. Such places abound.

One that is easily accessible is found in the Marin Headlands north of the Golden Gate Bridge near San Francisco. Here there are parallel ribbons of red chert that run across the landscape. Each ribbon is about an inch thick. The total thickness of the chert, from sea level to the top of the Marin Headlands, is several hundred feet. Simple arithmetic shows that the number of individual, repetitive ribbons must be in the tens of thousands.

In West Texas, along the highway that runs south of Guadalupe Peak, the highest point in the state, road construction has cut through thinly layered rocks. Each layer is about a tenth of an inch in thickness and consists of a couplet of white and brown laminae. Study the white ones. Most have delicate features, almost at the microscopic level, an indication that each white layer required a considerable amount of time to accumulate. These layers have actually been counted. There are 261,162 of them. Again, a vast amount of time must have passed for the entire sequence of thinly layered rocks in West Texas to have accumulated.

At Fossil Butte in southwestern Wyoming there are exposures of deeply weathered rocks where the individual layers are paper thin. Each layer represents a distinct episode of sediment accumulation. Each one is a tick of a geologic clock. How many ticks are here? More than six million.

But how much time has passed for the tens of thousands of ribbons in the red chert near San Francisco to form, or the hundreds of thousands of white and brown laminae in West Texas, or the stack of millions of paper-thin layers in southwestern Wyoming? How long did it take for the Grand Canyon to form? Or for a mountain range to rise? These are the types of questions that were asked during the nineteenth century when geology was developing as a science. And different ways were proposed to answer them.

Charles Darwin provided one of the answers. In his book *On the Origin of Species*, first published in 1859, he suggested that current rates of erosion could be used to estimate how long it took a broad river valley to form. He chose the Weald in southeastern England as an example.[2] The region had once been dominated by a large dome of chalk. The chalk had eroded over time so that two high cliffs, North Downs and South Downs, now formed the boundaries of the valley. Darwin assumed that the rate of erosion of those walls was the same as the rate of erosion at the nearby cliffs at Dover, where the chalk walls were eroding away at a rate of one inch per century. The two prominent chalk walls at the Weald were twenty-two miles apart. Hence, according to Darwin, it must have taken about three hundred million years for the Weald to form.

John Phillips of Oxford University looked at the problem in another way: How long had it taken for all of the layers of sediment to accumulate? He estimated that the total thickness of all of the strata in the world was at least ten miles. If the average accumulation rate was an inch a year—taken from the rate that he thought sediment accumulated along river deltas—then a few hundred million years were required for all of the sedimentary layers to form. Irish geologist John Joly asked how much time had passed for salt to have accumulated in the oceans, assuming that the oceans were originally freshwater, and that all of the salt had come from rocks being washed off the continents. He estimated it had taken eighty million years. The most highly regarded estimate of geologic time was made by William Thomson, also known as Lord Kelvin, who asked how long it had taken the Earth, assuming it had formed in a molten state, to cool down to the solid mass it is today. He eventually decided it had happened in about twenty million years.

Each of these methods had a serious shortcoming: Long-term rates of erosion or of sediment accumulation are hard to estimate, the accumulation of salt has been episodic, and the radioactive decay of uranium, thorium, and potassium keeps the interior of the planet warm. But all of the methods did have something in common: They all indicated that the Earth must be very old. But exactly how old? How could absolute geologic ages be determined? The key to this, as told in the previous chapter, was the regularity and reliability of the radiogenic dating of rocks. And that changed everything.

Much is made of the great age of rocks—and, by implication, the very long history of the Earth, a history that is often referred to as *deep time*, because it is a critical point to understanding how the Earth works and is key to understanding the planet's past. The idea of deep time is on par with the ideas contributed by Copernicus that the Earth is not at the center of the universe, by Darwin that all life-forms are descended from common ancestors and, hence, are genetically related, and by Sigmund Freud, who showed that it was not rational thought but the unconscious that controls much of human behavior. Each of these ideas, including deep time, removed ourselves from a place of favored position and of self-importance to one in which we are one aspect of the world around us.

Admittedly, the length of the time periods in geology are so incredibly long—often measured in millions or billions of years—that they are difficult to comprehend. And so analogies are often used to convey a sense of what such great immensities of time represent.

One common analogy is to imagine taking a photograph of the Earth every year during each of the 4.51 billion years of its existence, then making a time-lapse movie of the result. If the movie is shown at a standard rate of thirty frames per second, then it would take five years to watch the entire movie. The formation of the Acasta Gneiss would appear about six months after the movie starts. The end of the dinosaurs would come about three weeks before the movie ends.

Another one is to imagine walking along a timeline of Earth history, each stride taking you one step back in time. If one step equals a thousand years, then after four steps one would have passed through all of written history. After walking the length of a football field, one would be back in the time when ancient humans and Neanderthals shared the planet.

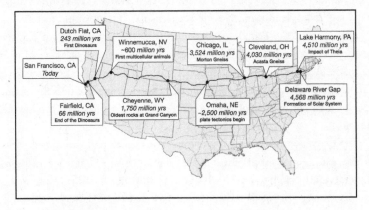

Timeline of the Earth's history projected along Interstate 80.

More specifically, imagine beginning the walk in San Francisco and following an eastward route along Interstate 80. After crossing the Bay Bridge and reaching the community of Richmond, one would be as far back in time as when the most recent ice cap started to form over Antarctica. Just before Fairfield in California would be the time when the age of dinosaurs ended. Continuing to go back in time, halfway between Sacramento and Reno, Nevada, near Dutch Flat in California, the first dinosaurs would appear.

Close to Winnemucca, also in Nevada, one would find the first multicellular animals. Only one-tenth of Earth history has yet been traveled. For the remainder of the walk, the only life on the planet would be microbes that lived in the sea.

By Salt Lake City, an early episode of mountain building would be in progress—the Grenville Orogeny—one with remnants that are seen today in the rocks of the Adirondacks in New York, the Ozark Mountains in Arkansas, and Enchanted Rock and the Llano Uplift near San Antonio in Texas. By Cheyenne, Wyoming, the oldest rocks at the Grand Canyon will have formed. Somewhere near Omaha, Nebraska, one would see the beginning of plate tectonics.

From there it is still a long walk to Chicago, where one would encounter the mass of molten rock that solidified and became the Morton Gneiss. In Cleveland the rocks of the Acasta Gneiss form. The impact of Theia and the proto-Earth occurs when one arrives at Lake Harmony in the Pocono Mountains in Pennsylvania. At the Delaware Water Gap on the border between Pennsylvania and New Jersey the solar system would form.

The point of these analogies is to illustrate the vastness of our planet's history compared to everyday experiences. And there is another point to be made: Almost all of the evidence of our planet's past is missing, erased through weathering and erosion, by the tremendous forces that raise the mountains, and by the cycling of great volumes of material back into the Earth's interior.

To offer yet one more analogy, reading the Earth's past through its rocks is like reading a book with most of the pages missing. That it is possible to understand so much of the Earth's past, to read the "book" of the Earth's history through its rocks, is owed to the fact that every "letter" of the book is unique, that every layer of rock, every stone, every grain of sand has a distinct history. It is because there are so many individual histories that a history of the Earth can be deciphered and a coherent story can be told.

And yet, there is more to lament. The rock record—the "letters" of the "book" of our planet's history—is not evenly distributed across time. For the first five hundred million years, nothing remains except for a hundred or so tiny zircons. For the most recent five hundred million years, almost half of all of the rock that formed during that period, at least on the continents, still exists, mostly as great layers of sediments. As one looks back in time, there is less and less evidence. And so the distant past will never be known as well as the more recent past.

But there is a startling lack of the oldest rocks. Very few rocks between the time period of the Acasta Gneiss and the Morton Gneiss have survived. It is estimated that less than a *millionth* of all of the rocks that existed between 3.6 and 4.0 billion years ago still exist. What happened? Why does such a minuscule amount remain? What happened to the Earth so that this earliest part of the rock records—much more than would be expected if only weathering and erosion had taken place—has been erased?

◆

Look at the Moon through binoculars or a small telescope. The surface is covered with craters. Each one was formed by a high-velocity object slamming into the Moon. Now look carefully at the surface.

Three are two distinct types of terrane. The dark areas have fewer craters and so they must be younger. These are vast lava flows, roughly equivalent in chemistry and mineralogy to the flows that erupt on the seafloor of the Earth or in places like Hawaii, but much, much larger. By comparison, the bright whitish areas—the lunar highlands—are saturated with craters. There are so many craters that they lie atop one another. This is the old surface of the Moon, the part that solidified from the lunar magma ocean. It was once thought that the intense cratering seen in the lunar highlands dated back to a time soon after the magma ocean solidified: that is, when the Moon's crust formed. That is not quite what happened.

Rocks from six different sites on the Moon were collected by the Apollo lunar missions. After they were dated, it was found that almost all of the rocks could be put into one of two age groups. The older ones dated back about 4.4 billion years to when the lunar crust formed. The others, which represented the bulk of the rocks collected by Apollo, had ages that ranged from 3.85 to 3.95 billion years. It was soon suggested that these younger rocks must represent a time in lunar history, hundreds of millions of years after the lunar crust had formed, when the lunar surface was bombarded by an intense swarm of meteors, a swarm so intense that, as recorded in the lunar highlands, almost all evidence of earlier impacts was wiped out. In short, this later fusillade of meteors had reset the age of much of the rocky surface of the Moon, an event that was dubbed the Late Heavy Bombardment.

At first, many lunar scientists were skeptical that such a catastrophic event had occurred. They argued that the rock samples returned by Apollo were highly biased. All six landing sites were on the nearside of the Moon and all six sites were near the lunar equator. But, since Apollo, more than a hundred meteorites in museums have been identified as lunar rocks based on the similarity to the mineral content of the Apollo samples. Those museum samples would have come from all over the Moon, including the lunar farside. And the museum samples also showed a clustering of ages at around 3.9 billion years.

But why did such an intense bombardment happen hundreds of millions of years after the Moon formed? Why were the lunar highlands peppered with so many craters? An explanation was a long time in coming, but a reasonable one can now be offered.

At the end of the Grand Tack the main objects of the solar system were not quite in their current orbits. Jupiter and Saturn were close to their current orbits, and the four rocky inner planets were still adjusting their orbits. But a significant change was yet to come to Neptune and Uranus, the two gas-rich planets beyond Saturn. There was also much more rocky debris in the asteroid belt between Mars and Jupiter than there is today. It was that combination—the sudden adjustments of the orbits of Neptune and Uranus and the clearing out of most of the debris in the asteroid belt—that probably accounts for the Late Heavy Bombardment.

The explanation is based on the Nice Model, developed by astronomers at the Observatoire de la Côte d'Azur in Nice, France. They proposed that, after the Grand Tack, the main objects in the solar system had not yet settled in stable orbits. In their calculations, both Jupiter and Saturn shifted slightly. That caused larger movements of Uranus and Neptune and sent much of the debris in the asteroid belt into orbital chaos.

Their calculations also show that this chaos started about 3.9 billion years ago and lasted about a hundred million years. Tens of thousands of small objects were thrown out of the asteroid belt and down into the inner solar system, pounding the four rocky planets and the Moon. Then the solar system stabilized. And for the next four billion years the orbits of the main planets and their moons have remained essentially unchanged.

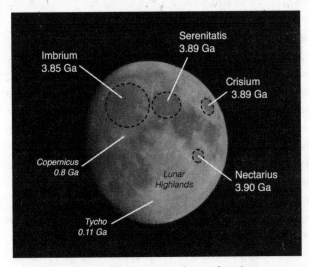

Ages of lunar basins. The locations and ages of two large craters,
Copernicus and Tycho, are also shown. (Ga is a gigayear = 1 billion years)

Look again at the Moon. Almost every impact crater you can see formed at this time.[3] If you now look at the Moon without binoculars or a small telescope and see a face, the right eye of the face is a dark circular feature known as the Imbrium Basin. That basin formed 3.85 billion years ago. To the west of the Imbrium Basin are three other dark circular basins with similar ages: Serenitatis, 3.89 billion years; Crisium, also 3.89 billion years; and Nectaris, 3.90 billion years. All of these basins formed during the Late Heavy Bombardment.

And if the Moon was getting pummeled by huge impacts at this time, so was the Earth. Its larger size and greater mass meant that it would have been hit by many more large objects, forming many more large impact basins. But, as of yet, no one has found clear evidence of the Late Heavy Bombardment on the Earth. No one has found an impact feature on the Earth that dates back to this early period.

The oldest impact feature yet identified on the Earth is the Vredefort Dome impact structure in South Africa. It is a feature about 200 miles in diameter and it formed 2.023 billion years ago. The next oldest is the Sudbury structure north of Lake Huron in Ottawa, an impact crater about 150 miles in diameter that formed 1.849 billion years ago. Both happened much later than the Late Heavy Bombardment and both are much smaller than the Imbrium or other basins on the Moon.

A much older impact feature *might* exist along the western coast of Greenland. Named for a nearby fishing village, the Maniitsoq structure is an ancient circular feature that seems to have involved the intense crushing and heating of rock. Conclusive evidence of an impact origin—such evidence would include the recovery of minerals that could form only under the extremely high pressures and temperature of such an impact—is yet to be presented. But the structure, whatever it is, is about three billion years old. In which case, if it does prove to be an impact feature, then it would be the oldest and one of the largest yet discovered on the planet.

Notwithstanding the case pending about the Maniitsoq structure, there is evidence of much earlier impacts hitting the Earth. Though the craters have not been found, the debris that would have been thrown up and fallen back to Earth has been discovered.

A dozen or so such blankets of impact-produced debris have been found in South Africa and in Australia. They occur as discrete beds, at most a few feet in thickness, and contain small glassy beads. The glassy beads are the key. They represent material that was vaporized by the great amount of energy produced by the impact and thrown into the air. The vaporized material then cooled rapidly as it fell back to Earth, forming the small beads. The oldest such bed of vaporized beads formed 3.472 billion years ago, still a few hundred million years younger than the Late Heavy Bombardment, but an indication that rocks of substantial size were bombarding the Earth and doing so on a regular basis.

During the Late Heavy Bombardment, the surface of the Earth was covered with craters. It is estimated that a new large crater, say, a hundred miles or more in diameter, formed every hundred thousand years or so. By comparison, the rate at which such craters form today is about one every hundred *million* years.

And the high cratering rate would have influenced how the Earth's surface evolved. A large impact might have induced volcanic activity. It would have vaporized part of the oceans, injecting a large amount of water vapor into the atmosphere and temporarily changing weather patterns. It would have sent a burst of intense heat through the atmosphere and across the surface, killing any life-forms that may have existed. And that would have put a bottleneck in how early life could have evolved.

◆

If one climbs up through the rock record, moving from older rocks to younger ones, one finds that animal fossils appear suddenly and in great abundance above rocks that seem barren of life. This sudden appearance is known as the Cambrian Explosion because it occurs throughout the world at the base of Cambrian-age rocks. This sudden burst of life on the planet has been known for almost two centuries. It was certainly known to Charles Darwin, who puzzled over it greatly because he considered it to be the most difficult problem facing his theory of evolution.

In *On the Origin of Species*, Darwin admitted that, "the question of why we do not find rich fossiliferous deposits belonging to these assumed earlier periods prior to the Cambrian system, I can give no satisfactory answer." He continued. "The case at present must remain inexplicable; and may be truly urged as a valid argument against the views here entertained." But, if his theory was right, then, he insisted, the development of life must have been gradual. And evidence of life-forms would be found in rocks that predated the fossil-rich Cambrian Period—that is, in Precambrian rocks.

Much effort was made during the next hundred years to address what has often been referred to as "Darwin's dilemma." Among those who searched was William E. Logan, director of the Geological Survey of Canada, who reported in 1865 that he had made such a discovery. In fact, he was so convinced that he had found tiny fossilized shells in limestone along the Ottawa River west of Montreal—in limestone that was definitely Precambrian in age—that he formally named his discovery *Eozoön canadense*, the "dawn animal of Canada." But his discovery was firmly discounted when similar-looking "fossils" were found in blocks scattered around the volcano Vesuvius. A fossil shot out of a volcano was something few were prepared to accept.

In 1883, Charles Doolittle Walcott, who later served as a director of the United States Geological Survey and as a secretary of the Smithsonian Institution, found what he thought were specimens of Precambrian life at the Grand Canyon. But his discovery was also widely dismissed, one critic writing that it required a "demand upon imagination." Nevertheless, today Walcott's specimens are considered animal fossils, the first Precambrian organisms ever found.

The person who set the course for the modern study of Precambrian life was Stanley A. Tyler of the University of Wisconsin. In 1954 he reported that he had found evidence of early life in a layer of chert along the northern shore of Lake Superior. This layer, known as the Gunflint Chert, runs for nearly a hundred miles from the base of the Gunflint Range in northeastern Minnesota to a place near Thunder Bay in Ontario. When examined under a microscope, the chert reveals itself to be made up of interlocking minerals of quartz. If one studies the chert as carefully as Tyler did, one will see many tiny particles between the minerals. These are in a variety of shapes. Some resemble spheres, others rods, or filaments. These microfossils, as Tyler would prove, are the petrified remains of bacteria that lived in an ancient ocean.

The microfossils of the Gunflint Chert lived 1.88 billion years ago, which make them much older than the Cambrian Explosion, which occurred, in a relative sense, much more recently—a mere 541 million years ago. Tyler also realized that many of the microfossils that he was seeing through his microscope were remarkably similar to microorganisms that are alive today. That meant they were as complex as any bacteria living today. And so they could not represent the earliest form of life on the planet. Earlier, less complex forms must have existed.

Today the earliest complex form of life is found in the Dresser Formation in Western Australia. These, as in the Gunflint Chert, are microfossils. And life was, indeed, flourishing in the Dresser Formation as early as 3.5 billion years ago. Which again means that some form of more primitive life must precede that found in Western Australia.

In a region in West Greenland, formally known as the Isua Supracrustal Belt, melting ice has recently revealed a small section where rocks with ripple marks have been found. Those ripple marks were made in sand that was disturbed by rhythmic wave action at the edge of an ancient ocean. It is the oldest known occurrence of such wave action. And though these rocks do not contain microfossils, they do contain *biomarkers*, chemical traces of life. And so, though the actual life-forms are not visible, there is strong circumstantial evidence that life existed on the Earth at least as old as these particular rocks, which have an age of 3.8 billion years.

This happens to be close to the end of the Late Heavy Bombardment, and may not be a coincidence. Impact cratering during the Late Heavy Bombardment may have erased all earlier evidence of life. And so direct evidence may never be found. But there is evidence that some form of life must have existed on the Earth before that period. And it is based on a new field of study known as molecular phylogenetics, which searches chemical sequences in modern forms of life to determine evolutionary history.

The key is determining sequences of chemical units in long biological molecules, such as DNA, RNA, and proteins.[4] In DNA and RNA, the chemical units are the molecular fragments—the nucleotides—that connect the two helical strands of DNA and RNA and that determine the genetic code of an individual. In proteins, the units are the sequence of amino acids that comprise the protein. The evolutionary history of life is revealed by the similarity of the patterns of these various chemical units. The more similar the patterns are between two types of organisms, the more closely the two are related. Remarkably, there are also patterns that are shared by *all* life-forms. And it is those patterns that reveal the earliest evolutionary history of life because such patterns could be shared by all life-forms in only one way: Those patterns must have come from an ancestor that was common to all life that is living today.

The identity of this earliest ancestor came as a surprise. Of all of the life-forms alive today, it was most closely related to hyperthermophiles, microorganisms that live in extremely hot environments. The first hyperthermophiles were discovered in 1965 in the hot springs of Yellowstone National Park. Since then, dozens of species have been identified, some found in the deep sea around volcanic vents where the great depth has raised the boiling point of water to 280° Fahrenheit (about 140° Celsius).[5]

Following the clues provided by molecular phylogenetics, this earliest ancestor of all life-forms must have also lived in extremely hot, watery environments. But such an environment cannot be where life originated because the long molecules that are essential for life—DNA, RNA, and proteins—could not have been assembled and remained stable in such an environment. These ancient thermophiles must have had a protective membrane, such as hyperthermophiles do today. Which means they were already complex life-forms, and again, some form of life must have existed before them. But why did they survive? It was the extreme conditions of the environment in which they lived.

The Moon-forming event with Theia would have vaporized much of the early Earth, leaving the early planet sterile. If any living cells or complex molecules existed before the event, they would have been vaporized by the event and would not have survived. That is not true of the meteor impacts during the Late Heavy Bombardment. While many were substantial, none would have had enough energy to sterilize the planet. At most, individual impacts may have vaporized the upper few hundred feet of the oceans where much of early life might have been living. But they would not have extinguished life-forms that lived in the deepest parts of the oceans in water that was already extremely hot.

And that is what molecular phylogenetics suggest. All life-forms on the planet today—including us—are descended from a common ancestor that was similar to modern hyperthermophiles. The Late Heavy Bombardment—ultimately caused by the last orbital adjustments of Jupiter and Saturn and of Uranus and Neptune and to the scattering of rocky debris out of the asteroid belt—frustrated an early development of life, eliminating all early life-forms except those that lived in what became a protective environment of extremely hot water in the deep sea.

◆

But that still leaves unanswered the question of the origin of life. It just pushes the question back to a time before the Late Heavy Bombardment, to before 3.8 billion years ago. And there are so few rocks of that age that exist and can be examined. But examined they are. And in great detail.

Because even microfossils are unlikely to have survived from before the Late Heavy Bombardment, those scientists who pursue the question of life in the very earliest period of the Earth's history have had to look for other evidence. They look for chemical traces that very early life-forms may have left. In particular, they look for a depletion of the isotope of carbon-13 relative to carbon-12 in the very oldest rocks.

It has been known since the mid-twentieth century that living organisms incorporate less carbon-13 than carbon-12 in their systems, the slightly heavier atoms of carbon-13 requiring an expenditure of a bit more energy in metabolic reactions. And so a depletion of carbon-13 in an ancient rock could be a sign of ancient biological activity.

Such a depletion had been found in a small rocky outcrop in the extreme northern part of Labrador where the rocks have been determined to have an age of 3.96 billion years. A speck of carbon found inside a zircon recovered in Western Australia also has a carbon-13 depletion. It has yielded an age of 4.1 billion years. And the greatest age of any rocky sample yet collected with a depletion was taken from the eastern shoreline of Hudson Bay in Quebec. That potentially record-setting sample has an age of at least 3.77 billion years and could be as old as 4.28 billion years.

All of these claims of evidence of life before the Late Heavy Bombardment are controversial. Yet it is intriguing to wonder: What was the origin of life? And where did it originate?

To be clear, while the details of life's origins remain elusive, the general outline of how life originated is gradually emerging. The origin of life is now in the realm of plausible, accessible chemistry and is no longer considered a miraculous or rare event. It may have been inevitable, part of planetary evolution. And that makes the question of the origin and evolution of life a geological issue.

It is certain that life could not have originated in the vastness of the early ocean because such a large body of water would have diluted any concentration of small molecules, making the assembly of larger, more complex ones impossible. And so pools of water where small molecules could concentrate must have existed. Among the areas that seem the most promising are those where there was a cycle of wet and dry conditions, the alternation between wet and dry concentrating small molecules even more. But even in those conditions the assembly of larger molecules would have been slow unless there was a mechanism that could speed up the assembly. And such a mechanism could have been supplied by clay minerals, which have a flat layered structure, and which would have formed naturally in a wet-to-dry environment. Furthermore, laboratory experiments show that when clay minerals are irradiated by high-energy ultraviolet light—such as the light that would be coming directly from the Sun—the flat surface of clay minerals becomes electrically charged, thus attracting small molecules. And from that long molecules could be assembled.

All of this is to illustrate that a great deal of thought, laboratory work, and field work have gone into answering the question of the origin of life. And, from this, some scientists have concluded that the environmental conditions on the early Earth may have been too severe to have allowed life to form during the few hundred million years between the Theia impact and the Late Heavy Bombardment. The rate of impact cratering was high. The Sun was not yet stable, and so the Earth's surface was probably receiving high levels of lethal radiation. The early ocean was so large that intense storms may have been common, ripping up whatever land existed and churning up the shallow seas. And so a place where there was a more benign environment, one more suited for the origin of life, was needed. Such an environment existed then on Mars.

Mars is substantially smaller and has much less mass than the Earth, and so the rate of impact cratering has always been less. It is half again as far from the Sun as the Earth, and so the intensity of lethal radiation would have been reduced by a factor of two. It has probably always had a less-dense atmosphere than the Earth, and so storms would have been less severe. And, based on the findings of the numerous spacecraft that have landed on the planet, Mars did have bodies of liquid water on its surface early in its history—and today there are clay minerals.

Supporting the idea that life may have come from Mars is the fact that, in laboratory situations, some microorganisms have been shown to survive high shock pressures, which means they would probably have survived both being launched from one planetary surface and crashing into another one.

All of this makes Mars a possible place for the origin of life. Those who designed and operated NASA's *Curiosity* rover, which landed on the red planet in 2012, concluded that early Mars was a possible "well-spring of life" where "most or all of the major ingredients required to support life" probably existed early in that planet's history.

So it is reasonable—though far from proven—that Mars might be the birthplace of life. And that life was brought to the Earth by one or more meteorites that originated on the red planet that came blazing through our early atmosphere and that landed on the Earth's surface, which would later develop conditions that were suitable for life's continued development. Maybe. And if that is true, then, in the broadest sense, we were originally Martians.[6]

◆

If a spacecraft from another star system arrived at some random time in the history of the solar system, it would probably have found life on Earth. And if it had arrived sometime before the last several hundred million years, it would have found that a particular form of life dominated the planet. That dominant form would have been the stromatolite, which, because of its simple structure, might seem inconsequential in the planet's history. But stromatolites are just the opposite. They are one of the most influential forms of life ever to inhabit the Earth.

For a long time stromatolites were known only from the fossil record. They were first described in 1825 by John Honeywell Steele, a physician from Saratoga Springs, New York. Steele wrote a letter about his discovery to Professor Benjamin Silliman of Yale College for publication in the *American Journal of Science*, the nation's first scientific journal. In the letter Steele mentioned a patch of ground two miles west of Saratoga Springs where there were strange concentric circles outlined in the rock.[7] He described them as "mostly hemispherical" and that they varied "in size from half an inch to that of two feet in diameters." They had alternating bands of dark and light layers. The dark layers were thin and contained particles of an unknown origin that were about the size of mustard seeds. The light bands were thicker and composed of siliceous and limy sand. Steele did not speculate on how the circles might have formed—or what they were.

In 1847, James Hall of the Rensselaer Polytechnic Institute in Troy, New York, studied the same circular features and, waiting more than twenty-five years to write about them, suggested they might be some type of fossil life. Then, in 1908, Ernst Kaldowsky, director of a museum of mineralogy in Dresden, Germany, reported similar-looking features in rocks in the Harz Mountains of northern Germany. And he gave them a name, *Stromatolith*, which was translated into English as "stromatolite," which means "layered stone."

A major advance in the study of stromatolites came in 1914 from Walcott while he was doing work in the Big Belt Mountains of Montana. Here he found stromatolites in abundance. He collected samples and studied them under a microscope, reporting that he could see features that were reminiscent of a common type of bacteria, once called blue-green algae but now more commonly referred to as *cyanobacteria*. Others would confirm his work. And that would become the key to understanding the importance of stromatolites. They are the most abundant microfossil in the rock record and have a wide range of ages. And they have now been found in many places in North America—and around the world.

Stromatolites have been found in the Snowy Range of the Medicine Bow Mountains of southeastern Wyoming, age 1.7 billion years; along roads cut in Glacier National Park in Montana, age 1.5 billion years; at the bottom of the Grand Canyon in Arizona, age 1.2 billion years; along the Keweenaw Peninsula of Michigan, age 1.1 billion years; in the Van Horn region of West Texas, age 1.0 billion years; in California in the Mojave Desert, age 650 million years; and in Capitol Reef National Park in Utah, age 180 million years. One of the most impressive and easily accessible exposures is at the western end of the Champlain Bridge in Gatineau, Quebec, age 450 million years. The oldest ones are in Western Australia, age 3.5 billion years. And the size of an exposure can be amazingly large. Those in the Belcher Islands in Hudson Bay cover more than a thousand square miles and are several hundred feet thick.

The past dominance of this curious form of microbial life led scientists to wonder: Might stromatolites still be living somewhere on the planet today? The first living examples were found in 1933 on a tidal flat along the shoreline of an island in the Bahamas. Since then, several additional sites have been found. They exist at the bottom of Pavilion Lake in British Columbia, at Lake Bacalar on the Yucatán Peninsula, and along what is occasionally a dry stream bed within Anza-Borrego Desert State Park in southern California. But it is at Shark Bay in Western Australia where they have been found in profusion, where there are fields of stromatolites that can be found over an area of hundreds of square miles. That fact that stromatolites still survive is due to the harsh conditions of the environment at each of these locations. The water is either extremely hot or contains a high amount of salt or is extraordinarily acidic. It is such extreme conditions that have kept other organisms away that would compete with or prey on the stromatolites.

Because there are living examples, we know how stromatolites grow. The cyanobacteria produce a mucus that forms a thin sticky mat that traps small particles floating in the water. Cyanobacteria also have the ability to pull atoms of calcium and molecules of carbon dioxide directly out of the water and use them to produce calcium carbonate (limestone), which cements the small particles together, forming a hard layer. The living cyanobacteria then moves upward so they are not buried. Over time, this cycle of sediment capture and growth creates the banded structure common in stromatolites, a structure that is often referred to as "cabbage heads," and that gives rise to various forms, including domes, cones, and upright cylinders. Admittedly, the growth rate is slow. Those at Shark Bay grow about a tenth of an inch a year.

So one can imagine sailing across an ancient ocean and looking down into the shallows and seeing what must have seemed to be an endless field of stromatolites. That, by itself, illustrates how important this life-form was in early Earth history. But they had an even more important role to play, one that influenced both subsequent biological evolution and the geological future of the planet. It stems from a basic fact: Of all of the life-forms that ever existed on the planet, or that exist today, only cyanobacteria have the ability to produce oxygen.

◆

The earliest forms of life probably derived energy from chemical sources. One possible source of energy was sulfur that was energized and had accumulated around volcanic vents. Some forms probably extracted hydrogen and carbon dioxide that were dissolved in ocean water and were combined to produce methane, which was then released into the atmosphere. But chemical sources of any kind could provide only small amounts of energy. The evolution of new forms would be driven by the harnessing of more energy. And that meant making use of the energy in sunlight.

Photosynthesis is the use of sunlight to combine chemicals to store energy. Its origin is perhaps the most widely discussed yet one of the most poorly understood events in the history of life. To complicate matters, there are several types of photosynthesis, but only one type that produces oxygen. And it has evolved only in cyanobacteria.

In cyanobacteria, the energy of sunlight releases hydrogen from water molecules, then adds the hydrogen to molecules of carbon dioxide to make organic compounds. The net result is that molecules of oxygen are produced. To be clear, plants and algae duplicate the process—and, hence, produce oxygen—but they do so by having cyanobacteria trapped inside their cells. At some time, during the long history of life, cyanobacteria were incorporated into other cells. As those other cells grew and divided, so did the cyanobacteria inside of them. And the ancestors of the ingested cyanobacteria can be seen today. They appear under microscopic examination of plant and algae cells as tiny internal structures known as chloroplasts. It is inside the chloroplasts of plants and algae that photosynthesis takes place. Chloroplasts also have their own DNA, which is different from the DNA of the main cell. An ancestral link between chloroplasts and cyanobacteria has now been firmly established by numerous studies that have shown that the DNA of chloroplasts is identical to that found inside cyanobacteria living today.

It is unknown when the first molecules of oxygen were produced by photosynthesis. What is known is that oxygen was being produced at least as long ago as 2.7 billion years because a particular type of deposit known as a *banded iron formation* started to accumulate on the ocean floor.

Underwater volcanic vents are constantly spewing out a venerable zoo of chemicals into the oceans. This was first seen in action in 1977, when the research vessel *Alvin*, carrying three men, descended to the deep ocean floor near the Galápagos Islands. It was there that the now famous "black smokers" were discovered, vigorous dark plumes rising from the ocean floor. Water samples were taken and analyzed and they showed that most of the "black smoke" rising from the vents was dissolved iron. And so it must have been the same in the early ocean where, for hundreds of millions of years, iron, which is soluble in water, continued to accumulate. That is until a significant amount of oxygen had been produced. Then, when there is oxygen in the water, iron readily precipitates and a deposit of iron oxide forms on the ocean floor, resulting in an ocean floor covered with a layer of rust.

And so it was 2.7 billion years ago. And the production of oxygen continued, in fits and starts, until about 1.8 billion years ago. By then, there were massive amounts of iron oxide on the ocean floor. These deposits appear today as the seemingly countless thin red bands of iron oxide separated by equally thin bands of silicate-rich shales and cherts, hence the name banded iron formation. And they are found on every continent. In North America, they are the rocks that comprise the iron-rich regions around Lake Superior and extend into Ontario, Michigan, Minnesota, and Wisconsin. Ninety percent of all the iron that has ever been mined comes from banded iron formations, from rocks formed during the Archean Eon.

The concentration of iron is so high in these formations that iron can be extracted at low cost and in abundance. And it is those two attributes that have allowed iron and its alloy, steel, to be used so widely. Iron or steel beams are required to construct a building taller than a few stories. Girders and cables made of iron or steel are essential in the construction of bridges that span wide rivers or canyons. Iron or steel is found in automobiles and most home appliances. They are used in the manufacture of offshore oil platforms, pipelines, railroad engines and railroad tracks, and in much else. And so it is worthwhile to pause and realize that this key component of our industrialized society comes from the action of a singular form of life—cyanobacteria—that existed for more than a billion years before there were any animals or plants on the planet.

And yet, the production of banded iron formations is not the most significant influence that cyanobacteria has had on the planet. Once most of the iron had precipitated out of ocean water, which happened about 2.4 billion years ago, oxygen began to accumulate in the atmosphere. We know this because many new types of minerals—oxidized forms of copper, nickel, and uranium, as well as other metals—suddenly appear in the geologic record. And they could have formed only if an oxygenated atmosphere existed.

But the amount of oxygen was not yet very high. Today's atmosphere is about 20 percent oxygen. The atmosphere 2.4 billion years ago was 1 to 2 percent. Nevertheless, that small amount was enough to have a major impact on the Earth's subsequent history, one that produced the Great Oxygenation Event. And the consequences would be many.

One of the immediate consequences was an abrupt cooling of the planet. Carbon dioxide and methane are both greenhouse gases, which means they have the ability to trap the Sun's radiation in the atmosphere. This warms the Earth's surface to temperatures higher than would be possible without greenhouse gases. The amount of carbon dioxide in the atmosphere was already decreasing sharply as cyanobacteria became more abundant. These microscopic organisms pulled more and more of this greenhouse gas out of the atmosphere and, through photosynthesis, poured more and more oxygen into the atmosphere. And the increased oxygen in the atmosphere reacted chemically with the methane, reducing the amount of that greenhouse gas.

The result was a sudden plunge in surface temperature, a sudden shift to a cold climate. Liquid water could no longer exist on the planet. The oceans froze and the entire planet went into a deep freeze.

Evidence of this ancient frozen world can be seen today. It is found along a section of the Trans-Canada Highway that crosses through southern Ontario.

◆

The earliest ice age for which there is firm evidence is known as the Huronian Glaciation, first recognized in 1907 by Arthur Philemon Coleman of the University of Toronto, who was searching for ore deposits, as was typical of most geologists who worked in the nineteenth and early twentieth centuries. In his quest, he was given what he described as "a few scratched stones" that had been found in a silver-mining region near Sudbury in Ontario. The stones sparked an immediate interest because he recognized the scratches to be akin to glacial striations, grooves made when the slow movement of glaciers cause rocks to grind and slide against each other, leaving sets of parallel lines. Because the "few scratched stones" had been found in deposits that were very old, Coleman realized that this was evidence of a very ancient ice age, one that had occurred much earlier than a recent period when sheets of ice had covered parts of North America. Furthermore, he suggested that this ancient ice age had been more extensive than the recent one. Subsequent work has proven him right on both counts.

The recent period of ice ages—which will be discussed later in this book—is known as the Quaternary Glaciation. It began about three million years ago and consisted of a series of advancing and retreating ice sheets. The most recent sheet retreated a mere twelve thousand years ago after reaching halfway between the North Pole and the equator at its greatest extent. By comparison, the ancient glaciation discovered by Coleman lasted two hundred million years and occurred between 2.42 and 2.22 billion years ago. Evidence for it has now been found in many places around the world, including in South Africa, Russia, Finland, and Australia, which means it was a global phenomenon. But the best evidence is found in Canada, in rocks immediately north of Lake Huron, hence the Huronian Glaciation.

These rocks comprise the Gowganda Formation, which can be seen in road cuts along the Trans-Canada Highway between Sudbury and Sault Sainte Marie. Many road cuts reveal a striking stone composed of both rounded and angular pebbles and boulders, some of pink granite, surrounded by hardened gray sand and mud. The stone is a *tillite*, a telltale sign of the action of a glacier. But it is neither the tillites or the possibility of finding "scratched stones" that brings serious students of our planet's past to southern Ontario. Most come to find the feature that has made the Gowganda Formation world famous: *dropstones*.

As glaciers melt, lakes form along their margins. In the winter, the lakes are frozen, but, in the summer, they are almost free of ice. This produces a cycle change in the sediment that is deposited on the lake bottom: fine clay and mud in the winter, coarse sand and silt in the summer. Each layer represents one year of sediment accumulation.

The Gowganda Formation has many such layers, most only a fraction of an inch thick. At some locations, hundreds of layers are exposed. And if one looks closely at the hundreds of thin layers, one will find places where fist-sized stones have disturbed the normally flat-lying layers, making it look as if the stone had dropped down from above and caused the underlying layer to bow downward. That is exactly what has happened. These are the famous dropstones of the Gowganda Formation. And they are known to form in only one way: by the release of stones from a melting iceberg onto a lake bottom.

This episode of global freezing is known as Snowball Earth, a period when the entire planet was encased in ice. Geologic evidence shows that this condition persisted for two hundred million years, then the planet warmed again. Evidence for the reversal can also be seen in southern Ontario.

The rugged hills and long peninsulas along the north shore of Lake Huron are composed of a hard, erosion-resistant rock known as a *quartzite* that is part of the Lorrain Formation. The quartzite is mostly tightly compacted white sand. It accumulated by erosion, after the planet warmed and liquid water started to flow. The quartzite of the Lorrain Formation sits directly on top of the friable glacial deposits of the Gowganda Formation. It is the contrast between the hard quartzite and the underlying glacial deposits that gives rise to the spectacular scenery in Killarney Provincial Park in Ontario. The high-standing, white ridges of the La Cloche Mountains south of Sudbury are exposures of the erosion-resistant Lorrain Formation. The valleys between the ridges, many of them containing lakes, are where the easily eroded rocks of the Gowganda Formation lie.

This transition, seen from the highway that passes through Killarney Provincial Park, marks a time when the Earth was half its present age. It indicates a time when the planet underwent a major transition, so much so, that the beginning of the Huronian Glaciation—the base of the Gowganda Formation—is the beginning of a new geologic eon, the Proterozoic, literally "a time of early life."

Life-forms will become more complex and more diverse and they will be much more abundant. Just as important, tectonic plates will have finally formed and will be moving slowly and steadily. And, from that, the first major mountain ranges will form, as well as the first supercontinent.

3

The Children of Ur

Early and Middle Proterozoic Eon: 2.420 to 0.720 Billion Years Ago

I t could be said, without too much exaggeration, that Staten Island is one of the geologic crossroads of North America. It is there, in a triangular area that measures at most twenty miles long and seven miles wide, that one finds each of the three major rock types: igneous, sedimentary, and metamorphic. And each of the rock types on Staten Island formed at very different times.

The igneous component lies beneath the northwestern corner of the island. It is the southern extension of the same magmatic event that formed the rocks that are exposed along the northern face of a great cliff face known as the Palisades that runs through the New Jersey side of the Hudson River. That event, with its pulses of molten rock, occurred about 190 million years ago. The southern and eastern sectors of the island are underlain by sedimentary rocks, mostly silt and sandstone with the occasional fragments of dinosaur bones. The bones indicate that these rocks are the youngest of the trio of rock types on Staten Island and have a reliable age of seventy million years. And at the core of the island is a greenish-gray metamorphic rock that is a familiar sight to commuters who ride along the Staten Island Expressway between Slosson Avenue and Clove Road. This rock is actually a piece of the Earth's mantle that was squeezed and heated and infused with water and subsequently forced upward and now lies exposed. The squeezing and the heating and the infusion occurred about five hundred million years ago.

It is the close proximity of these three different rock types with three very different ages that poses a challenge. Before the second half of the twentieth century, geologists could not explain the geologic history of an area as seemingly simple as Staten Island because they were hampered by a long-held tenet of geological thought. It was then widely assumed—and those who tried to counter it were harshly criticized—that masses of rocks formed essentially in place, that the various blocks that comprise the Earth's surface might have moved up or down to create mountains or valleys, but they had not moved horizontally by any significant amount. To propose otherwise was preposterous. The view today is completely different.

Notwithstanding the awe one feels when looking up to or down from a high mountain range, vertical movement of the Earth's surface is a *minor* component of the total movement. It is the horizontal sliding of great blocks that has determined our planet's history, at least for the last 2.5 billion years. And so it is for a place as small as Staten Island, where the collision of large blocks hundreds of millions of years ago led to a part of the Earth's mantle being squeezed up to the surface, then to the pulling apart of large blocks that led to the magmatic activity.

It was this revolutionary change in geological thought that fundamentally changed how we view the Earth and its history and, by implication, how we came to understand the geological history of North America, how we learned that the continent is primarily an amalgamation of very large landmasses that have been shoved together.

But to understand that history—both of the Earth as a whole and North America in particular—it is necessary to pause and delve slightly deeper into how the Earth works today.

◆

Look at a map of the world. There is a natural division between land and ocean, between the continents and the ocean basins, a division that exists because the continents stand higher than the ocean basins. That division is a direct consequence of a difference in mineral compositions of continental and oceanic rocks.

The most abundant rock on the planet is *basalt*, a dark igneous rock that forms the floor of the ocean basins. Basalt is composed mostly of two minerals—olivine and pyroxene. Both are rich in iron and magnesium. By comparison, the continents have an average composition close to that of granite, also an igneous rock, which is composed of quartz and feldspar. Quartz is silicon dioxide. Feldspar is a range of minerals that are rich in aluminum, sodium, and calcium. That difference in mineral compositions accounts for why the continents stand higher than the ocean basins. It is simply a matter of buoyancy. Quartz and feldspar are less dense than olivine and pyroxene. And so the continents are less dense than the ocean basins. As a result the continents are buoyed up higher on the Earth's hot, pliable interior.

Look again at a map of the world, this time paying attention to the outline of the continents along the Atlantic Ocean. The bulge of Brazil seems to fit nicely into the great bight of southwestern Africa, like two pieces of a jigsaw. If North America is moved southward and rotated slightly clockwise, the coast from Florida to Nova Scotia will run close to the one from Liberia to Morocco. The coastal contours of Labrador and Greenland can be fitted against Spain, the British Isles, and the coast of Norway. Could this be a coincidence?

The Atlantic coastlines of the continents fit together like
pieces of a jigsaw, suggesting the continents were once together
as a single supercontinent and have drifted apart.

Abraham Ortelius, a Flemish cartographer, is usually credited with producing the first modern atlas, which he published in 1570. Among the more than four dozen maps showing various regions and political states, his atlas includes a single large map of the entire world. It is centered on the Atlantic Ocean and shows what is now known to be an exaggerated protuberance of Brazil and an equally exaggerated indentation of the coast of western Africa. He must have spent years contemplating what this might mean because it was not until 1596 that he proposed that the conformity of the two opposing coastlines was not an accident. He suggested that North America and South America had been "torn away from Europe and Africa . . . by earthquakes and floods." As to when this had occurred, he turned to the ancient Greek philosopher Plato and his story of Atlantis and suggested that the great upheaval that had destroyed Atlantis had also torn the continents apart.

And that is where the matter rested for three hundred years. Several notable people, including the English philosopher Francis Bacon and the world traveler and naturalist Alexander von Humboldt, made the same observation that Ortelius had, but they did not advance the idea further. Not until 1893 did anyone suggest that there was additional evidence that South America and Africa had once been connected.

In that year, Austrian geologist Eduard Suess reported that identical fossils of the extinct fernlike tree *Glossopteris* had been found in both Argentina and South Africa. But the idea that the Earth's surface could not move horizontally any significant amount was too much for him to overcome. And so he suggested that a land bridge, now submerged, had once existed between the two continents. Suess seemed to busy himself inventing continents that popped up like corks, then sank like anchors. But he did not address the question why the coasts of Brazil and Africa seemed to fit so snuggly against each other. Twenty more years passed. Then, after considering Seuss's claim about *Glossopteris* being found on opposite sides of the Atlantic, and adding much additional evidence, one man was convinced that South America and Africa had once been adjacent and had since drifted apart.

This renegade in scientific thinking and orthodoxy was Alfred Wegener, a German meteorologist and arctic explorer. In 1908 he was between planning for arctic expeditions, working as a tutor at the University of Marburg, when, while browsing through the university library, he found a scientific paper that listed fossils that had been found on both sides of the Atlantic. Among the fossils listed were, of course, *Glossopteris* and a crocodile-like dinosaur Mesosaur. Intrigued by the list, Wegener began to search other scientific papers looking for additional evidence that the opposite sides of the Atlantic had once been connected.

Wegener discovered that the sequence of rocks in the Appalachian Mountains of North America were similar to the sequence of rocks in the Scottish Highlands in Scotland and in the Caledonian Mountains in Scandinavia. He also realized that the great coal deposits in the eastern United States were the same age—from the Carboniferous Period—as those in northern South America and western and central Europe.

He wrote a book, *Die Entstehung der Kontinente und Ozeane (The Origin of the Continents and the Oceans)*, which was published in 1915. In it he dismissed Suess's idea of sunken land bridges as an explanation because such a land bridge would have included all of the Atlantic Ocean. Instead, he proposed another outlandish idea: All of the continents had once been connected and, since then, they had moved apart slowly, plowing their way through the ocean basins in a manner similar to how an icebreaker plows through sheets of arctic ice.

His idea about the continents drifting was dismissed immediately. Critics argued that the plowing of the continents through the ocean basins was physically impossible—which it is. More importantly, geologists simply refused to believe that Wegener had provided proof that the continents had been connected. Moreover, he and his idea were ridiculed. "A fairy tale," pronounced geologist Bailey Willis of Stanford University of the idea of continental drift. He accused Wegener of having "delirious ravings." He suggested that the Marburg tutor was in possession of a "mystical German mind" that dealt with things "not apparent to the senses nor obvious to the intelligence." Mostly, though, other geologists simply ignored Wegener's idea.

In 1930, on his fourth expedition to Greenland, Wegener died in a blizzard while attempting to deliver food to two colleagues working at a remote outpost. Would-be rescuers later found his body and buried it in the ice. Thirty more years would pass before his idea that the continents had shifted their positions on the planet by great distances would be widely accepted.

The story of the development of the theory of plate tectonics in the 1960s and its acceptance by the scientific community by the 1970s is one filled with many crucial insights made by scores of scientists working at dozens of institutions in many different countries. One of the key moments came in 1962 when Harry Hess, head of the Department of Geology at Princeton University, published what he called "an essay in geopoetry"—apparently he was prodding his colleagues to think imaginatively—in which he proposed that the floor of the Atlantic Ocean was spreading open from a central range of undersea mountains. Those mountains are now known as the Mid-Ocean Ridge. Furthermore, according to Hess, as the seafloor was spreading open along the Mid-Ocean Ridge, it was also working like a conveyor belt and carrying the continents along.

In 1965, John Tuzo Wilson at the University of Toronto expanded on Hess's proposal and suggested that the outer shell of the Earth was divided into rigid plates that were in constant motion, sliding horizontally over the Earth's surface. It was the pulling or pushing or sliding along the boundaries between plates that was responsible for most of the earthquake activity, where most of the active volcanoes were located, and where mountains were now rising.

And in 1968 Jason Morgan, also at Princeton University, gave a general account of how the entire system worked. New oceanic crust was forming along the Mid-Ocean Ridge where the seafloor was spreading apart. Because new crust was forming, old crust was being destroyed along the great trenches on the floor of the oceans where oceanic crust was sliding back into the Earth. Where two plates were simply sliding against each other, major fault systems formed, such as San Andreas.

This system of spreading and destroying and sliding of the Earth's crust could explain the observations of continental drift compiled by Wegener. South America and Africa were once connected and they did move away from each other. Furthermore, it could be estimated how fast the two continents were moving.

New oceanic crust forms because hot material is rising from the mantle along the Mid-Ocean Ridge. As the crust forms, it cools. And it cools in such a way that the magnetic minerals contained within it align themselves in the direction of the Earth's magnetic field. Because the magnetic field occasionally flips—the North Magnetic Pole becomes the South Magnetic

Pole and vice versa—over the millions of years that the floor of the Atlantic has been forming, a series of long magnetic strips have formed on the floor of the ocean. Knowing when reversals in the magnetic field had happened and knowing the distance of each strip from the Mid-Ocean Ridge, it is easy to determine how fast the Atlantic Ocean has been spreading open. It is on the order of an inch or so each year. And that is also true for the motions of the other plates.[1] But how to measure such slow rates?

The first measurement of slow-plate motion was done in the mid-1980s. It made use of radio telescopes that could receive radio waves from distant objects in space. The telescopes used those distant objects as reference points that could show how much the ground beneath each radio telescope had moved. It was a tedious process and required years to accumulate enough data to do the calculations. Finally, in 1986, Irwin Shapiro of the Harvard-Smithsonian Center for Astrophysics announced the results. Over the previous several years a radio telescope in Massachusetts had moved steadily away from a radio telescope in Sweden at the average rate of about an inch each year.[2]

Such measurements of plate motion are now routine and make use of Earth-orbiting satellite systems, such as the Global Positioning System (GPS). The motion of tectonic plates is now undeniable. The Atlantic Ocean is getting wider. And the tectonic plates—and the continents that are attached to them—do move.

◆

But when did plate tectonics begin on the planet? That question is best answered by looking for key evidence in the geologic past that could only have been left by the movement of tectonic plates.

One of the features of the theory of plate tectonics is that a massive amount of oceanic crust is pulled down and slides into the Earth, a process known as subduction. This process of subduction is remarkably efficient. It is estimated that about 99 percent of all of the oceanic crust that has ever formed has slid back into the Earth. But there is a small amount that did not. Instead, it was scraped off and pushed up onto the edge of continents and is readily recognized because it forms a distinct suite of rocks, ones that often have the texture of snakeskin. This suite of rocks is known as an *ophiolite*, coined from the Greek words for snake and rock.

An ophiolite is a section of oceanic crust. At the bottom is mantle material, *peridotite*. In the middle is an igneous rock that never reached the surface and so cooled slowly, and that has the composition of basalt, *gabbro*. And at the top in the form of lava flows and other volcanic features is *basalt*.

This trinity—peridotite-gabbro-basalt—is what one looks for when searching for evidence of past plate collisions and subduction. The greenish-gray rock that forms the core of Staten Island is part of an ophiolite.[3] One of the most complete ophiolites exposed anywhere in the world is found at Gros Morne National Park along the western coast of Newfoundland. Another one that is readily accessible and that shows the full suite of rocks is at Point Sal on the California coast north of Santa Barbara. All three of these—Staten Island, Gros Morne National Park, and Point Sal—have ages of a few hundred million years. A considerably older ophiolite is exposed near Payson, Arizona, about ninety miles northeast of Phoenix, and it, too, displays all three rock layers. And it has an age of 1.73 billion years, which means that plate tectonics was certainly in full operation by that time in Earth history.

The oldest ophiolite in North America is among a group of rocks known as the Purtuniq Complex along the northern coastline of Quebec. Its age is two billion years, making it one of the oldest ophiolites on the planet. But the oldest one yet found is in China, the Zunhua-Wutasisha Ophiolite, which runs for a few hundred miles and is located east of Beijing. It formed 2.55 billion years ago, which is a notable age because the oldest known ophiolite would have formed just before the Great Oxygenation Event and the Huronian Glaciation. And that may not be a coincidence.[4]

Major evolutionary changes in biology occur when there are dramatic changes in the environment. And the beginning of plate tectonics would certainly have been one of those dramatic changes. Crustal material was being cycled into the Earth in a regular manner by subduction. And because subduction was occurring, the nature of volcanism changed with more explosive eruptions. And that would have changed what volcanic gases were being emitted, which would have changed the chemical composition of the atmosphere. Which, in turn, changed the chemistry of the oceans.

But there is a broader view that must be taken. Before plate tectonics became a planet-wide phenomenon at around 2.5 billion years, there was an even more profound event on the planet. And evidence for it has been found inside diamonds.

◆

Peer inside a diamond. A well-cut stone displays a quality known as brilliance, the different facets reflecting light in a myriad of interesting ways. But brilliance is not the only factor that determines the value of a diamond. Others include color and size. And then there is clarity, or as those who deal in the sale and purchase of diamonds know it, the degree of flawlessness.

There are two types of flaws inside diamonds. One is due to irregularities in the arrangement of atoms, which gives rise to visual features known as feathers. The other is the presence of what appears as tiny light or dark spots known as inclusions. Both can break the heart of a diamond miner because such imperfections greatly lower the value of a diamond. But, for geologists, inclusions are an unexpected gift. They are the ultimate time capsule because they hold material that is the most pristine sample of the Earth's interior as it was long ago.

Diamond is a mineral form of carbon. So is graphite. They differ in how the carbon atoms are arranged. In graphite, the atoms are arranged in sheets that can slide over each other, which is why graphite has a slippery feel. In diamonds, the atoms are arranged in a structure reminiscent of the cross bracing one sees in the design of many bridges. That is what gives a diamond its exceptional hardness.

Graphite is the stable form of carbon at room temperature. A diamond is only stable at temperatures between about 1,600° and 2,400° Fahrenheit (about 900° and 1,300° Celsius) and at pressures that correspond to depths greater than about 90 miles (about 150 kilometers). And so it is only within the temperature range and at those depths, or pressures, that diamonds can form.

There are two more important points to make about diamonds. Based on radiogenic dating, most diamonds formed between 3.5 and 1.0 billion years ago, which makes them the oldest pieces of the Earth that most people will ever see or ever handle. Also, when a diamond does form, small amounts of the surrounding mantle material gets inside the diamond. These are the inclusions.

And so trapped inside these perfect containers are tiny bits of the Earth's mantle as it was billions of years ago. The study of thousands of diamonds has shown that these tiny bits are of two types. Most are peridotite, which is the rock that comprises most of the upper mantle. The others are a form of basalt, *eclogite*, which is a volcanic rock that has been cycled back into the Earth and compressed by the high pressure. Peridotitic diamonds have the full range of ages from 3.5 to 1.0 billion years. Eclogitic diamonds have been produced inside the Earth only since about 3.0 billion years and most since 2.7 billion years, signaling that a major change was beginning to happen inside the Earth hundreds of millions of years before the onset of global plate tectonics, hundreds of millions of years before the formation of the oldest ophiolite. And that major change, because it involved large volumes of the Earth's interior and because details are still being worked out, is known, quite informally, as the Super Event.

But how might the Earth have worked before plate tectonics and the so-called Super Event? That would depend on how the planet was losing its internal heat. Such heat originated from the Earth's formation and from the continued release of energy by the decay of radioactive elements, mainly uranium, thorium, and potassium. Because the Moon was much closer there was also a substantial contribution from lunar tides. And so, in the distant past, the Earth's interior was much hotter than it is today. And that would have required the heat to be transferred from the interior to the surface in a different manner. Today, and for the last 2.5 billion years, the main transport of heat has been through plate tectonics; that is, through the *horizontal* movement of the surface and the interior. Before three billion years ago, the heat inside a much hotter Earth would have been transported primarily by *vertical* movements of the interior, by a global dynamic known as lid tectonics.

In lid tectonics, the outer part of the Earth consists of a single stagnant plate, a plate that, because it is still hot and pliable, is too weak for subduction to happen. Instead, hot mantle material rises up and, in places, punches through the stagnant lid and transports molten material—and, hence, heat—from deep in the mantle up to the surface. The continued eruption of this molten material causes a large platform to form. As more and more molten material is erupted, the platform gets heavier and heavier and sinks downward, eventually collapsing back into the mantle.

But that cannot be the full story. Neither the 4.030-billion-year-old Acasta Gneiss nor the 3.542-billion-year-old Morton Gneiss in Minnesota can be made in this manner. And so much is yet to be learned about how the early Earth worked and how the earliest rocks were produced.

But that something happened on the planet between the regime of lid tectonics—or whatever type of tectonics dominated the early Earth—and that of plate tectonics is undeniable.[5] The inclusions in diamonds tell us that something major happened. So does the volume of continental material that exists today. During the period between 2.78 and 2.63 billion years ago—a mere 150 million years—more than half of all of the continental material that exists today was created.

But how did it happen? And what was the Super Event? There is much speculation. It has been described as a "catastrophic overturn" or a "flushing" of the mantle. It certainly involved large volumes of water, presumably from the oceans, that mixed with the mantle to produce granitelike molten rock that then solidified into much of the continental crust that we see today.

The immediate result of the event was the creation of scores of small continental masses scattered across the planet. As the first tectonic plates began to form, they moved these small masses around, causing many of them to assemble into a dozen or so larger masses. These larger masses are known as *cratons* because they are the stable cores of all of the continents, made of highly deformed, mainly metamorphic rock that is often covered by layers of younger sediments.

◆

Eight cratons came together to form what is the core of North America. The largest is the Superior Craton. Today it underlies much of Ontario and Quebec, a substantial part of Manitoba, the eastern halves of both Dakotas, most of Minnesota and northern Wisconsin, and forms essentially the entire Upper Peninsula of Michigan. Another was the Slave Craton, which lies in the far north and includes the area around Great Slave Lake. Beneath several Canadian provinces—Northwest Territories, Nunavut,

Manitoba, Saskatchewan, and Alberta—are two more cratons, Rae and Hearne, named for two arctic explorers. The Medicine Hat Block lies beneath Montana and the Wyoming Province lies beneath Colorado and much of Wyoming. The smallest of the eight is the Sask Craton, which now lies between the Hearne and Superior cratons. And the eighth is the Nain Craton, also known as the North Atlantic Craton because, unlike the other seven, it has fragmented since being added to the others. Pieces of the Nain Craton are found in Newfoundland and Labrador, Greenland, and across the Atlantic Ocean in Scotland.

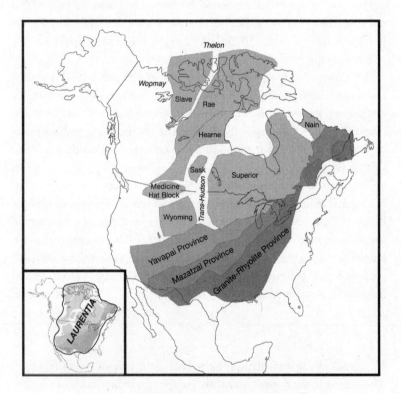

The eight cratons and three geologic provinces that make
up an early North America called Laurentia.

These eight cratons began to assemble two billion years ago and the assembling took three hundred million years. The same was happening elsewhere on the planet, small numbers of cratons assembling to form the cores of other continents; notably, at least four cratons came together to form Australia and at least three to form what is now northern Europe.

As the cratons were moved around and assembled into the various continental cores, they collided and became welded to each other. And, from that colliding and welding, the first massive mountain ranges were formed.

That is not to say that mountains did not exist before then. Mountain ranges had certainly been created when huge meteors had struck the Earth early in its history and thrust up large sections of the crust to form rings of mountain ranges, similar to those seen on the ancient surface of the Moon today. And there were probably massive volcanoes that formed at the top of the heat pipes that fed molten rock to the surface during the period when lid tectonics dominated the planet. But such early mountains were probably not towering peaks. The Earth was much hotter then—and the crust and the mantle much more pliable—and did not have the strength to support the weight of a great mountain range.

And so the Proterozoic Eon is probably when the first great mountain ranges formed. And they formed when cratons were still growing from smaller masses and when the cratons themselves were colliding and assembling into continental cores.

The first episode of mountain building recognized on North America is the Algoman Orogeny,[6] named for the small town of Algoma in Wisconsin that lies about midway along the trace of where this very old mountain range once stood. Most of the Superior Craton had already been assembled, but there was a small continental mass, known as the Minnesota River Valley Terrane,[7] that rammed into the Superior Craton. The collision forced the Earth's surface to fold and buckle and pushed up a mountain mass that once stretched from the middle of South Dakota to the Upper Peninsula of Michigan and into the Sudbury region of Ontario.

The lofty peaks of the Algoman Mountains, among the first major ones on the planet, eroded away long ago. Evidence that they ever existed is found in the rocks that were produced by the intense heat and pressure generated by the collision. A belt of schist and granite that runs through Voyageurs National Park in Minnesota is a product to the Algoman Orogeny. The ages of zircons recovered from the belt show that the collision began about 2.7 billion years ago and lasted about fifty million years.

There were, of course, other collisions, especially between cratons, and that built other early mountain ranges. The Slave and the Rae cratons collided 1.9 billion years ago and produced the Thelon Orogeny in what is now northern Canada. The oddly named Wopmay Orogeny, named for a Canadian flying ace of the First World War, Wilfrid Reid "Wop" May, also involved the Slave Craton. The largest of these collisions and early building of massive mountains involved the Hearne and the Wyoming cratons colliding with the much larger Superior Craton. The resulting mountains were comparable to the Himalayas of today. And though evidence of the collision and mountain building—this episode is known as the Trans-Hudson Orogeny—are a challenge to find because so much of the evidence has been overprinted by later events, there is a notable relic.

It is found as the granite in the Black Hills of South Dakota. The pressure and heat of the Trans-Hudson Orogeny melted some of the mantle, and that molten rock rose up into the crust, where it slowly solidified. It remained buried for almost two billion years. Then a more recent orogeny, one that would lift a major section of western North America, pushed up a block of the once-molten rock, now a granite, causing the overlying layers of rocks to dome upward. Then erosion stripped away the overlying layers exposing a structure that looked like a giant blister. And at the center of the blister is the granite, the light gray, erosion-resistant rock that the four faces at Mount Rushmore are carved into.

At the time of the Trans-Hudson Orogeny there was another surge in the production of continental material. Though not as large as the Super Event of the Archean Eon, this second surge did produce a significant amount of the continental crust that exists today. This surge in production is important in the history of North America because four great masses were produced, each one primarily volcanic in origin and each one, in succession, was to collide with and weld to the cratonic core.

The first of the four was the smallest. It produced a range of peaks that are notable because deeply eroded remnants of them can be seen today. They exist as two ridges, no more than a few hundred feet high, that run through northern Wisconsin and across the Upper Peninsula of Michigan. These hills are the Penokee Range. And they give to the few residents who live in this sparsely populated region the right to brag and point to those hills and say that they live near one of the world's oldest mountain ranges.

The Penokean Orogeny happened about 1.8 billion years ago. The mountains it produced ran at least from New York to South Dakota and, perhaps, as far as Arizona. And though what is left of those mountains is not impressive topographically, the amount of erosional debris is.

Weathering and erosion broke down most of the original minerals, leaving only highly durable quartz, which, as grains of sand, was moved by rivers and spread across a broad coastal plain. The grains became tarnished with a film of iron oxide that colored the normally white quartz to various shades of pink and red. The grains accumulated into thick deposits, which compressed into a hard quartz-rich rock known as the Sioux Quartzite. It is this hard rock—this debris from the original Penokee Mountains—that forms the series of waterfalls at Sioux Falls in South Dakota. This was a place, so the Sioux Quartzite reminds us, that was once close to a coastal shore of an early North America.

The other three landmasses came from the same direction and impacted against the edge of an early North America in the same way. The second one was the Yavapai Province. Soon after its arrival—here "soon" means in less than a hundred million years—a third landmass, the Mazatzal Province, arrived. Both are named for native people of the American Southwest who live in an area where rocks of these two provinces are well exposed. The rocks of the Yavapai run from northern Arizona to Wisconsin. Those of the Mazatzal run from southern Arizona at least as far as Michigan and may include rocks found as far north as Newfoundland and Labrador.

The fourth and largest of the landmasses to be added was the Granite-Rhyolite Province, the name betraying its volcanic origin. Rocks of this province run through northern Texas and comprise most of the basement rock beneath Louisiana, Mississippi, Arkansas, Tennessee, Kentucky, Indiana, and Ohio. They are also found in southeastern Ontario.

Collectively, these four large landmasses added an area to the continent that was up to a thousand miles wide and nearly three thousand miles long. More than two-thirds of what is now North America had been assembled. But the continent was still so different from the one known today that a different name has been given to it.

It was Austrian geologist Eduard Suess—the man who had proposed that land bridges had once connected South America and Africa—who proposed the name. It was 1909, and he had just published the third volume of what would become a four-volume work entitled *Das Antlitz der Erde*

(*The Face of the Earth*). It was in volume three that he noted that the oldest parts of North America were the broad region around Hudson Bay, the vast interior and northern sections of Canada, and the central part of the United States—what we know as the eight cratons and the four large landmasses that were added later. Suess also noted that the oldest rocks on the continent that were easily accessible—at least, accessible to most geologists of the early twentieth century—were in the mountains north of the St. Lawrence River near Montreal. This is the Laurentide region of Quebec. It is from that name that he derived a name for ancient North America: Laurentia.

◆

Laurentia was not the only ancient continent named by Suess. He identified four others. There was Fennoscandia, which comprised what is now the Scandinavian Peninsula and Finland. Today that has been expanded to include most of northern Europe and the ancient continent known as Baltica. And there was Angaraland, which is modern-day Siberia. Enough was then known about the rocks of Antarctica that Suess proposed that this mostly ice-covered region represented yet another ancient continent. And then there was Gondwana, which had since broken apart, but had been composed of South America, the southern half of Africa, and of India and Australia. He chose the name because the fernlike fossil *Glossopteris* had been found in all four places, but it was especially abundant in the Gondwana region of central India.

Unfortunately, Suess could not shake himself loose of the idea that the continents were fixed and had been connected by land bridges. It was up to Alfred Wegener, working a generation later, to try to release geology from the idea of stationary continents. In that, as already recounted, he failed.

But his mark remains today in one of the names he proposed. He suggested that Seuss's ancient continents of Laurentia, Fennoscandia, Angaraland, and Antarctica had once abutted against massive Gondwana and that the four had since drifted away. Wegener gave a name to this former grand assemblage. He called it Pangea, literally "all of the land."

Then, decades later, as the theory of plate tectonics was being developed, it was realized that some of the ocean basins, such as the Atlantic Ocean, must have gone through cycles of opening and closing. Thus, it was concluded that the continents must have also gone through cycles of dispersing, shifting around, and reuniting. A superlative name was given to those single grand all-inclusive landmasses that must have assembled occasionally. They were called supercontinents.

That supercontinents must have existed before Pangea was an idea that was accepted immediately. Also several smaller landmasses that were *almost* supercontinents. Gondwana was one example, composed of all of the major continental masses except Laurentia and Fennoscandia. There was Pannotia, or Greater Gondwana, which had included almost all of the continental land-masses immediately before Pangea formed. And Laurasia, which had formed soon after Pangea broke up and had been composed of *almost* all continental masses. It was even proposed that there had been a supercontinent during the Archaean, smaller than Australia is today, known as Ur. And it, presumably, had been broken up by the Super Event and from its fragments many of the cratons formed, making these cores of all modern continents the children of Ur. But, if a supercontinent must include all major landmasses, then only three supercontinents have existed in the Earth's history.

Pangea was the most recent. The earliest was Columbia. Evidence for it was first established after it was discovered that similar rocks exist along the eastern edge of the Columbia Basin in Washington State and along the eastern coast of India.[8] The intermediate supercontinent is Rodinia, the name derived from the Russian infinitive *rodit* that means "to beget" or "to grow." A Russian word was chosen to recognize the major contribution Russian geologists have made to the study of this second supercontinent. The infinitive *rodit* was used because the breakup of Rodinia was followed immediately—the breakup of the supercontinent begat—by the appearance of the first multicellular animals on the planet.

But when did these three supercontinents form? The answer, as expected, comes from zircons.

Zircons form when molten rock cools and solidifies. The more molten rock is produced, the more zircons are produced. And more molten rock is produced when a lot of continental crust forms—as during the Super Event—or when continental masses collide and mountains rise. The last is the key to timing the creation of the supercontinents. When there was more mountain-building activity, more molten rock was produced and, hence, more zircons would have formed.

Radiogenic dates have been determined for more than a hundred thousand individual zircons. And, when they are lined up according to age, there are five distinct peaks in zircon ages. The oldest is at 2.7 billion years, the Super Event. The next is at about 1.8 billion years, the formation of Columbia. Then, 1.1 billion years ago, when Rodinia formed. A peak at 0.6 billion years is the formation of Gondwana and one at 0.3 billion years is the supercontinent Pangea.

The arrangement of the various continental masses that came together to form Pangea is well established. What is today North America, South America, and Africa were nestled together. Antarctica, Australia, and India were along what is now the eastern coast of Africa. And northern Europe and most of Asia were connected to the other masses through the northern part of North America and Greenland.

The exact arrangements of the continental masses when Rodinia and Columbia existed is still problematic. Each year more evidence is uncovered and the masses get nudged around. But two things are clear. Laurentia was at the center of both Rodinia and Columbia, surrounded by the other continental masses. And, as the assembly of Rodinia was almost completed, a continental mass known as Amazonia, the heart of modern Brazil, crashed into Laurentia, causing one of the longest mountain ranges in the history of the planet to form.

Known as the Grenville Orogeny, named for the quaint municipality of Grenville in southern Quebec, where rocks crushed and folded by this event are readily seen, this episode of mountain building began 1.2 billion years ago and lasted about three hundred million years. Its effects are found from Labrador to Texas and continue into Mexico. Among the many rocks that formed at that time and have since been lifted and exposed by erosion are the Long Range Mountains of northwestern Newfoundland and Labrador, the Adirondacks of New York, Old Rag Mountain and Hogback Mountain of the Blue Ridge Province in Virginia, the Nashville Dome in central Tennessee, the Ozark Mountains of Arkansas and Missouri, and the Llano Uplift of central Texas. Of the last, this includes the great dome called Enchanted Rock near San Antonio, rock that was formed and solidified from a molten state during the Grenville Orogeny. It also happens to be the oldest rock in Texas, 1.082 billion years old.

And, thus, the last major piece of Rodinia was now in place. But even as Amazonia was pushing against Laurentia and the mountains of the

Grenville Orogeny were forming, Rodinia was already starting to break apart. And evidence for this can be seen as a deep geologic scar that runs across what is now the center of North America.

◆

The five great lakes of North America—Superior, Michigan, Huron, Ontario, and Erie—are contained within basins that were formed, in the main, some eighteen thousand years ago when a mile-thick sheet of ice slowly inched its way southward and scooped out the land. But at Lake Superior there is an additional, much older element that contributed to the outline of that basin, one that reaches back 1.1 billion years to the time of the Grenville Orogeny.

The outline of Lake Superior is distinct from the other four lakes in two ways. First, the western half of Lake Superior is significantly deeper than the other lakes. And, second, the western half of Lake Superior narrows to almost a point at the western end of the lake at the port city of Duluth, Minnesota. Both the greater deepness and the narrowing are due to a giant rift that formed in the Earth's surface, a rift that extends far beyond Lake Superior and is known, formally, as the Midcontinent Rift.

The Midcontinent Rift runs along an arc from Oklahoma to Minnesota and Michigan then to Alabama. Lava flows that fill the rift (shown as gray) are exposed only in the northern half, readily seen at several waterfalls and on Isle Royale. Farther south the flows are covered by later sediments. The location of the Morton Gneiss in Morton, Minnesota, is indicated.

The Midcontinent Rift extends outward from Lake Superior as two major arms. One arm runs southwest and passes through Minnesota and Iowa and into Kansas and probably reaches as far south as Oklahoma. The other runs southeast through Wisconsin and Michigan, through Ohio, and probably extends to Alabama. Surface evidence of the rift can be found only in Minnesota, Wisconsin, and Michigan. Elsewhere it is buried by sediments. But that it does exist and that the trends can be followed is due to the huge amounts of lava that were erupted when the rift formed and that now fill it. The lava is both more susceptible to magnetism and denser than the surrounding rock, and so the rift, though most of it is buried, can be followed by mapping how the gravity and the magnetic field vary across the region.

The thick piles of dark lava erupted from the Midcontinent Rift can be found in many places. At Taylor Falls near the border between Minnesota and Wisconsin, the St. Croix River cascades down a series of steps. Each step is an individual lava flow. At Big Manitou Falls in Wisconsin water of the Black River drops more than a hundred feet down a sheer cliff of ancient lava flows, making Manitou Falls the highest waterfall in Wisconsin and the fourth highest east of the Rockies. The picturesque views at Palisade Head and Shovel Point in Minnesota and on Isle Royale in Michigan are due to these flows. The Keweenaw Peninsula of the Upper Peninsula of Michigan is a wide arm of the same volcanic rock, here nearly three miles thick and where hundreds of individual flows are exposed.

The opening of the Midcontinent Rift lasted about twenty million years. It occurred during a lull in the collision between Amazonia and Laurentia, during a brief respite in mountain building during the Grenville Orogeny. Then the movement resumed, the mountains continued to grow, and the Midcontinent Rift was prevented from growing and severing Laurentia into two halves.

The Grenville Orogeny ended about a billion years ago. Rodinia was complete. Then after the passage of another hundred million years or so, activity internal to the planet began to break Rodinia apart. The various continental masses reversed their motions and moved to new positions, each one creeping outward and away from Rodinia's central core, away from Laurentia.

◆

It was during the billion years between the formation of the first supercontinent, Columbia, and the breakup of the second, Rodinia, that two-thirds of what is now North America was assembled. That, in itself, makes this an important time in the history of the continent. But the same billion years is also notable for a lack of biological change.

Columbia never truly broke apart and never fully dispersed. Instead, the various continental masses shifted around and re-amalgamated into Rodinia. And that meant the climate remained stable. And so did the chemistry of the oceans.

Without environmental change, there was no pressure for evolutionary change. The life-forms that dominated the planet when Columbia formed were almost the same ones that dominated the planet when Rodinia broke up. Those scientists who search for and study fossils have found a remarkable sameness for this period, so much so that one has remarked, in a clever adaptation of Churchill's famous phrase: "Never in the course of Earth's history did so little happen to so much for so long." The prolonged period of biological stability has been given an informal name: the Boring Billion.

The wide dispersal of continental masses after the breakup of Rodinia did change the climate. And from that, the chemistry of the oceans did change. New life-forms evolved, some radically different. The first plants and the first animals would appear. Their existence would, in turn, modify both the climate and the oceans, so that the geological and biological histories of the planet would become more tightly interwoven.

In short, as the name implies, Rodinia did beget a very different Earth.

4

Gardens of Ediacaran

Late Proterozoic Eon: 720 to 541 Million Years Ago
Cryogenian and Ediacaran Periods

A fossil is any preserved impression or remains of a once-living thing from a very long time ago. The most familiar ones are those that were once bones—mineralization has changed them into rock—such as those of mastodons or dinosaurs or many of the other odd and curious creatures of the past that grace the halls of many museums. There are fossilized shells, from both the land and the sea. Imprints left in rocks by leaves or by the footprints left by dinosaurs or other animals are also a type of fossil. So are the burrows made by animals that were living underground long ago, hiding from predators or digging and searching for food. But, technically, stromatolites are *not* fossils. They comprise microscopic life that were trapped and bound together by a gelatinous ooze—which, in a strict sense, makes them *organo-sedimentary structures*—but to those who buy or sell fossils, they list them with the rest.

There is often an effort to find and admire the oldest of something. The oldest dinosaur bone is 230 million years old and was found in Argentina. The oldest shells are tiny ones that have been found in thick sections of limestone and gray shale on Mount Slipper along the Yukon-Alaska border. The oldest among them formed 810 million years ago.

But what are the oldest organism remains large enough to be readily viewed and held in one's hand? That distinction belongs to a coiled form of marine life known as *Grypania*. The oldest such fossilized relics have been recovered from an iron mine near Ishpeming, Michigan, and have an age of 2.1 billion years. The coils are about the size of a penny. If the larger ones were unwound, they would stretch for about two inches. Exactly what type of life *Grypania* represents is still debated—it is unclear if it's a colony of bacteria or of algae—which means that no one has been able to relate them to anything living today, and so they may not have any living relatives.

Grypania spiralis, the oldest known macroscopic fossil (2.1 billion years old), is from the Negaunee Iron Formation of Michigan's Upper Peninsula. Coils are about an inch in length. (Photo courtesy of James St. John)

That such an organism existed two billion years ago means that life was starting to become more complex than a simple assemblage of loose cells.[1] But the further development of life stopped when the supercontinent Columbia began to form. There was a great pause in the evolution of life when no major new forms of life—no life-forms more complex, at least in general appearance, than *Grypania*—would appear until after the breakup of Rodinia. Then several bursts of biological innovation would occur, including the appearance of the first "hard structures"—that is, the tiny shells of Mount Slipper. This great pause in the evolution of life—and, in general, in the evolution of the Earth, with no major changes in climate or in ocean chemistry—is known as the Boring Billion.

But what caused the Boring Billion to end?

◆

Supercontinents have extremely arid climates because most of the land is far from the sea. Water-filled storms can move only partway across the land before the water is dropped as rain, leaving the remainder of the land dry. It is only after a supercontinent has broken up and the individual landmasses have dispersed that a wet climate returns.

And when a wet climate does return, the rain, slightly acidic from the carbon dioxide in the atmosphere, falls and dissolves the rock minerals. The dissolved minerals are swept to the sea, and, because they are usually rich in phosphates and nitrates, they stimulate rapid microbial growth.[2] Then, as microbial growth increases, the amount of carbon dioxide pulled out of the atmosphere increases. And less carbon dioxide means a cooling of the planet.

During the breakup of Rodinia, the change from a single large landmass to many individual landmasses was so dramatic that the cooling of the Earth became extreme. Huge fields of snow and ice began to form at each pole. The snow and ice developed into ice sheets more than a mile thick. The sheets expanded, moving toward the equator from both poles. A point was reached when so much sunlight was being reflected by the ice sheets that little of the Sun's heat was being absorbed by the Earth. The planet went into a deep freeze. It was a second Snowball Earth.[3]

W. Brian Harland of the University of Cambridge was the first to suggest that the Earth had once been encased in ice. He made his suggestion in 1964 and based it on the fact that he and others had found glacial deposits with ages of about seven hundred million years on almost every continent. Furthermore, Harland's work showed that some of those glacial deposits had formed on land that was near the equator, indicating that sheets of ice must have reached from both poles to the equator. But there was a problem with his suggestion. If the planet had once been completely encased in ice, then it would have been highly reflective. And it would have reflected essentially all of the Sun's light. So what had made the ice melt and changed the planet into the nearly ice-free condition it is in today?

There was much ruminating about Harland's work. Most geologists discounted his suggestion, unable to overcome such a strong objection. Then in 1992 Joseph Kirschvink at the California Institute of Technology offered a solution.

He realized that, though the planet was covered entirely by ice, the internal part of the planet was still working. The mantle was still churning, the tectonic plates were still moving, and the volcanoes were still erupting. The volcanoes were pumping more and more carbon dioxide into the atmosphere. Furthermore, because the planet was encased in ice, sunlight was not penetrating through the ice and reaching where marine life still existed—probably concentrated around a few local hot springs—and so carbon dioxide was not being pulled out of the atmosphere by photosynthesis. Instead, it was steadily increasing.

Paul Hoffman of Harvard University took the ideas of Harland and Kirschvink a step further. He had been studying glacial deposits in Namibia that were the same age as those studied by Harland elsewhere. He noticed that the glacial deposits in Namibia were always overlain by a thick series of carbonate rocks that, from their chemistry and their structures, must have formed in warm seas. How could the Earth's climate have changed so fast from a deep freeze to hot conditions? Then Hoffman realized that was exactly what would have happened if Kirschvink was correct.

Because it enclosed the entire planet, a global ice sheet would not begin to melt until there was a significant rise in temperature. And there would not be a significant rise until a large amount of carbon dioxide had re-accumulated in the atmosphere. Then, once the melting had started, it would be catastrophic.

The ice would retreat quickly. As it did so, more and more land and ocean would be revealed. Because a land or an ocean surface is much less reflective than ice, more and more sunlight would be absorbed by the planet, which would make the temperature rise even faster. That and the huge amount of carbon dioxide in the atmosphere would heat the Earth to an extreme hothouse condition. The amount of carbon dioxide dissolved in the oceans would also increase dramatically. And that would result in the rapid precipitation of carbonates in warm, shallow seas.

This close association of ancient glacial deposits capped by carbonate rocks has now been found throughout the world. In North America it can be found at Oxford Peak south of Pocatello, Idaho, at Mineral Fork in the Wasatch Mountains of Utah, and at Kingston Peak south of Death Valley in California.[4]

This period of global glaciation—this second episode of a Snowball Earth—marks the end of the Boring Billion. It is a geologic period known as the Cryogenian, "the beginning of the cold." Eventually the climate did stabilize. An environment not yet seen on the planet became established. And radically new life-forms appeared.

◆

The port city of St. John's on the island of Newfoundland has two distinctions. First, it is the easternmost city in North America. Second, according to residents, St. John's is the oldest British settlement in the New World. Of the latter, there is some contention. Most historians assign that distinction to Jamestown in Virginia, which was established in 1607. But the residents of St. John's are quick to correct this. They maintain that English fishermen had set up camps at St. John's during the century before Jamestown, camps that were not recognized as official settlements by the British government until the 1630s.

The city has another, less well known distinction. It was in St. John's in 1872 that local attorney Elkanah Billings found what is regarded today as the oldest fossilized evidence for animals. Based on the general shape of the few specimens he found, and as a tribute to the island where he found them, he named them *Aspidella terranovica*, "the little shield from the New Land."

Aspidella terranovica is a trace fossil. It is an impression that has been left in the rock by a living organism, similar to the mark you can make by pressing your hand into mud or wet sand. At St. John's, and at several other locations along the eastern coast of Newfoundland, this earliest-of-all animals is found in black shale. The impressions themselves are small, appearing as ovals about the size of a dime with a narrow outer ring that surrounds a small shield-like dimple. Billings, who would later become the first paleontologist employed by the Canadian government, never doubted that *Aspidella terranovica* was an early form of life. But this contemporaries did. The problem was the age of the black shale.

The absolute age-dating of rocks was more than a generation into the future. But the *relative* ages of different rock layers had been well established. It was based of the laws of succession, outlined first by Danish naturalist Nicolaus Steno in the seventeenth century. These laws were based on sensible ideas, such as younger rocks sit atop older rocks. And any disturbance within a rock layer, such as a crack, had to have happened after that layer had formed. In the early nineteenth century British geologist William Smith added to these ideas with his law of fossil succession. Simply put: The history of life had proceeded through distinct phases, and each phase had left distinct fossils. In other words, if two rock layers had identical fossils, then those two layers had formed at the same time. Moreover, if two layers had different fossils, then, knowing how the succession of fossils changed with time, one could determine which of the two layers was older; thus it was possible to determine the *relative* ages of the layers.

Much work in geology in the early nineteenth century went into establishing how the succession of fossils had changed with time. From that, the geologic column was established, a listing of the various layers of rocks and fossils according to their relative ages. The column is usually displayed vertically with older events nearer the bottom and increasingly younger events higher up, so that rising through the geologic column is a visual climb up a rock face with distinct layers, where each higher layer is younger than the ones below.

Because much of the initial work in formulating a geologic column was made by British geologists, many of the names reflect places or people in the British Isles. The Cambrian Period is derived from *Cambria*, a variant of the Celtic name *Cumbria* for Wales. The Ordovician and the Silurian are named for ancient Welsh tribes. The Devonian is named for Devon County in England and the Carboniferous is an adaptation of "the Coal Measures," an old British term for the extensive coal deposits found in England and Wales.

Even as the first attempts were being made at construction of a geologic column that could be applied to the entire world, it was clear that there was something missing in the oldest rocks. They were completely devoid of any traces of former life. In comparison, rock layers of the Cambrian age or younger were often filled with fossils. The change was so stark—as already mentioned, Darwin commented on this and saw it as a challenge to his idea about natural selection—that rock layers older than the Cambrian Period were often referred to as primordial strata. They have also been called the Azoic, "a period of no life," or simply the Precambrian.

And that is how Earth history was envisioned when Billings made his discovery of *Aspidella*. Using Steno's laws of succession—today more properly called the laws of stratigraphy—it was clear that the black shale was Precambrian. And so no forms of life could exist within it.

This idea continued to dominate geology well into the twentieth century. In 1946, an Australian mining geologist name Reginald Sprigg found imprints in rocks that looked like casts from a jellyfish. He was working in the Ediacara Hills of South Australia, where he also found other types of imprints. Most were unlike any forms of life alive today. He made lists and created a catalog. In every case the rocks with these strange imprints were clearly of Precambrian age. He insisted that they must be evidence of Precambrian life, but few would agree with him. It was another decade before opinions would change and the world of geology would open itself and accept the evidence of Precambrian life.

It was the summer of 1965. Tina Negus, then fifteen years old and already showing a passion for nature, persuaded her parents to take their annual vacation to Charnwood Forest in northwestern England. It was here, hiking through the forest, that Negus spotted a fernlike impression on one of the craggy rocks. She told her geography teacher about her discovery. The teacher consulted several books and, finding the appropriate section, read that the rocks of Charnwood Forest were much too old—they were Precambrian in age—to contain evidence of past life. The teacher informed the student, insisting that Negus could not have seen a fossil imprint.

A year later Roger Mason, also fifteen years old, and two friends were climbing the walls of an abandoned quarry in the Charnwood Forest when one of the friends recognized a fernlike impression in the rock. Mason made a rubbing of the discovery and showed it to his father. The father, a minister, showed it to a geology professor at the local university. The professor and the father returned to the site with Mason and convinced themselves that the young man and his friends had, indeed, found a genuine fossil. The professor published a description of the fossil in a scientific journal, naming the discovery *Charnia masoni*, after the place where it was found and after the person who had brought it to his attention. Months later, at Adelaide University in Australia, another professor read the description and realized it was similar to some of the impressions Sprigg had found in the Ediacara Hills. The second professor went on to agree that *Charnia masoni* had been a living organism and to suggest that it might have a living relative in the sea pen, a marine animal that resembles a plump, old-fashioned quill pen with a bulbous structure at the base that is used by the organism to anchor itself to the seafloor.

Thus was the beginning of the slow acceptance of the idea that life had existed before the Cambrian Period, that the discoveries that had been made by Mason and Negus and by Billings and Sprigg did represent early forms of life. The importance of these early life-forms to the Earth's history was fully acknowledged in 2004 when members of the International Union of Geological Science, an august body of scientists who come together and decide such matters, added a new time interval to the geologic column. The Ediacaran Period would be the geologic period between the end of the second Snowball Earth and the beginning of a sudden burst and great diversification of life that would begin the Cambrian Period. The name comes from the Ediacara Hills, where many of the best fossils have been found. And the organisms themselves, whether their fossilized impressions are found in Australia or elsewhere, are known as *Ediacarans*.

◆

Ediacarans have been found at only a few dozen places in the world. Much searching is often required to recover even one good specimen. And much persistence and much patience are needed to discover something new. In 2014, a single specimen of a new species was found on a desolate hillside north of Death Valley. This single specimen, worm-like in appearance, measured a tenth of an inch long and a hundredth of an inch wide. Those who search for Ediacarans are rewarded by the unusualness of their discoveries, not by the size of what they find.

There are four sites that are particularly good in that they have an abundance of exposed specimens and that they cover a substantial area. One is the namesake in the Ediacara Hills of the Flinders Ranges in South Australia. Others are in Namibia and around the White Sea region of Russia. A truly remarkable site is close to where Billings found his first specimen along the eastern coast of the island of Newfoundland.

Billings made his initial discovery from outcrops of black sandstone and shale near the intersection of Prescott and Duckworth Streets in downtown St. John's. Today these streets are covered by brick buildings and asphalt, but there is a location a short distance to the north at the intersection of Duckworth and Holloway streets where black shale can be found. And several faint impressions of *Aspidella terranovica* can be seen.[5] But far south of the city, they can be found by the thousands.

From St. John's southward to Aquafort, a distance of about fifty miles, there are a dozen or so roadside exposures. The best one is at Ferryland, a quaint village of a few hundred people immediately north of Aquafort, where excellent specimens of *Aspidella* can be seen.

Continuing south and following a more adventurous coastal route, one passes Cape Race Lighthouse. It was here that a telegraph operator received the distress calls from the *Titanic* in April of 1912 and relayed them to other ships and to other land locations. Beyond the lighthouse, still holding close to the shoreline, in a few more miles one arrives at the southernmost point of this section of the island of Newfoundland. This is Mistaken Point, a windswept promontory barren of trees and grass where only moss and lichen survive. The name comes from the fatal errors made by sailors who mistook it for nearby Cape Race, a guidepost to a safe harbor at St. John's, and ended up shipwrecked on the rocky coast. Hundreds of shipwrecks have happened there.

Mistaken Point is also famous in geology. It is here along a strip of rugged, naturally eroded coastal cliffs that tens of thousands of fossil impressions have been found and cataloged. They range from the barely visible to those that are bigger than a truck tire. Radiogenic age-dating shows that these are the oldest Ediacarans. This was one of the spots on the Earth where "life first became big." And it happened 565 million years ago.

But what type of life-forms were the Ediacarans? The most common ones are shaped like spindles. Some took the form of long, feather-like fronds with vertical stalks that held them to a seafloor. Others sprawled across the ocean bottom. *Charnia masoni* is here. So is *Aspidella*.[6] Those at Mistaken Point lived at the beginning of the age of the Ediacarans. More complex forms evolved later.

Cloudina, which grew up to a few inches in length, looked like a series of paper cups stacked on top of each other. It is one of the most widely distributed of all of the Ediacarans. The first examples were found in Namibia. A few have been found in North America in British Columbia, Nevada, and California. *Swartpuntia* had one of the smallest distributions. It, too, was first found in Namibia, and examples have been uncovered in California and in North Carolina. *Swarpuntia* has features that make it seem related to the frond-like specimens, but in general appearance it looks like an upright weather vane with three vertical fins.

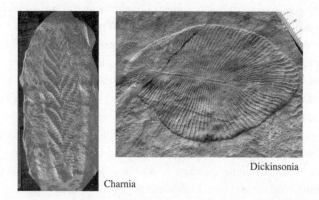

Dickinsonia

Charnia

Trace fossils of two Ediacarans: the feathery Charnia (photo courtesy of Smith609) and rubber raft–like Dickinsonia (photo courtesy of Verisimilus).

These strange-looking organisms, all soft-bodied, seem as if they came from another planet and, in a way, they did. The Earth was different hundreds of millions of years ago. The atmosphere had a much higher concentration of carbon dioxide. It was filled with more dust and the sky may have appeared red instead of blue. The ocean was sulfuric and more acidic. These were mostly sessile organisms, either attached to the seafloor or freely floating and moved about by ocean currents. The ones at Mistaken Point were in water that was too deep for sunlight to reach, and so they did not get energy by photosynthesis and were not plants. Instead, their flat body plans suggest that they absorbed nutrients directly from the surrounding water. That would explain why they seemed to be composed of networks of small tubes and of disks. And if there is a poster child for this strange assortment of life-forms, one that displays many of the oddities that seem unique to the Ediacaran Period, it would be the saucer-shaped, raised-ridged organism called *Dickinsonia*.

Some have likened it to a flattened segmented worm. Others see its general form as similar to the crown of a giant mushroom. It has been suggested that it represents an early jellyfish. It has the general appearance of a poorly constructed air mattress that is not fully inflated. It has an obvious midline with raised ridges radiating out from it. It almost has bilateral symmetry, but not quite. Its two ends are distinct, but which one might be a head and which might be a tail is impossible to determine.

It grew up to four feet in diameter. Hundreds of specimens have been found, mostly in South Australia and in the White Sea region of Russia. Only one possible example of *Dickinsonia* has been found in North America, a diminutive specimen smaller than a fingernail with what appears to be eight nearly equal-sized ribs attached to a midline. Whether this is an immature example of *Dickinsonia* or something entirely different has yet been determined.

Given that so many specimens have been found and examined closely, and that they come in so many different sizes, it is surprising that no progress has been made in determining how this organism grew. It might have grown outward from the midline. Or it might have grown from the outer ends of each of the raised ridges. It is also an open question whether it could propel itself or whether it was at the mercy of ocean currents and tides to be moved around. It certainly had no feature that might have anchored it to the seafloor. All of this raises an important question: Exactly what type of organism was *Dickinsonia*? Was it an animal?

From careful studies of their fossil imprints, Ediacarans seem to lack any indications of appendages. There are no examples of a mouth or a gut. There is nothing obvious that links them to animals. And that is true of *Dickinsonia*. However, a chemical known as cholesteroid—the ancient equivalent of cholesterol or, to put it bluntly, ancient fat—has been found associated with a specimen of *Dickinsonia* found in Russia. That discovery pushes the opinion that *Dickinsonia* was some type of animal because all animals today, except sponges and a few mollusks, produce such chemicals and no other life-forms do.

There is an animal characteristic that Ediacarans do lack, though. After the examination of thousands of specimens, the writing of many long and highly detailed descriptions, the classification of these varied and highly unusual organisms into more than a hundred different species, no one has yet uncovered any evidence of predation. No eye, antenna, bristle, tooth, claw, or jaw has been found. There is also no clear evidence of tracks or burrows that could have been made by one type of Ediacaran searching for other types, intent on devouring them, or made by those that might have been trying to escape or hide. Instead, there were great fields of Ediacarans such as *Charnia masoni*, *Cloudina*, and *Swartpuntia* that numbered in the thousands, perhaps in the hundreds of thousands, each one collecting nutrients from the surrounding sea and growing and reproducing without the threat of being consumed.

In a nod to the biblical Garden of Eden, this biological world of no predators and, hence, no prey has been called the Garden of Ediacaran. It lasted twenty million years. At its close there were organisms that seemed to have some of the features of animals—and of predators. An example is *Kimberella*, found in South Australia and Russia.[7] It both resembled modern-day slugs and seemed to leave marks, or perhaps tracks, similar to those made by slugs and snails today. But it was probably grazing off the microbial mats that still covered vast regions of the shallow seas. And so predation may not yet have begun.

But species do change. New ones do form and others do disappear. Understanding why and when they change has been one of the main pursuits of biology for more than a hundred years. And has been one of the main factors in the geological history of the Earth. For that reason, before continuing, it is necessary to go back and consider in more detail the history of life up to this point.

◆

Decades ago when I was sitting in an introductory biology class, I was told that there were two general types of cells: prokaryotes and eukaryotes. Today biologists know there are three.

The eukaryotes are the same as they were before. These include large single-celled microorganisms such as algae and amoebas and the cells of yeast and mold. All of the cells of our bodies are eukaryotic cells. So are the cells of every animal and every plant that has ever lived.

Every eukaryotic cells has a nucleus that contains the genetic material, molecules of DNA. All also have mitochondria. And most plant cells have chloroplasts.[8] And both mitochondria and chloroplasts contain strands of DNA that are distinct from the strands of DNA found in the nucleus.

A prokaryote, as I was told years ago, was every cell that was not a eukaryote. Prokaryotes included all of bacteria. But, fifty years ago, the ability of biologists to classify bacteria was limited by the methods available to study microscopic organisms. They could be examined under a microscope to reveal the general shapes or they could be grown in laboratories to see what substances they consumed. Then in 1977, Carl Woese of the University of Chicago changed everything when he developed a way to determine the sequence of long biological molecules.

Today the sequencing of DNA is routine. It reveals, among other things, how closely two different organisms are related. The more similar the DNA is between two different organisms, the closer is their evolutionary relationship. It has confirmed that human beings are more closely related to cats than to crabs. It has also revealed, somewhat surprisingly, that yeast and mold are more closely related to animals than to plants. But such detailed sequencing was not possible in the 1970s. And so Woese had to rely on a different method, one that was much cruder than today's DNA sequencing, but that was still based on the sequencing of long molecules.

He chose RNA, a molecule that is found in all cells and that is shorter than DNA, and so easier to sequence. He took RNA from both eukaryotes and prokaryotes and confirmed that these two types of cells have very different RNA sequences, that they are evolutionarily far apart. He also showed that there were two fundamentally different types of bacteria. And that these different types of bacteria were as different from each other as each one was from the eukaryotes.

From such distinct differences, he proposed that there were three fundamental types of cells, which he called the "three domains of life." There were eukaryotes and bacteria. And there was a third: archaea.[9]

As Woese originally envisioned the distinction, bacteria were the former prokaryotes that were "typical" bacteria, which included the oxygen-producing cyanobacteria and some of the microorganisms that cause deadly diseases, such as pneumonia and cholera. Archaea were "unusual" bacteria because the examples he analyzed had been collected in extreme environments, such as the hot springs of Yellowstone and the high salt concentration of the Great Salt Lake. Later work would show that archaea are quite common. They include the microorganisms that live in our intestines and, because of their particular type of metabolism, release methane that is responsible for the occasional burst of unacceptable behavior (aka, a burp).

How might bacteria and archaea be related? Did one descend from the other or did they arise separately? How long have they been distinct?

These and similar questions are still being pursued. But one thing is clear: It was the merging of the two domains, bacteria and archaea, that led to the eukaryotes.

Broadly speaking, most of what is contained within the cell wall of a eukaryote—the cytoplasm—originated as an archaea. And within the cytoplasm are chloroplasts and mitochondria that have their own DNA as well as their own cell walls. These chloroplasts and mitochondria were originally bacteria.

Chloroplasts are cyanobacteria. The source of mitochondria has not been as easy to trace, which suggests mitochondria may have had multiple origins. One possible living ancestor might be the bacteria *Parcoccus denitrificans*, which is commonly found in the soil.

Today bacterial and archaeal cells are often seen, through microscopes, to be engulfing and digesting each other or to be infecting and destroying each other. And so it must have been in the past. But, long ago, for an unexplained reason, an engulfment of a bacterial cell by an archaeal one or the infection of an archaeal cell by a bacterial one resulted in both cells surviving and establishing a mutually beneficial relationship. The archaeal cell protected the bacterial cell from further harm and the bacterial cell supplied energy for the surrounding archaeal cell to function. In any case, the result was a newer and larger and more energy-rich type of cell: the eukaryote.[10]

But when did this merging happen? When did the first eukaryotes form?

Algae have been found in the Gunflint Chert, which has an age of 1.88 billion years. If the macroscopic fossil *Grypania* found at the iron mine near Ishpeming, Michigan, proves to be a eukaryote—there is still much debate about this—then the first appearance of eukaryotes can be pushed back to at least 2.1 billion years.

The long period of geological and biological stability known as the Boring Billion was yet to come. But when that period of stability ended by the breakup of the supercontinent Rodinia and a second episode of a Snowball Earth followed, another turning point happened in the history of life. The Ediacarans appeared. And they, as all large life-forms are today, were eukaryotes.

As the work done by Hoffman showed, the second episode of Snowball Earth was not simply a global deep freeze. It was a period of rapid and wild oscillations of climate, times when the entire planet was encased in ice, then in a hothouse condition, then ice, then hothouse. These oscillations happened at least four times. The bacteria and the archaea that then existed—and that dominated the planet—not only survived but were unaffected. The eukaryotes were not.

The higher energy needs of eukaryotes meant that they were more affected by the cycles of extreme climate changes than either bacteria or archaea. As such, those cycles would have imposed an intense survival filter that resulted in a series of genetic bottlenecks. During these bottlenecks, most of a population would die off, leaving a group of survivors whose genetic diversity was highly skewed from the original gene pool. Then, if the new population managed to recover and grow to huge numbers again, it, too, would suffer from another cycle and, again, many individuals would die, skewing the genetic diversity even more.

It is probably through such rapid cycles in population numbers and the repeated skewing of genetic diversity—a "bottleneck and flush" as it is called—that radically new life-forms develop, and not simply through the slow and steady accumulation of many mutations.[11] But this does not tell in what direction evolution will occur. That is told by the specific changes that happen to the environment. And one change that seems to have controlled much of evolution is the amount of oxygen in the atmosphere.

To be clear, oxygen is not essential to all life-forms. It is required to be at high levels only for those that have a high level of metabolic energy. And there is evidence that the amount of oxygen in the atmosphere did increase substantially after the end of the second and latest Snowball Earth, though why it increased is still debated. The amount of oxygen in the atmosphere was far less than it is today, but it apparently was enough to reach a tipping point, one in which an accumulation of many cells would benefit; that is, one in which multicellular organisms would have an evolutionary advantage.

◆

That the end of Snowball Earth had something to do with the rise of the Ediacarans is undeniable. The four cycles of icehouse/hothouse lasted

150 million years. Less than five million years after their end, the earliest known Ediacarans, those at Mistaken Point, appeared.

The reign of Ediacarans lasted twenty-five million years. And through those years, new forms of life appeared, at a slow pace, transitioning from the frond-like *Charnia* to the quilted-appearing mattresses of *Dickinsonia*. And then these mostly immobile filter feeders of the sea disappeared.

The abrupt disappearance could have been caused simply by the evolutionary appearance of the first predators, the first organisms that could search out and consume other organisms. One of these early agents of death was *Protohertzina*, the first terror with teeth, which appears just a million years before the end of the Ediacaran Period. It was only a tenth of an inch in size, but had a dastardly set of razor-sharp teeth that could tear and shred other organisms apart. It was from *Protohertzina* and similar early animals that the first food chain was born, that organisms were feeding on other organisms that fed on plants that derived their energy by the combination of air and sunlight.

Renewed environmental stress may also have contributed to the demise of Ediacarans. If so, it has not yet been clearly seen in the rocks of the Proterozoic Eon, in part, because so many rocks of this longest-of-all geologic eons have been erased. The cycles of icehouse/hothouse with the melting of the ice and the roaring and scouring of huge water floods certainly contributed to the erasure. There was a great scouring of the landscape, one that can be found at many places across North America, though is most starkly and impressively displayed in the walls of the Grand Canyon.

5

The Great
Unconformity

Early Paleozoic Era: 541.0 to 443.8 Million Years Ago
Cambrian and Ordovician Periods

Go to the Grand Canyon. Go to the Grand Canyon to look. To sit or stand or walk around or follow one of the trails that descend into the canyon. Walk at least partway into the canyon. Then return another day and look again. The scene will be different. The change may be subtle or dramatic, but it will be different.

Within a single glance, one can view more of the planet's history than is possible anywhere else in the world. It is one of the world's great natural wonders, "the most sublime spectacle on the earth" according to one of the first geologists who saw the canyon.

The first impression one has about the Grand Canyon is the immensity of the chasm. It is nearly three hundred miles long, up to twenty miles wide, and, in places, more than a mile deep.[1] The steepness of the walls and the unusual landforms attract immediate attention. Clarence Dutton, one of the first to study the geology of the canyon in the latter half of the nineteenth century, said he was shocked when he first saw the Grand Canyon because it did not conform to what he regarded as beautiful or noble in nature. He initially saw it as "grotesque" and "oppressive." But familiarity changed that. And the change came suddenly once he became aware of the "grace and dignity" of the canyon. And then there were the colors. To the new visitor, they seem subtle. Dutton initially described the colors as "tawdry and bizarre," but again, after familiarity and after seeing the canyon under many different light conditions, he described and many tones and shades as "modest and tender."

I recommend seeing the Grand Canyon first from the South Rim at Lipan Point at the eastern end of the canyon. From there, looking west, one can see the Colorado River as it enters the deepest part of the canyon. The Colorado River is narrowest here, pinned between the darkest and the steepest walls of the canyon. These are hard rocks made of metamorphosed schist and igneous granite. These are the rocks of the supercontinents of Columbia and Rodinia. These are the rocks that formed during the Boring Billion.

Immediately above the schist and the granite, still viewing from Lipan Point, one's eyes climb up to patches of a crumbling red rock. These red rocks are in layers that are tilted. These are sedimentary deposits that accumulated in large basins that had formed along the edge of Rodinia as this supercontinent was being torn apart.

Above the rocky remnants of Columbia and Rodinia and above the layered and tilted red rocks that record the demise of Rodinia and continuing all the way up to the canyon rim is the famous layer-cake assemblage of thick horizontal beds that has made the Grand Canyon a favorite of artists and photographers and hikers and sightseers. With one exception, these thick, flat-lying rocky units represent episodes when seas advanced across Laurentia. These transgressions of the sea left behind thick deposits of ocean sediments. The most prominent one is the Redwall Limestone, which can be seen all around the canyon as a sheer vertical cliff several hundred feet thick. The exception to the ocean sediments is near the top of the canyon wall. It is the Coconino Sandstone. It is composed of ancient sand dunes that formed on land during an extremely arid period in the history of Laurentia, long after Rodinia had broken apart and after a new supercontinent, Pangea, had formed.

But this is getting ahead of the story. The stories of the Redwall Limestone, the Coconino Sandstone, and much else will be told later. For now, it is necessary to descend down into the canyon to the Colorado River and reminisce about the person who made the first recorded passage through the Grand Canyon—and the geologic discoveries that he made.

He was John Wesley Powell. Powell had been an officer in the Union Army during the Civil War who lost his right arm during the Battle of Shiloh when a minié ball passed through it. For most people, the loss of an arm would have limited them, but for Powell, as one biographer has written, "it affected Wes Powell's life about as much as a stone fallen into a swift stream affects the course of a river." All of his life, he was enthusiastic about everything he did—and that included a study of the Colorado River.

On May 24, 1869, Powell and eight men set out in four large wooden boats to begin a descent of the Green and the Colorado Rivers. The stepping-off point was Green River Crossing in the Territory of Wyoming, then the farthest west the tracks of the Union Pacific Railroad had reached. It took them three months to reach the stretch of the Colorado River that is visible from Lipan Point. By then, their rations were meager. The men were living off musty flour, spoiled bacon, dried apples, and coffee. It was here that they readied themselves for the most perilous part of the trip, a passage through the Grand Canyon.

It was on August 18, their fourth day in the canyon, that their progress slowed considerably. During the previous three days they had made much progress running the various rapids and doing little portaging. But the fourth day was a day-long portage, around especially large rapids. Powell took advantage of the slowed progress to climb partway up the canyon walls and survey the geology at close range.

"I climbed so high," he later recounted, "that the men and boats are lost in the black depths below, the dashing river is a rippling brook." The first part of his ascent was agonizingly slow as he climbed up steep hard granite. Then, about a thousand feet above the river, his pace quickened as he moved easily over "a rust colored sandstone and greenish yellow shale." The sudden change in rock type immediately caught his attention. It meant a disruption in the rock record. It meant that a major change had occurred in the deposition and erosion of the rocks. Later, when writing about this, he described the sandstone and shale as "broken, ragged, non-conformable rocks, in many places sloping back at a low angle." It was the first mention of a key feature in the history of North America.

How important the disruption might be and how far beyond the Grand Canyon it might extend could only be answered after years of fieldwork, something that Powell did not have time to do. Much of his career would be consumed by administrative work in Washington, DC, as the director of two government agencies—the United States Geological Survey and the United States Bureau of Ethnology at the Smithsonian Institution. But he knew a man who could do the work—Clarence Dutton, a career military man who had been detailed by the United States Army to work for Powell on a geologic survey of the western United States.

Those who knew both men described Powell as flamboyant and Dutton as reserved. Powell's descriptions of the Grand Canyon are from the river's edge. Dutton's are from the rim. Powell provided the first general view of the canyon. Dutton was the first interpreter, seeing the many details and weaving them into a whole. When one sees the canyon in broad daylight, one often thinks of Powell and his passage through the canyon. But when it is sunset and one marvels at the scene, one is seeing the Grand Canyon through Dutton's eyes.

From 1876 to 1881, Clarence Dutton worked every summer on a survey of the geology of the Grand Canyon and the broad surrounding area. Powell specifically directed him to the question of the "non-conformable rocks." From that work, Dutton would identify "a vast body" of ancient strata that had been "flexed and faulted" then "enormously eroded."

As Dutton first saw—and as generations of geologists after him continue to confirm—there is a major disruption of the rock record here. One that was eventually found all across the continent. In many places, one can literally place a finger where there is a transition from the rocks of the ancient Earth and those of a much more recent planet. One can stand astride a boundary where one foot is placed on rocks that are billions of years old and the other foot on those that are a few hundreds of millions of years old.

A boundary as important as this one deserves a name. And Dutton gave one to it. The name marks both the simplicity and the drama represented by the boundary. It is known as the Great Unconformity.

◆

In geology, an *unconformity* is a boundary between two dissimilar rocky units. The distinction may have been made by erosion, one or more layers of older rock stripped away, then, after a passage of a considerable length of time, additional layers laid down on the eroded surface. Or the distinction may simply represent a pause—again one of a considerable length of time—when no new rock layers were being deposited, then deposition resumed. In any case, when one looks at an unconformity, one is looking at a gap in time, a punctuation in the geologic record.

The most important unconformity in the history of geology—here "important" means influencing geologic thought—is the one at Siccar Point along the southeastern coast of Scotland. There, in 1788, Scottish farmer and noted naturalist James Hutton took two friends to show them an abrupt change between two rocky units. The lower unit, a gray sandstone, rose up out of the sea as a set of almost vertical beds that were cut off abruptly by the horizontal beds of an overlying red sandstone. According to Hutton, the beds of the gray sandstone had formed horizontally, then were tilted into their present vertical positions by some force acting inside the Earth.[2] The gray sandstone was then eroded, leaving a truncated, flat surface, and the beds of red sandstone were deposited on top of it. In all, the deposition of the gray sandstone, its upheaval and tilting and truncation, and the deposition of the red sandstone must have taken a very long time. But how long? Hutton had no means to answer that question. But we have the means to do so today.

It is now known that the gray sandstone at Siccar Point was deposited on the floor of a deep sea 425 million years ago. The overlying red sandstone was deposited over a land surface by streams and rivers 345 million years ago. Thus, the gap in time between them—the amount of time it took to raise and tilt the gray sandstone, erode its surface, and deposit the red sandstone—was eighty million years, a considerable length of time by any standard. But it pales in comparison to the gap in time found in the deepest part of the Grand Canyon.

There are several unconformities in the walls of the Grand Canyon. There is one between the Redwall Limestone and the underlying Muav Limestone, in this case, a gap in time of about a hundred million years. But the one of immediate interest, the one that is found deep in the canyon and was first noticed by Powell in 1869 and named by Dutton in 1882—the Great Unconformity—represents a much, much longer gap in time. It is between the horizontal beds of the whitish Tapeats Sandstone and the dark Vishnu Schist, ancient sand and mud that has been metamorphosed and that has streaks with bright red blocks of Zoraster Granite running through it. The gap in time between the Tapeats Sandstone and the Vishnu Schist/ Zoroaster Granite is 1.2 billion years.

For most people, it is a challenge to descend several miles down a trail and into the Grand Canyon to place a hand on the Great Unconformity. Fortunately, it is a continent-wide feature. There are many places, much more convenient—though nowhere as spectacular as the Grand Canyon—where the Great Unconformity can be found.

In Las Vegas, Nevada, a mile east of the intersection of Lake Mead Boulevard and Hollywood Boulevard, State Highway 147 passes directly over the Great Unconformity. There is a convenient pullout and a small sign that explains the geologic importance of the site. The sign instructs travelers to follow a trail that passes along the base of nearby Frenchman Mountain and in doing so walk along the exposed trace of the Great Unconformity. As one makes the walk, to the left, in the direction of Frenchman Mountain, are the whitish rocks of the Tapeats Sandstone. And to the right are the dark rocks of the Vishnu Schist shot through with the red blocks of the Zoroaster Granite. It is exactly the same as in the deepest levels of the Grand Canyon. The gap in time is 1.2 billion years, a period that represents one quarter of the Earth's existence.

In Colorado, the Great Unconformity can be found in Unaweep Canyon, in Black Canyon of the Gunnison National Park, at Manitou Springs, and within the Red Rock Amphitheatre near Denver. It is found at Camp Creek near Dillon in Montana, in the Shoshone Canyon in Wyoming, in Ruby Canyon in Utah, and in the Franklin Mountains near El Paso, Texas. It is exposed at the St. Francois Mountains on the Ozark Plateau in Missouri and just south of Glen Lock Dam near Chippewa Falls, Wisconsin. In the northeastern United States, it is found in the Adirondack Lowlands, south of Ticonderoga, New York, where the Potsdam Sandstone, the geo-equivalent of the Tapeats Sandstone, sits atop a much older rock.[3] The contact between them is the Great Unconformity.

And there are many more sites. On the Manitoulin Islands along the northern shore of Lake Huron. Along Highway 401 near Kingston, Ontario. On Somerset Island in arctic Canada. And it has been found on other continents, including near Cape Town, South Africa, and Petra, Jordan. It is a global feature that represents a long period of deep erosion of all the continental masses.[4]

Estimates have also been made of the amount of material that must have been removed. Again, the indestructible zircons are brought into play. Isotopes of oxygen and hafnium are used as indicators of erosion and sedimentation rates. And they indicate that, across most of the continental landmass that then existed, *a few miles* of material was eroded in the billion-year gap in time represented by the Great Unconformity.

By any measure, it was a truly deep scouring of the continents. But what process, what set of conditions, could have removed so much material from all of the land surface? The age-dating of zircons indicate that this period of continent-wide erosion ended about six hundred million years ago. That is near the end of the Cryogenian Glaciation, the end of the second Snowball Earth.

In hindsight, it all makes sense. For tens of millions of years, great sheets of ice moved across the entire surface of the Earth. They then retreated. And they did this at least four times. And each time, as the sheets advanced and retreated, they scraped and bulldozed and pulverized the land surface. Then, when the great ice sheets melted, massive floods of water washed huge amounts of material from the continents and into the oceans, creating a great gap in time in the geologic record.

To be sure, a similar thing must have happened two billion years earlier at the end of the first Snowball Earth. And that did produce a major change in the Earth's history, one that is marked by the beginning of a new geologic eon, the Proterozoic. The Cryogenian Glaciation and its aftermath also produced a major change in the Earth's history, one that is also marked by the beginning of a new geologic eon—the one we live in today—the Phanerozoic, literally, the time of "visible life."

◆

And that introduces an inconsistency that confuses. If the Phanerozoic Eon is the time of "visible life," why are the Ediacarans excluded from it? Why did the Phanerozoic Eon begin after most of the Ediacarans disappeared? In part, the reason goes back to when geology started as a science and how the various geologic periods were originally defined.

In 1835, Adam Sedgwick of Cambridge University, after conducting an exhaustive survey of the countryside of England and Wales, announced that the oldest—meaning the lowest—strata that he could find that contained fossils was a series of rocks in the mountains of central Wales. Included among those were *trilobites.*

Trilobites were animals that could scurry and feed and, possibly, hunt other animals. They had a well-defined head and tail. They had eyes, often bulging out from their bodies. Each one had a mouth. And a segmented body with an external hard skeleton and sets of paired appendages. This complexity made Sedgwick and his contemporaries wonder: Shouldn't earlier forms of animal life have existed that were less complex and more primitive in appearance? Because he was unable to find such evidence, Sedgwick eventually concluded that none had existed. Instead, because complex life had appeared suddenly, its creation must have been miraculous, the work of a supernatural being. Not everyone agreed.

Charles Darwin was one of the dissenters. Though he had no proof, Darwin was convinced that more searching of the geologic record would reveal that a more primitive form of life did exist before Sedgwick's Cambrian-age fossils. And decades of work—and new techniques, involving better microscopes and the ability to analyze isotopes—would prove him right. Ediacarans were an earlier form of life. So were the microfossils of the Gunflint Chert. And the stromatolites. And yet, in defining the beginning of the Cambrian Period—and the Phanerozoic Eon—we are tied closer to Sedgwick than to Darwin in that we have excluded the Ediacaran.

Ediacaran fossils are scarce. Consequently, there are large uncertainties in how to correlate them around the world, or even across a continent. Their appearance, as it is currently understood, was not abrupt enough to define the beginning of a geologic period. There is no place where one can go and place a finger on a rocky outcrop and say that this spot marks the beginning of visible life.

Moreover, it is still not clear what type of life Ediacaran represent. They were not plants. Some *might* have been animals. If so, there is no clear evidence that Ediacarans were ancestors of modern animal life.

Most were immobile, living their lives fixed to a rock or to a bed of sand and passively feeding off microbes or other nutrients contained in the water and that passed through their elaborate filtering structures. But some could move. Some Ediacarans did leave trails, showing that they could squirm or crawl or somehow pull themselves around. But how far they might have been able to move using their own mobility is still an open question. What is clear is that when they did move, their movement was confined to the surface of the ocean floor. They did not dig down and penetrate, to any significant degree, into the muck and mud on the ocean floor.

And that is important. Such penetration into the muck and mud can be seen in the fossil record. It means that whatever organism could burrow into the ocean floor could probably also transport itself considerable distances. That is the key to defining the beginning of the Cambrian Period. It is the distinction used by members of the International Commission on Stratigraphy to determine when the Phanerozoic Eon began.

The overall purpose of the commission is to define the boundaries of geologic time by deciding exactly where in the world each stratigraphic boundary is to be found. From that, a detailed vertical chart is made—one that is familiar to anyone who has riffled through a geology textbook or has read one of the guideposts at the Grand Canyon—that shows, beginning with the oldest ones at the bottom, all of the geologic time periods over the entire history of the Earth. But, first, the stratigraphic boundaries have to be defined.

To do that, members of the commission meet at semiregular intervals to discuss and debate and, finally, to agree on a specific location for each boundary. It is a never-ending task because the boundaries are always getting better defined as improvements are made in the age-dating of rocks and as more fossils are discovered. But that is still the main purpose: to define a specific location for each boundary, a Global Stratotype Section and Point (GSSP) or, as it is known in geology, a "golden spike."

The golden spike for the beginning of the Cambrian Period was decided in 1992. It is at a point partway up a low cliff at Fortune Head near the tip of the Burin Peninsula on the island of Newfoundland, a location that is readily accessible by foot. Those who travel there will find a plaque that tells how to find, within a sequence of rocks that are well exposed, the exact position in the geologic record that marks the beginning of the Cambrian Period. According to the plaque: "It is located at 2.3 metres above the base member 2 of the Chapel Island Formation and is exposed on the coast, 250 metres northeast of this plaque." If one is still has difficulty, there are usually other visitors nearby who may be of help.

But to understand why this specific site was chosen as the Cambrian golden spike requires a trained eye. According to the detailed description

written by members of the commission, the GSSP for the beginning of the
Cambrian Period is defined as the moment in time when the small animal
Treptichnus pedum first appeared. No mineralized fossil, no imprint, and
no bodily remains of *Treptichnus pedum* have ever been found. But that it
existed—and that it had a wormlike character—is evident in the tiny burrows it made. The burrows are distinct, being sinuous and having loops as
the creature made repetitive movements through muddy sediments, probably in search of food, producing a pattern reminiscent of a small fan or a
small piece of twisted rope.

It is from that moment in time, when *Treptichnus pedum* began to probe
into the sediment, ingesting some of it, withdrawing back into the main
part of its burrow, then probing, expelling undigested parts of its food from
another body part, that marks the beginning of the Cambrian Period. And
that makes it the beginning of the Phanerozoic Period, which more cor-
rectly should be known as "when the Age of Animals begins."

Then, if one works upward through the rock record, moving forward in
geologic time, one will find the mineralized remains of complex creatures.
If one puts an absolute time to the first appearance of *Treptichnus pedum*, it
occurred 541 million years ago. Then tens of millions of years passed—tens
of millions of years in which small wormlike creatures evolved and diversified into complex animals with external skeletons that could scurry and
swim and hunt and are now found in abundance as skeletal fossils.

But, in following Darwin's idea, intermediate forms of animals must have
existed that were between the wormlike creature *Treptichnus* and those that
were more complex. It took decades of searching, but such intermediate
forms were eventually found. They were first discovered in a short section
of black shale that is located high on the side of a steep ridge in what is
today Yoho National Park in the Canadian Rockies of British Columbia.

◆

Pilgrimages are made today by both professionals and nonprofessionals to
see the oddly wonderful animal life-forms that preceded us all. It is, then,
strangely appropriate that to make such a trek requires a long, arduous
hike, a personal undertaking that adds to the feeling of reverence that many
people have for the site. It is located near Mount Burgess. The rocks that

contain this remarkable trove of the earliest animal life-forms are known as the Burgess Shale.

Given the importance attached to the Burgess Shale, it also seems appropriate that its discovery has a slight mythical tone. It was late August 1909, the end of summer fieldwork in this part of the world. Charles Doolittle Walcott, head of the Smithsonian Institution and an expert on Cambrian Period fossils, was leading a small party of assistants, which included his wife, Lura Ann Walcott, down a steep slope covered in loose rocks. Each person was riding a horse. Mrs. Walcott's horse stumbled and slid down the slope, turning over a large slab. Mr. Walcott came to her aid. After confirming she was all right, he examined the slab, noticing that it had several unusual markings. He split the rock open with a sledgehammer. Inside was a great treasure: a collection of strange and, at the time, totally unknown fossilized imprints of ancient life.

Walcott returned in 1910 and established a quarry on the flanks of this steep ridge. He worked at the quarry almost every year until 1924. At that time, at age seventy-four, he had collected more than sixty-five thousand specimens. He was still describing and classifying them when he died in 1927.

More than two hundred thousand specimens have now been collected, representing more than a hundred different species. The Walcott Quarry measures about ten feet high and is less than a city block long, capped on either end by metamorphosed rocks that contain no fossils. And so many of its best treasures have probably already been discovered. But searches are underway to find similar exposures in the Canadian Rockies, to add to the collection that Walcott originated.

What makes these specimens so rare is that most are of soft-bodied organisms, so delicately preserved that one can see tiny features of antennae, bristles, appendages, even the contents of the guts of a few individuals, revealing what those creatures were consuming hundreds of millions of years ago.

The extraordinary preservation was due to the unusual circumstances of the burial of these animals. At the time this region was close to tropical. A steep cliff existed just off the coastline. (This can be seen in the rocks that surround Walcott Quarry.) Most of the animals whose remains have been found in the Burgess Shale lived on or at the base of the cliff, well below the level where ocean waves churned up debris. Occasionally, an avalanche of mud would slide down from near the top of the cliff and carry animals that lived on the side of the cliff down into deep water. The fine particles of mud would then slowly settle down through the water, entering every small opening, precisely preserving the soft parts of the body. Moreover, because these soft part came to rest a few hundred feet below the surface of the sea, at a depth that, in a Cambrian sea, was devoid of oxygen, bacteria could not exist and consume and destroy the soft-body parts.

Intense interest in the fossils of the Burgess Shale is not due solely to their great age or to the preservation of soft-body parts. It is also due to the bizarre appearance of many of the creatures. There is *Wiwaxia*, which looks like a pinecone lying on its side with two rows of spikes running along its back. *Hallucigenia* was named for its dreamlike or hallucinogenic appearance. It had tentacles that were probably used to propel itself and to search for food. It also had spines along its back to protect itself from predators. There is *Opabinia* with five eyes and a long nose that probably passed food to its mouth. Fewer than twenty good specimens of *Opabinia* have been found and each measures less than three inches in length. By comparison, *Anomalocaris* was a giant. It had two appendages at the front of the head that probably grabbed food and fed it to a round mouth that had plates that pointed inward, looking, somewhat bizarrely, like a pineapple ring. The largest specimen of *Anomalocaris* recovered from the Burgess Shale is twenty inches in length.

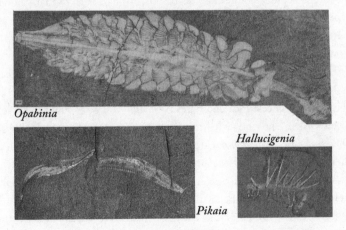

Trace fossils of animals from the Burgess Shale. *Opabinia* (photo taken by Charles Doolittle Walcott), *Pikaia* (photo taken by Charles Doolittle Walcott), and *Hallucigenia* (photo courtesy of the Smithsonian Institution).

And then there is *Pikaia,* a small wormlike creature that had what appears to be a notochord that ran the length of its body, suggesting that it might be an early ancestor of all animals with a backbone, including ourselves.

Since Walcott's discovery, fossilized remains of early animals similar to those in the Burgess Shale have been found at several other sites. In North America they have been found in the Wellsville Mountains of the Wasatch Range in Utah and in the hills immediately west of the small town of Pioche in Nevada. Additional ones are in Greenland and Australia. The most impressive discovery of such early animals was made in southern China in 1984. Among the specimens recovered from China was one of *Anomalocaris* that is more than six feet in length.

The animals of the Burgess Shale—and these are definitely animals, many have eyes and appendages and a mouth and a gut—are so diverse and so bizarre in appearance that it was initially thought that most were unrelated to any animals living today. Stephen Jay Gould of Harvard University was one of the strongest proponents of this idea. He dwelled on the outlandish anatomy of many specimens, arguing that the Burgess Shale contained the most diverse forms of animals to ever appear in the Earth's history. He referred to them as "weird wonders" and thought that most represented evolutionary dead ends, animals with fundamental body types that appeared then subsequently disappeared, never to be seen again. His view has now been tempered by the recovery of many more specimens from the Burgess Shale and from the other sites.

Detailed examinations show that most of these early animals can be assigned to animal groups that we see today. As mentioned, *Pikaia* is probably an early ancestor of chordates. *Wiwaxia* is now thought to be related to mollusks. The ones with the most bizarre appearance—*Hallucigenia, Opabinia,* and *Anomalocaris* are good examples—seem weird to us because they are represented by obscure animal groups, such as the velvet worms, small arthropods that linger in the dark under the sea, and that look like a cross between a centipede and a caterpillar.

Because the fossils of the Burgess Shale were buried quickly and en masse, they provide a snapshot of the range of animal life that then existed. That it was highly diverse is undeniable. But that brings up the same question that was asked about the sudden appearance of the Ediacaran, though, this time with more flourish: What forces drove such incredible diversity at the beginning of the Cambrian? In a time interval of a few tens of millions of years, animal life-forms expanded from wormlike *Treptichnus* to complex trilobites. In fact, all of the major body plans of animals seen today can be traced back to this short interval.

It was the evolutionary equivalence of an explosion, an event that is known as the Cambrian Explosion.

◆

Much has been written about the biological innovations that suddenly appeared during the Cambrian Explosion. Among the most striking were the first appearance of hard shells, the first animals with eyes, and the first animals that were predators. And these developments were probably related.

The Cambrian Explosion seems to have been something akin to an arms race, one that was a constant one-upmanship between predator and prey. The Ediacarans were notable for their lack of defenses and of sensory organs and for their immobility. Those animals that evolved during the Cambrian Explosion abounded in defenses and sensory organs and were highly mobile.

Eyes gave predators the means to spot and track and capture prey. Eyes gave prey the ability to see predators and take evasive action.

The development of hard shells is especially noteworthy because it shows how the development of a new characteristics can be beneficial in unexpected ways. During the Cambrian Explosion, there was a dramatic increase in the concentration of calcium in seawater. (This will be explained in a moment.) Because high amounts of calcium can be detrimental to a cell's functions, those cells that could concentrate the calcium further and excrete it would be more prolific and have a better chance at survival. Those excretions came in the form of calcium carbonate (limestone), which built up around the cell and, in animals, would be used to form hard shells—a protection against predators and against the occasional poundings by wave action.

But the prey-predator arms race—the development of shells and of eyes—was more likely an outcome of the Cambrian Explosion rather than a cause. The cause that would have led to these biological innovations would have come from the external environment. And one of the major changes that occurred to the external environment during the Cambrian Explosion was dramatic increase in sea level.

Look again at the Grand Canyon. Immediately above the Great Unconformity are three thick horizontal units. Each one formed on an ocean floor. The lowermost one is the Tapeats Sandstone. Above it is the Bright Angel Shale. And above the shale is the Muav Limestone. That sequence—sandstone, shale, limestone—means that, as time progressed, the depth of the ocean in which these sediments were being deposited was getting deeper and deeper.

The Tapeats Sandstone has features known as ripple marks that formed in shallow water as waves washed back and forth near a sandy beach. The Bright Angel Shale is composed of silt and mud, both much finer than sand. Silt and mud can remain suspended in water longer than sand and can be carried farther from shore and into deep water. The Muav Limestone, because it is almost devoid of sand and silt and mud, must have formed in very deep water, far from the shoreline, too far for rivers to sweep debris out to those distances.

This continuous sequence of rock formations—Tapeats Sandstone, Bright Angel Shale, Muav Limestone—records a time when, on this part of the planet, the ocean water was getting deeper and deeper. In fact, by studying these three rocky units beyond the Grand Canyon, one can determine that the shoreline was moving west to east; that is, that the ocean was *transgressing* from west to east across Laurentia.

And as the shoreline was moving eastward, wave action was pounding higher and higher onto continental land, exposing more and more fresh rock and releasing more and more ions, such as calcium, into the ocean. This led to the development of hard shells, which was a factor in the prey-predator arms race.

But why did sea level rise so dramatically during the Cambrian Period? The amount of rise—at the Grand Canyon—is several hundred feet. Usually, when one sees evidence of a rise in sea level, one immediately thinks of the melting of thick sheets of ice or of ice caps. But ice had disappeared from the planet long before the start of the Cambrian Period.

Another way to raise sea level is to reduce the volume of the ocean basins. This could happen if the rock that comprises the ocean floor becomes hotter and, hence, more buoyant, as when new ocean ridges form. And new ocean ridges were forming during the Cambrian Period as continental masses were shifting and a new landmass called Gondwana started to form. But would this be enough to account for several hundred feet of sea-level rise?

There is another explanation. It is based on the idea that the entire planet shifted—like turning a ball in your hands—causing the positions of the North Pole and the South Pole to change, or to *wander.*

Though controversial, the idea is not as outrageous as it may originally sound. Huge shifts in the positions of poles—that is, *true polar wandering*—have occurred on other objects in the solar system. The geologic histories of the Moon and Mars and several other objects—Ceres, the largest asteroid; Europa, a major moon of Jupiter; Enceladus, a moon of Saturn where more than a hundred geysers are spewing ice and gas into space—indicate that each of these objects has changed its orientation in a major way since it formed. And so the Earth might be expected to have also undergone such a major reorientation.

For the Earth, the evidence relies on measuring the direction of ancient magnetic fields in continental rocks, a technique that requires considerable care and an extraordinary amount of expertise. By determining the direction of the ancient fields, the position of a continent in latitude can be determined in the distant past. Such an effort has shown that, at the beginning of the Cambrian Period, Laurentia, which had already split from Rodinia, was near the South Pole. Thirty million years later, it was straddling the equator.

That was an exceptionally high rate of motion. It means that the entire continent moved northward at a rate of a mile every five thousand years, and that Laurentia was shifting its position by more than a foot a year. That rate is three to ten times faster than the rates that continents move today. And all continents during the Cambrian Period were moving at such high rates. Either something had accelerated all plate motions for thirty million years or some new phenomenon was giving the appearance of rapid plate motion. Careful examination of the magnetic directions of Cambrian-age rocks on all continents show that it was the latter.

All of the continents, including Laurentia, appear to have moved rapidly during the Cambrian Period because, during those thirty million years, the entire planet rotated onto its side. The only way this could have happened was by a sudden change in how a large mass was distributed on the surface of or inside the planet.[5]

For Mars, true polar wandering occurred after there was an eruption of massive lava flows and after four giant volcanoes formed in an area known as the Tharsis Province. For the Moon, it was the filling of several deep impact basins on the nearside with dense lava—seen today as the dark areas on the Moon. For the Earth, it was something that happened internally.

Exactly how this would have happened is a matter of debate. One possibility is that, during its final assembly, Rodinia was surrounded by a ring of subduction zones. At each zone was a dense slab of ocean crust that was sliding slowly into the Earth. The slabs accumulated beneath Rodinia. Eventually—this is where there is a great deal of debate—the accumulation of many dense slabs became unstable. This dense mass started to slide faster and deeper into the Earth in a *mantle avalanche*. That changed how mass was distributed inside the Earth and caused the Earth to rotate onto its side in an episode of true polar wandering.

During the Cambrian Period, the Earth rotated about 60°. The position of Laurentia shifted from being near the South Pole to being near the equator. That caused flooding of hundreds of feet of seawater as the continent encountered the great bulge of water that lies along the Earth's equator.

As sea level rose, the position of the shoreline moved from west to east across Laurentia, leaving behind this notable sequence of rocks: Tapeats Sandstone, Bright Angel Shale, and Muav Limestone. As the sea transgressed, it eroded more and more of the continental land surface, releasing large volumes of sediments and increasing the concentrations of chemical elements, such as calcium, in seawater (which lead to the development of the first hard shells). By the end of the Cambrian Period, essentially the entire continent was flooded. The only major section of dry land was a roughly oval-shaped region over the Hudson Bay region of Canada and a broad arch that ran across the center of the United States and into Mexico. It was along one edge of this dry land that the Burgess Shale formed.

Eventually, the sea did retreat. As it retreated, the action of sea waves again scoured the landscape and left another erosional mark, another unconformity. This one is at the top of the Muav Limestone and separates it from a much younger limestone, the Temple Butte Formation. The Temple

Butte is much less impressive than the Muav, so it is often overlooked by those who hike into the canyon.

The sea retreated and sea level dropped because an ice cap formed over the South Pole. The ice cap grew because there was a general drop in global temperature and because most of the continental masses that had dispersed when Rodinia broken up had now reassembled into Gondwana. Because ice is more likely to form over land than water, the land-covered South Pole was now ideal for a new polar ice cap.

This new period of glaciation was not as severe as an episode of Snowball Earth, but it was severe enough to cause an extinction of animals. And that extinction is noted by geologists as the beginning of a new geologic period—the Ordovician.

◆

In the aftermath of that extinction, the Ordovician Period became, in many ways, a second explosion of life. The creatures that now inhabited the planet—still all marine creatures—would grow larger and burrow deeper than those of the Cambrian Period. They would develop more elaborate ways to hide from predators and more aggressive ways to hunt for prey.

The animals of the Cambrian Period were generalists, capable of surviving in a wide range of environments. Those of the Ordovician Period were specialists, animals that had adapted themselves to narrow habitats. And that made the animals of the Ordovician Period more diverse. One can see this great diversity by going to Cincinnati, Ohio, a city that is built on one of the most fossil-rich regions of the world.[6]

It has often been said that if all of the fossils were removed from beneath Cincinnati, the city would drop several hundred feet. Fossils are everywhere. They literally tumble out of hillsides. Road cuts are favored places to find fossils. But they are being dug out by the ton where excavations are underway for new buildings. And, as a matter of routine, they are often scooped up by the shovelful by people who are digging in their backyards to plant new gardens or to erect new fence posts.

In fact, it is a challenge *not* to find a rock in or around Cincinnati that contains a fossil. Clam-like brachiopods are the most common. Bryozoans are especially prized because they look like slender tree branches. The first corals to appear on the planet did so during the Ordovician Period. And they are found in great abundance in and around Cincinnati. The most prized ones are the horn corals that have wrinkled surfaces and resemble a bull's horn in size and shape.

And there are trilobites. Lots of trilobites. A trilobite had a hard exoskeleton that it discarded as it grew, much as lobsters do today. An individual might discard more than two dozen shells, which explains why so many trilobite shells have been found.

The largest trilobite was found just fifty miles north of Cincinnati near Dayton, Ohio, in 1919 by workmen who were digging an outlet tunnel for a dam. The specimen is fourteen inches long and ten inches wide. It is on display at the Dayton Museum of Natural History and serves as the museum's logo. In 1985, in recognition of this record-breaking specimen—and after an enthusiastic letter-writing campaign by Dayton schoolchildren—this species of trilobite, *Isotelus maximus*, was proclaimed the official state fossil by the Ohio State legislature.[7]

But why did a second explosion of life happen? In part, it involved cycles of bottleneck and flush: The glaciation cycle at the end of the Cambrian Period severely reduced the size of animal populations, causing some animal species to become extinct. Then the few surviving populations increased dramatically, adapted to new environmental conditions, and produced new species. This would happen repeatedly in the Earth's history. But the Ordovician Period was different.

Almost all of the fundamental body plans of animals were in place by the end of the Cambrian Explosion.[8] But the diversity of animals by then was low. That changed during the Ordovician Period when, over a period of about thirty million years—the same length of time as the Cambrian Explosion—there was a dramatic increase in the variety of animal life.

The Great Ordovician Biodiversification Event, as it is called, produced many more types of brachiopods, bryozoans, cephalopods, graptolites (tiny colonial animals that lived in skeletal tubes), corals, and trilobites. And many of these were more complex than their Cambrian ancestors.

The trigger for the great surge in biodiversification is still debated. The ice cap over the South Pole at the end of the Cambrian Period had melted and much of the planet was covered by seawater. What dry land that did exist probably existed as archipelagoes. And there were rapid changes in sea levels during the Ordovician Period, radical changes to the shoreline environment that caused marine animals that lived in shallow water to adapt, adding pressure for them to evolve.

A barrage of meteors also hit the planet during the Ordovician Period. Whether the impacts were spaced out over millions of years or occurred simultaneously cannot yet be decided from the evidence. Among the craters that were produced are ones near Decorah, Iowa; Ames, Oklahoma; Calvin, Michigan; and Menomonie, Wisconsin. So is the large crater known as Clearwater East in northern Quebec and the small one that forms the Slate Island along the northern edge of Lake Superior. How an increase in meteor impacts could have increased animal diversity is unknown.

In short, there is no lack of global disturbances that might have caused the Great Ordovician Biodiversification Event. The physical environment did change—and cycles of bottleneck and flush were happening. But the physical environment provides only the conditions that are conducive to evolutionary change. There still has to be a biological mechanism for rapid change to occur. And that was a stumbling point for biologists for many years.

That is, until a rapid mechanism was found in Hox genes.

◆

Evolution by natural selection, as envisioned by Darwin, predicted a gradual change in the diversity and complexity of organisms. This was supported by Mendelian genetics in which the traits of an adult were passed on to an offspring through discrete inheritance units known as genes. Each new population of offspring differed only slightly from their parents. According to Darwin, a major evolutionary change required a long succession of ever-changing populations. But that is not what is seen in the fossil record.

The fossil record shows that populations of organisms maintain long periods of stability interrupted by episodes of rapid change. This idea was formalized in 1972 by paleontologists Stephen Jay Gould and Niles Eldredge in the concept of *punctuated equilibrium*.

Gould and Eldredge based the concept on their studies of trilobites and snails, two animals groups that are preserved in large numbers in the fossil record. By 1983, punctuated equilibrium received additional support when a genetic mechanism was uncovered.

That same year Hox genes were discovered by research groups working independently in Switzerland and in the United States. These are the genes that control the basic body plan of an individual.

Genes are part of the DNA code that is used to make the proteins that make up all organisms. If the DNA just made proteins randomly, an organism would be just a jumble of cells and proteins. In order for organisms to grow into a particular shape, the DNA must contain instructions to regulate the growth and shape of the organisms. And those instructions are the Hox genes.

Hox genes act like master switches during embryonic development. They bring reality to the catchy phrase that all biology students learn in introductory classes: Ontogeny recapitulates phylogeny. That is, the developmental stages of an embryo (ontogeny) go through the successive adult stages in the evolution of its remote ancestors (phylogeny). For example, a human embryo seems to go through stages in which it is similar in appearance to a fish, an amphibian, then a reptile. Hox genes explain why all vertebrates, including humans, have gill slits and tails early in embryonic development.[9]

Lowly sponges have just one Hox gene. Most invertebrates have fewer than a dozen. Human beings and most vertebrates have a few dozen. The sets of Hox genes in vertebrates are similar to one another and to those in invertebrates, which indicates an ancient origin.

One group of Hox genes switches on the formation of eyes in all animals that have them, including octopuses, insects, and humans. Another group determines the layout of the basic body plan, the head-to-tail organization. Another determines whether an appendage on an insect will be an antenna or a wing or a leg, or on a vertebrate what types of vertebrae will form. Mutations in Hox genes can result in body parts and limbs being in the wrong place along the body.[10] Small changes in such regulatory genes are a major source of evolutionary change. And from this comes support for the evolutionary theory of punctuated equilibrium.

Evolutionary change does not happen steadily and gradually due to a steady stream of accidental genetic mutations. It is not the appearance of new genes that causes morphological changes. Instead, such changes occur in the evolution of an organism when there are major environmental changes, and as a result, a new population of organisms seems to pop into existence in a geological instant—the "pop" producing the "punctuation" in punctuated equilibrium. As Gould and Eldredge envisioned it, the development of new species was similar to climbing a staircase. The climbing up one step happens because a major environmental change causes a different Hox gene to be expressed, and from that, a dramatic morphological change happens in a population.

Moreover, it was the reproduction of extra sets of Hox genes that produced segmented bodies. The loss of some Hox genes might explain the sudden appearance of animals with bilateral symmetry. In short, sudden evolutionary change indicated by the fossil record—which was a major concern that Darwin had—did not occur because a large number of new genes suddenly came into existence. It occurred because of a change in a few Hox genes.[11]

◆

As mentioned, the animals of the Ordovician Period and those that came later were mostly specialists, not generalists. That made them more susceptible to environmental change and, hence, more likely to disappear. And that meant mass extinctions.

The first mass extinction defines the end of the Ordovician Period. It was another rapid cooling of the planet. The massive continental landmass of Gondwana still straddled the South Pole. When the cooling started, a large ice cap again formed.

This time geologists are confident they know the cause of the cooling. It was a mountain-building event that exposed steep and fresh rock faces to weathering, which pulled huge amounts of carbon dioxide out of the atmosphere, cooling the planet.

The mountains that were built grew along what is now the East Coast of North America. It was the beginning of the Appalachian Mountains.

6

An Ancient Forest
at Gilboa

Middle Paleozoic Era: 443.8 to 358.9 Million Years Ago
Silurian and Devonian Periods

I t can be a challenge, even when given more than a cursory look, to
distinguish between the two rock formations that lie beneath New
York City: the Manhattan Formation and the Hartland Formation.
Charles Merguerian of Hofstra University, who studied both for years,
has described the rocks of the Manhattan Formation as having "brown-to-
rusty-weathering" and containing the minerals biotite and muscovite. He
has described those of the Hartland Formation as having "brown-to-tan-
weathering" and containing biotite and muscovite, as well as plagioclase
and quartz. Thus, as should be evident, much of field geology is in the
discerning eye of the beholder. Whether or not one might possess a similar
ability, I suggest seeking out and examining these same two rock formations, which, by convenience, are readily accessible and in close proximity
to each other in Central Park.

Umpire Rock, one of the favorite places in the southern part of the park,
serves as a natural grandstand for those who gather in the spring and the
summer to watch games played on nearby baseball fields. It is one of the largest
exposures of rock in the park. It is part of the Hartland Formation. Its surface
is covered with great swirling lines that show how the rock was twisted and
folded when it was still deep inside the Earth and subjected to high

temperatures and high pressure. Several hundred feet to the northeast of Umpire Rock is a brick-covered octagonal building that houses a large carousel. Near there are rocky outcrops of the Manhattan Formation. It, too, has swirls that indicate that these rocks were once deep inside the Earth and once subjected to high temperatures and high pressure.

Because these two rocky outcrops are so similar in appearance, one is tempted to conclude that there is not much of a geologic story here. But that is wrong. The two formations originated in different places and under different conditions. The fact that they are now adjacent is a consequence of a major event in the geological history of North America.

It was the collision of two tectonic plates—one that included the continental mass of Laurentia and the other a mostly seafloor plate with scattered arcs of volcanic islands. The rocks of the Manhattan Formation came originally from the continent. They were sand and silt that had been weathered and eroded and washed into a nearby shallow sea. Those rocks of the Hartland Formation are different. They were originally muds that had accumulated slowly on the bottom of a deep ocean. The collision pushed the deep sediments of the Hartland Formation up against the sand and silt of the Manhattan Formation. As the collision continued, a large arc of volcanic islands rammed into Laurentia. Some of the rocks of the Manhattan and the Hartland Formations were forced upward to form mountains, while other rocks of these formations were shoved downward, where they were heated and compressed.

The collision began about 470 million years ago and lasted about thirty million years. This was the first *orogeny*, the first episode of mountain building, that would lead to the Appalachian Mountains of today. It produced Umpire Rock and the rocky outcrops near the carousel in Central Park. This was the Taconic Orogeny, named for the Taconic Mountains in eastern New York.

To be clear, the ancient mountains that were formed hundreds of millions of years ago during the Taconic Orogeny are *not* the same as the modern Taconic Mountains. Nor are any of the other nearby mountain ranges—the Hudson Highlands, the Berkshire Hills, or the Green Mountains. Each of these other ranges formed long after the Taconic Orogeny by other orogenies that contributed to the building on the Appalachian Mountains.

The mountains formed by the Taconic Orogeny were leveled long ago by erosion. And yet, much can be learned about them by following where the contact between the rocks of the Manhattan and the Hartland Formations runs today. The contact is known as Cameron's Line, named for Eugene Cameron of the University of Wisconsin, who, in the 1950s, was the first to follow and describe this discontinuity in the geologic record.

Cameron's Line has been found as far north as western Massachusetts. From there, it runs south, passing close to the towns of Torrington and Danbury in Connecticut. It passes beneath the eastern suburbs of White Plains, New York. Admittedly, exposures are difficult to find in these areas. But an excellent one is in the Bronx at Boro Hall Park (also known as Tremont Park) where a rocky outcrop of the Manhattan Formation lies just outside the right-field fence of a baseball field while an outcrop of the Hartland Formation is found near the left-field fence. The actual point of contact between the two formations is about thirty feet east of home plate.

From Boro Hall Park the line continues south. It loops through a section of Queens, then follows the East River—Cameron's Line is probably responsible for the alignment of this section of the river—makes another loop, this time through Manhattan, where it passes through the southern part of Central Park. It ends, as best as anyone has been able to follow it, on Staten Island, where there is an important clue as to how this line, and the ancestral Taconic Mountains, formed.

Recall that a block of green serpentinite lies in the center of Staten Island. It is the reason for the vegetative barrenness of Todt Hill, the highest point along the Atlantic seaboard south of Maine. Todt is a Dutch word that means "dead." Dutch settlers probably gave it this name because this and surrounding hills were devoid of trees. The reason for this is the chemical makeup of serpentinite. It is rich in chromium, nickel, and cobalt, which are toxic to plants. It came from a deep part of the ocean crust. That this block now sits on the surface and against a continental mass means it was transported a great distance. Similar blocks of serpentinite can be found elsewhere in the New York region.

A small block forms the promontory at Castle Point in Hoboken, New Jersey. Two other blocks lie along the edge of Long Island Sound at

New Rochelle and near Port Chester in New York. An excavation for a new water-supply tunnel unearthed a block of serpentinite beneath Brooklyn. Another was found a few years ago by workers digging a deep foundation for a building along Sixth Avenue in Manhattan. These all lie where they do today between of the collision that formed the Taconic Mountains.

Blocks of serpentinite can be found along a long line that begins at least as far south as Alabama, passes through the Carolinas and Virginia, through New York City, and continues northward to Newfoundland. These blocks are remarkably easy to find because the land surfaces above them—as at Todt Hill—are almost devoid of dense vegetation, appearing as open grasslands surrounded by forests. One of the largest of these is at Soldiers Delight Natural Environmental Area in Maryland. Another is at Nottingham Park Serpentine Barrens in southern Pennsylvania, the name betraying the barrenness created by the underlying serpentinite.

When seen through the discerning eyes of a geologist, this long alignment of serpentinite blocks is a clue that a long mountain range once stood there. And there are other equally compelling clues.

A thin line of dark volcanic rocks runs between Bridgeport, Connecticut, and Hanover, Vermont. This thin line was once a large volcanic island, probably similar to the volcanic islands of Indonesia today. The island had formed out at sea. The seafloor tectonic plate that was colliding with Laurentia carried this island up to and smashed it against the continent, helping to raise the ancestral Taconic Mountains. The island was deformed and compressed. The southern end of the island is exposed today at Webb Mountain Park in Connecticut, where the hard volcanic rock is a favorite of rock climbers. The center of the island lies close to Shelburne Falls, Massachusetts, where, in the downtown section along the Deerfield River, a dark lava erupted from the ancient island when it was still far out at sea.

Though evidence for the ancestral Taconic Mountains is compelling, one still wonders: If these mountains were so majestic, where is the debris that was carried away as the mountains were eroded? That answer, too, is easy to find.

Some of the debris was swept down and into the sea and is part of the Manhattan Formation. But there are two sides to every mountain range.

Debris was also swept down and across the continent's flooded interior. Here it accumulated as the Queenston Delta, a vast fan-shaped deposit of sedimentary rock that is best revealed today in numerous exposed cliffs found across much of eastern Ontario and western New York and across Pennsylvania and into Ohio.

The growth of the Taconic Mountains occurred during the Ordovician Period. So did their erosion. The leveling of these mountains occurred quickly, spurred on by the rapid chemical weathering of the many steep and few rocky surfaces. Carbonic acid in rainwater—formed by the dissolving of atmospheric carbon dioxide in raindrops—was the main agent that attacked and chemically broke apart those surfaces. As the weathering proceeded and as debris was swept away by water and wind, new rocky surfaces were exposed and they, too, were subject to the same chemical attacks. The net result was the rapid pulling of a large amount of carbon dioxide from the atmosphere. That, in turn, lowered the temperature, which caused a new glaciation and a mass extinction.

When the planet warmed, much of Laurentia was again covered by a shallow sea. As the warming continued, the seas again became tropical. Marine life proliferated; though this time, it was dominated by the growth of broad reefs—a key characteristic of the next geologic period, the Silurian.

◆

To see the Silurian Period, go to Chicago. The entire city rests on bedrock of that age, ancient reefs that today line much of the western shoreline of Lake Michigan. Reefs of a similar age cover much of eastern Indiana and western Ohio, but it is near Chicago, just a few miles south of the city center, that the most famous and most intensely studied Silurian-age reef in the world can be found.

The reef in question is the one revealed at the Thornton Quarry in Thornton, Illinois. Though the quarry is now closed, during the nearly century-long period that it operated, it provided most of the crushed stone and lime that was needed to construct the roads and the many buildings of the modern city. To remove such a vast amount of material, workers had to cut deep into the reef. They exposed entire vertical sections of the reef, which stand today as great vertical walls, some as high as three hundred feet. And within those walls, for those who take the time do to a detailed study, is a rare snapshot of the diversity of Silurian-age marine life.

There are corals and trilobites and clam-like brachiopods. Ancestors of modern squids and octopuses have been found, some nearly ten feet long. There are also records of great storms that pummeled the reef with crashing waves. The powerful waves ripped apart sections of the reef, leaving what appear today as great geologic scars in the quarry walls.

A study of the walls and the surrounding area show that the reef, at its greatest extent, was nearly two miles in diameter. It has a slightly oval shape. The pattern of shell fragments and sand grains show that, as it was growing, wind-driven waves were coming mostly from the southwest.

The reef revealed at the Thornton Quarry is just one of many. Most reefs cannot be distinguished. Most, as they grew, coalesced. And the result is huge platforms of Silurian-age reefs that have not yet been exposed, except for their top layers. But there is an exception. An unusual succession of reefs has been exposed in the walls of a long *escarpment*, or cliff, that runs intermittently for hundreds of miles, and that is known as the Niagara Escarpment, because a short section of the cliff is where Niagara Falls formed.

The western end of the Niagara Escarpment begins north of Chicago near the small town of Ashippun, Wisconsin. From there, it runs north and is responsible for the few prominent topographic features of western Wisconsin, including the Door Peninsula that forms the narrow piece of land that separates Lake Michigan and Green Bay. The escarpment continues north, then arches its way and runs east, forming the northern shorelines of Lake Michigan and Lake Huron. It is picked up at the headlands at Tobermory at the northern tip of the Bruce Peninsula in Ontario. It then runs generally south as far as the city of Hamilton, where the Niagara Escarpment bisects the city—this giant steplike feature is known locally as The Mountain—and where it gives rise to more than a hundred waterfalls and steep cascades, some descending more than a hundred feet.

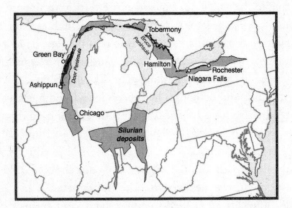

The Niagara Escarpment (shown in black) is a high cliff that runs for several hundred
miles from Ashipunn, Wisconsin, to Rochester, New York. It is responsible for
hundreds of waterfalls, including Niagara Falls. Silurian-age rocks are exposed in the
escarpment as well as across northwestern Illinois, eastern Indiana, and western Ohio.

At Hamilton, the escarpment takes a sharp turn to the east and passes
into New York, where it played a minor role in the history of the state—and
of the nation. It was the 1820s and the Erie Canal was being constructed.
One of the major challenges in construction of the canal was taking the
canal path up and over the Niagara Escarpment, a barrier that is a least
eighty feet high in this part of New York. Some investors grew nervous
and withdrew their support. But, after months of planning and the inven-
tion of a new way to dredge on a steep slope, canal engineers overcame the
obstacle. They designed what is now the famous Flight of Five: five locks,
closely spaced, that would raise or lower barges over the escarpment and
that gave rise to the city of Lockport, New York.

But the most famous section of the Niagara Escarpment is not at
Lockport nor fifty miles east at Rochester, New York, where the escarpment ends. Nor is it to be found among the many waterfalls and cascades
near Hamilton or along the shoreline of the picturesque Door Peninsula.
Instead, the most famous section is where the Niagara River drops over
the escarpment and forms Niagara Falls.

The reason for the falls at Niagara—and at hundreds of other places
along the escarpment—is the extreme hardness of the Silurian-age reefs

that formed in this region of the continent. They are composed mostly of dolomite, a magnesium-rich form of limestone. Dolomite is much harder than regular limestone (or calcium carbonate) and considerably more resistant to being abraded or crushed or broken apart than either shale or sandstone.[1]

At Niagara, the river drops over a cliff that is capped by dolomite. Beneath the dolomite are beds of shale and sandstone. As water cascades over the falls, it churns in a plunge pool. The churning sends up a spray that erodes the shale and the sandstone, which causes the beds to break away and removes support for the dolomite, which then crashes down as large blocks. The net result is that Niagara Falls is moving slowly upstream.

Originally, the Niagara River cascaded over the Niagara Escarpment at Queenston Heights near Lewiston, New York, seven miles downstream from the current falls. The repeated eroding and washing away of the shale and the sandstone and subsequent collapse of the dolomite caused the falls to retreat upstream. Today, the flow of water is managed, and the rate of retreat is a few inches per year. At that rate, it will take about fifty thousand years for the falls to retreat as far as Lake Erie, after which there will be no major falls along the Niagara River, only a series of steep rapids.

The retreat of the falls has left behind a gorge. Lining the top of both sides of the gorge is, of course, the hard dolomite. Immediately beneath it is a series of light-colored shales and sandstones that are also Silurian in age. And in the lowest part of the gorge, downstream of what is the second most famous section of the Niagara River—the Whirlpool—at the base of the wall of the gorge, there is a horizontal rocky unit with a distinctive red color. It is largely shale with some sandstone. This is the Queenston Delta, material swept down during the erosion of the ancestral Taconic Mountains. There is a feeling of comfort—and reassurance—in finding these rocks in the correct stratigraphic position beneath the Silurian-age reefs, to actually see the transition from the Ordovician to the Silurian periods.

It is also comforting to note that just twenty-five miles away, upstream along the Niagara River, at Buffalo, New York, the next transition of geologic periods can be found, from the Silurian to the Devonian. That transition is at the Peace Bridge, where riffles transition into slow-moving water.

◆

Do not be disappointed if it is difficult to find the transition from the Silurian to the Devonian Period in the moving water or in the surrounding rocks. The beginning of the Devonian Period is difficult to find anywhere in North America. The transition from the Silurian to the Devonian Period is much more pronounced in the British Isles, where the distinction was first made. It was originally established at the base of the Old Red Sandstone, a rock that can be followed across Wales and Scotland and across Norway. It is this transition that is so obvious at Siccar Point in Scotland that Hutton used it to convince his two friends that the Earth must be very old.

At one point, it was proposed that the extinction of graptolites should mark the end of the Silurian and the beginning of the Devonian Period. But these tiny animals that lived in skeletal colonies were found as fossils in rocks with a great range in age, and so that was abandoned. Today the beginning of the Devonian Period is defined by the first appearance of a particular type of graptolite, *Monograptus uniformis*, and the type section (the golden spike) is marked by a bronze plaque attached to what has been designated as Rock Bed Number 20 of the Klonk outcrop northeast of the village of Suchomasty in the Czech Republic.

But from the perspective of North America, it is much harder to distinguish rocks of the Silurian and the Devonian Period. In many ways, the Silurian Period was a preamble to the Devonian Period. Marine life had recovered after the mass extinction at the end of the Ordovician Period. Tropical seas covered much of Laurentia. Reefs were growing. The first fish with jaws developed during the Silurian Period. The first plants moved onto dry land. During the Devonian Period, this would multiply into ecological communities of great complexity. The ocean would fill with fish. And the land would finally turn green.

◆

The Devonian Period was the Age of Fish. More specifically, it was the age of fish with jaws. It was a time period when large fish fed on small fish. And larger fish fed on the large fish. Sharks first appeared and became abundant and diverse. The familiar ray-finned fish of today—represented in modern form by salmon, cod, herring, and tuna—would arise and have only modest success during the Devonian Period, but would dominate the later seas. What was foremost in the Devonian seas—at least, foremost in terms of terror—were the top predators, the placoderms, the fish with movable jaws and armored skin.

Dunkleosteus was the king of the Devonian seas. It had a jaw like a guillotine and a had body covered with bony armor. It grew up to twenty feet in length and weighed more than a ton. It could chop sharks into chum. When it opened its great mouth, the suction pulled in prey. Then it bit down with a bite that rivals the iron-tight claps of modern alligators.

The first remains of *Dunkleosteus* were found in 1867 by amateur paleontologist Jay Terrell in the cliffs along the edge of Lake Erie near the town of Sheffield Lake. He named this animal the "Terrible Fish." In the 1930s, paleontologist David Dunkle of the Cleveland Museum of Natural History found additional specimens of the terrifying fish. In all, ten species have been identified. The largest was eventually named *Dunkleosteus terrelli* in honor of Dunkle and Terrell.

The Devonian seas contained the first ammonites, today highly sought by collectors for the tightly coiled fossilized shells. The Devonian was a period when reefs became more complex and included both the fragile twiglike rugose coral and the highly durable tabulate corals. Trilobites were still abundant. Eurypterids, forerunners of modern scorpions—which, during this time period, had long sharp spikes on their appendages that they probably used to stab prey—swam in the Devonian seas, and reached lengths of more than six feet. A more wicked-looking marine animal capable of swimming in short darts is hard to imagine. In short, the Devonian was a period when the seas were teeming with life. But that is only half of the story.

The other half occurred on land. After hundreds of millions of years of being confined to the watery parts of the planet, both plants and animals began to move onto the land.

◆

It is more difficult to live on land than in water. On land, gravity is forever holding one down, limiting one's mobility, causing one's limbs to sag. Temperature changes are more extreme on land than in water, which means that a land organism has to develop a way to maintain a stable temperature, never being too hot or too cold. And there is the constant threat of desiccation, of losing too much water to sustain metabolism or other essential life functions, while at the same time avoiding being drowned by a sudden onrush of too much water during a torrential downpour or a flood.

On land, nutrients had to be found either by hunting or by taking them up directly from the ground. They did not come floating by twice a day with the reliable flow of a tide. And then there were the deadly rays of ultraviolet radiation from the Sun. Such rays were energetic enough to destroy cellular structures and stop cellular functions, such as reproduction, unless there was a barrier that prevented such harmful rays from reaching the ground. Fortunately, such a protective barrier had been in place high in the atmosphere in the form of a layer of ozone that had existed since the Great Oxygenation Event, and that had become more protective as more oxygen was added to the atmosphere.

For some modern life-forms, such as arthropods, it looks deceptively easy to move from water to land and back again. But these creatures have had hundreds of millions of years to evolve tough external shells or skeletons that protect them from the extremes of changing environments. It is an evolutionary challenge. Vertebrates in particular made the transition from water to land only once, though they have managed to return to water several times, indicated by the range of aquatic birds and mammals, such as penguins, whales, and seals. For plants, the initial transition from water to land probably occurred long before it did for animals, but then hundreds of millions of years passed before much happened in their evolution.

The field of molecular genetics suggests that some type of primitive plant lived on land long before the Cambrian Explosion of animals, perhaps when the supercontinent Rodinia was still forming. But what these early land plants may have been—sheets of algae, perhaps—is still unknown. What is known is that the first firm evidence of land plants in the fossil record are in Ordovician-age rocks. These are fossilized spores of plants whose modern relatives are mosses and liverworts. These early plants never came to terms with the challenge of living on land. They avoided the issue of gravity by remaining small and squat. They ignored the problem of desiccation by drying out to a crisp when water disappeared, recovering and springing back to life when water returned.

The first true innovation for living on land came 420 million years ago when land plants developed internal tubular structures that could move water to different parts of the plant. The earliest such vascular plant was *Cooksonia*. It was as simple a vascular plant as can be imagined. It had no roots. Instead, it sprouted short horizontal stems. Its few branches were slender and leafless with Y-shaped forks. Each stem ended at a tiny pod that released microscopic spores. *Cooksonia* stood no higher than an inch or two above boggy plains, but that was enough to tower over ancient mosses and liverworts and to assume access to unshaded sunlight.

Reconstruction of *Cooksonia*, the oldest known vascular plant. It had a stem, but no leaves, flowers, or roots. Fossils indicate it stood as high as two inches (photo courtesy of Matteo De Stefano/MUSE).

The first specimens of *Cooksonia* were discovered in the British Isles and described in 1937 by British botanist William Henry Lang, who named them in honor of a collaborator, Isabel Cookson of Australia. Cookson had done much of the pioneering work in the study of land-plant evolution. Fossil imprints of *Cooksonia* have now been found world-wide, though only one area in North America has yielded specimens. Until 1973, only a single specimen of this early land plant had been discovered—at the Ridgemount Quarries near Fort Erie, Ontario. Since then, a few specimens have been found in the same general area, including a few near Buffalo, New York.

In essence, *Cooksonia* was nothing more than a few naked stems with tiny pods at the end of each stem. It was from this humble beginning that all plants would evolve. And the initial stages of evolution came quickly. In the world of plants, it was comparable to the Cambrian Explosion for animals.

New and complex body plans developed. From those, sophisticated life cycles would arise. There would be a constant race to grow taller, a continued pursuit for more sunlight.

At the beginning of the Devonian Period, it was a Lilliputian world of plant species. The tallest ones were still measured in inches. And plant species were still confined to estuaries and tidal flats. Sixty million years later, by the end of the Devonian Period, the planet would be covered in green. Plants would thrive far from rivers and streams. They would grow on hillsides. Much of the ground cover would be forests with trees more than sixty feet high. The roots from these trees would penetrate deep into the ground, changing the mode and rate of erosion, and thereby changing the way that rivers moved silt and sand across the landscape.

The change in erosion would pull great amounts of carbon dioxide from the atmosphere, plunging the planet, yet again, into a global glaciation—and eliminating much of life in another mass extinction. But before these dramatic changes occurred, three evolutionary innovations would be added to plants: leaves, roots, and seeds.

◆

It took thirty million years for these innovations to occur, to go from *Cooksonia* to plants with leaves, then roots, and, after a pause, seeds. Leaves were tiny at first, then got progressively bigger, dominating many ecosystems. But it took thirty million years. That seems a long time for stems to evolve into leaves and roots—and spores into seeds. And so scientists have steered away from thinking these innovations came from changes in the genetics of land plants. Instead, the long delay between the first vascular plant and the first leaves and roots was an environmental factor. And the arbiter of so much of the evolution of plants and animals is a rapid drop in the amount of carbon dioxide in the atmosphere.

Ironically, it was probably the success of early land plants such as *Cooksonia* that caused the levels of carbon dioxide to plummet. These small and leafless plants probably spread quickly across wet coastal plains. Such rapid growth would have pulled large amounts of carbon dioxide out of the atmosphere. Furthermore, coastal plains are the places that were most likely to be buried by debris being washed down off the continent. The burial of plants would have prevented the carbon in the plants from being released back into the atmosphere after the plant died. In any case, there was a dramatic drop in carbon dioxide levels just before the first appearance of land plants with leaves.

The oldest plant fossils with discernible leaves are the few rare specimens of *Eophyllophyton bellum* recovered from 390-million-year-old Devonian rocks in China. These miniature leaves are exquisitely preserved as thin films of carbon, black smears on rock surfaces less than a tenth of an inch in size. Even so, it is possible to see intricate networks of branching veins within the fossilized tissues. And, either simultaneously or soon after leaves developed, the next innovation in plant evolution came: roots.

Unlike leaves, roots are difficult to recognize in the fossil record, which explains why the evolutionary development of early roots is still poorly known. Roots lack the intricate outlines and geometric patterns of leaves, and so are often difficult to see in the fossil record. But there is a site where roots of early land plants are well preserved. It is at Miguasha National Park in Quebec.

This national park is a favorite for everything that is Devonian. Thousands of early fish fossils and early plant fossils have been recovered. But what is unique at this site is the record of early plant roots. They appear as tapering, clay-lined casts and molds in red sandstone beds. This was the beginning of a new style of erosion: The creation and penetration of roots not only broke up the underlying rock, but also accelerated soil formation, which produced a new type of habitat that would support the first animals that would venture onto land.

And soon after the first leaves and roots appeared, the first seeds made their debut. Seeds have a great advantage over spores in that they already contain embryos of plants and they carry their own food sources, the carbohydrates, proteins, and fats that are necessary for the initial development into a plant. Spores are little more than undifferentiated cells wrapped in protective coating. They require water for further development, and so plants that reproduce by generating spores are confined to be near sources of water. Seeds can be dispersed across the land and wait for the right conditions to germinate and develop into adult plants.

And so it finally came to pass, late in the Devonian Period, that seed-bearing plants covered the planet. And animals that came from the sea followed them.

The oldest-known air-breathing animal is a millipede that was found in 414-million-year-old rocks near Stonehaven, Scotland.[2] That means it lived about the same time as *Cooksonia*, and so animals seemed to have started to move onto land at about the same time that vascular plants developed. And as plants spread across the land surface—as leaves and roots and seeds developed—so did animals.[3]

Spiders and mites, centipedes and millipedes, possibly early forms of lobsters and crabs were the first to make the transition onto land. The four-legged tetrapods came later—but not too much later. *Tiktaalik*, discovered in Arctic Canada in 2004, was technically a fish, but had bony structures in its fins that allowed it to move as most four-legged animals do. *Ichthyostega* had lungs and limbs and probably moved through shallow water in swamps and, possibly, made forays onto dry land. Both lived 375 million years ago.

Whatever animals did venture onto land during the Devonian Period entered an entirely new ecosystem. Plants had been growing taller as they competed for more sunlight. Complex root systems were coursing through the ground as a means to anchor the tall-growing plants more securely to the ground. These two factors combined to produce what were—and what are still today—the highest concentration of biomass on the planet. These were the first forests.

◆

The first great forests on the planet were dominated by *Archaeopteris*, a plant with a treelike trunk topped by fernlike fronds. It grew to heights of eighty feet. Its fossilized remains have been found on every major landmass, and so forests of *Archaeopteris* probably once covered much of the planet.

Evidence of a forest of *Archaeopteris* is clearly indicated at Miguasha National Park. A steep sea cliff along the southern shoreline of the park has yielded an abundance of fragmented fossilized fronds and the occasional carbonized pieces of trunk. Intermixed with these fragments and pieces are fossils of Devonian fish, including placoderms, which suggests that storm waves occasionally inundated and tore across this section of ancient coastline, drowning the *Archaeopteris*, which then quickly regrew and established a new forest.

The first forests were of densely packed trees of *Archaeopteris*, a plant with fernlike leaves that grew up to eighty feet (twenty-five meters) in height (photo courtesy of Falconaumanni).

A much less spectacular Devonian-age outcrop—though one that is more accessible and where the collection of fossils is permitted—lies along a short section of Highway 120, west of the small town of Hyner in north-central Pennsylvania. This outcrop illustrates the happenstance nature of finding good sites because this particular site was uncovered by chance. It is the only exposure of rock of the Late Devonian Period within thirty miles. That it is well exposed today is due to a steep vertical cut made through the rock to construct a highway near a river. The location is known as Red Hill, a reference to a series of bright red river deposits exposed at the road cut. Here there is a wide range of fossilized flora and fauna of Devonian age, including the fragile remains of *Archaeopteris* intermixed with tiny, equally fragile remains of fossilized fish.

Though the first great forests to cover the land were of *Archaeopteris*—and so, in the eyes of many, *Archaeopteris* should be regarded as the world's first "tree"—there were isolated groves of much smaller plants that stood tall and that lived a few million years earlier than *Archaeopteris*. These groves were confined mainly to the edges of estuaries. The dominant plant in these groves was *Eospermatopteris*.

Unlike *Archaeopteris*, *Eospermatopteris* contained no wood. It had no leaves. The root system was more limited for *Eospermatopteris* than *Archaeopteris*. The trunk of *Archaeopteris* grew both upward and outward, and so *Archaeopteris* could have massive trunks a few feet in diameter. The trunk of *Eospermatopteris* grew mainly upward and, at most, was a foot or so in diameter, which meant it was limited to grow no more than about twenty feet in height. In form, *Archaeopteris* had a general resemblance to the giant conifers of today. *Eospermatopteris* would have had a general similarity to modern coconut palms.

But *Eospermatopteris* is older. And though few specialists regard it as a tree, it did grow in groves that, though limited in extent, should be regarded as the world's first forests.

The oldest forest of *Eospermatopteris* is in New York. Among the first specimens to be described were those collected near the small town of Gilboa in the 1860s and sent to James Hall, then director of the New York State Museum of Natural History. Hall identified the specimens as the trunks of an unknown type of plant. Little more was done to understand these specimens—or to collect more—until 1920, when excavations for a dam were started near Gilboa.

The plan was to build a dam that would create a water reservoir that would fill the valley where the town of Gilboa was located, the reservoir to supply water to the growing needs of New York City. In all, more than four hundred buildings were destroyed, either razed or burned, including three churches, two boarding houses, and many small stores and houses. Seven cemeteries and family burial plots were relocated. The new town of Gilboa sits immediately downstream of the dam. Those families that trace their histories back to the old town of Gilboa still voice resentment about the dam today.

As the excavations were underway, workers began to uncover more fossilized trunks. The New York State Museum of Natural History was contacted. It sent out one of its most experienced paleontologists, Winifred Goldring, to collect specimens.[4] She continued to work at the site until 1926, when rising water prevented further work. By then, she had collected more than forty new specimens. She gave them the name *Eospermatopteris*, from the Greek, meaning "dawn of the seed fern." She determined, from their bulbous bases, that they must have grown under swampy conditions. She also determined that at least three successive forests had grown at Gilboa, each one ultimately destroyed by the sea and buried. The bulbous bases had been secured to the ground with long strap-like roots that had anchored the plants in soft black muds.

After the dam was completed, the excavation sites were filled. They remained covered for the next ninety years. No additional study of the Gilboa forest would be made until 2010, when it was decided to reconstruct the dam and remove the quarry backfill. That gave a new generation of paleontologists a chance to study the world's oldest forest.

Hundreds of large sandstone casts of *Eospermatopteris* were recovered. More importantly, evidence for other types of trunk-based plants were found, suggesting that even the earliest forests were diverse ecosystems.

Then in 2019, a slightly older fossilized forest was found about twenty miles east of Cairo, New York. It, too, contained roots that indicated the presence of palmlike *Eospermatopteris*. But the ancient environment at Cairo had been drier than at Gilboa. There were many more root systems of individual plants and they were larger and more complex. Some were symmetrical, whereas others showed marked directionality. One root system, in particular, radiated from a central trunk that was a foot or so in diameter. There were more than a dozen primary roots running outward as far as thirty feet. From the great size and complexity of the root system, might this be an example of the woody, conifer-like *Archaeopteris*? Perhaps.

What is clear is that complex forests did exist at least as early as 388 million years ago. And that they had an initial heyday that lasted about twenty million years. Their end came when they fell victims to their own success.

Today forests are the dominant terrestrial ecosystem on the planet. There is, by far, more biomass in land plants than in all of the other components of the biosphere combined. In essence, forests are giant reservoirs of carbon. And when they grow—including when they first developed—they pull massive amounts of carbon dioxide out of the atmosphere, the reservoir of carbon consisting initially of trunks and roots and later leaves.

Furthermore, the development of root systems meant a new style of erosion came to the continents, one in which roots penetrated and broke apart the rocky surface. That led to the development of deep soils that could be swept away by wind and water.

And as the rate of erosion increased, so did the amount of nutrients that flowed down rivers and into the oceans. There was a blooming of microbial life, which drew down the amount of carbon dioxide in the atmosphere further. And when the microbial life died, they sank and decomposed, a process that used up most of the oxygen in the oceans.

The result was a dead zone. Anything that relied on oxygen—meaning fish, coral, and other marine animals—died. A dark sludge accumulated on the floors of the shallow seas. And then a third element entered to change the planet, and redirect the history of North America.

The planet was already cooling by the drawdown of carbon dioxide caused by the growth of forests and by the blooming of microbial life. The motions of the continents now enter the picture. Laurentia was still, more or less, an isolated continental mass lying close to the equator. As the Devonian Period was closing, two other isolated continental masses, Baltica and Avalonia, came crashing up against Laurentia. It was the second episode in the growth of the Appalachian Mountains—the Acadian Orogeny.

As those new mountains grew, they, of course, weathered and eroded, which pulled yet more carbon dioxide out of the atmosphere. And as the debris swept down the steep mountainsides and into the shallow seas, it blanketed the black sludge of oxygen-poor, carbon-rich material that was already accumulating at the bottom of those same shallow seas.

Massive carbon-filled, energy-rich deposits that have been so crucial to powering our highly mobile and industrial society were now starting to accumulate on the planet.

◆

The Appalachian Basin is one of the places where such carbon-filled, energy-rich deposits have accumulated. The basin runs from the Adirondack

Mountains in New York to Alabama, from the western flank of the Blue Ridge Mountains to the crest of the Cincinnati Arch and the Nashville Dome. When one thinks of carbon-filled, energy-rich deposits in the Appalachian Basin, one immediately thinks of coal, but a few tens of millions of years of evolutionary development would be necessary before Appalachian coal would begin to form. The black sludge that accumulated on the bottom of the oxygen-poor shallow seas of the Devonian Period became a different type of rock. It is a black shale.

Most of the black shale that accumulated within the Appalachian Basin is deeply buried. But there is an excellent exposure at the northern edge of the basin. It runs just south of Marcellus, New York, easily found on the southern side of Lee-Mulroy Road, south of the local fire station and south of the intersection with Slate Hill Road. The rock is jet-black and flaky. It is composed of tiny plants and animals and microscopic marine life that lived when microbial life was blooming in the shallow seas, when the seawater was depleted of oxygen, and so the tiny plants and animals and microbial life did not decay. The carbon within them was not released. Instead, the energy-rich molecules that comprised these forms of life are still entombed in the shale. That is where they have remained for nearly four hundred million years—that is, until a few decades ago. Those energy-rich molecules are now being removed by a technique known as hydraulic fracturing, or fracking.

The Marcellus Shale is the largest source of natural gas in the United States, a source of energy that is far more efficient and cleaner than either petroleum or coal. It lies beneath most of western New York and western Pennsylvania, eastern Ohio, and essentially all of West Virginia. The natural gas is trapped within the shale. To be removed, fluid that contains a variety of chemicals that improve the efficiency of the fracturing, and that protect the well from corrosion, is injected into the ground. Those chemicals have been contaminating groundwater, which has prompted the state legislators and the governors of New Jersey and New York to stop the use of hydraulic fracturing in those states. The technique is still used in Ohio and Pennsylvania, where the number of wells and the amount of natural gas being produced continue to increase.

◆

The formation of massive deposits of black shales, such as the Marcellus Shale, marks a milestone in planetary history. From this time forward in planetary history, energy-rich deposits of carbon in a variety of forms—as black shales or as coal, as tar sands or as petroleum—will form during discrete episodes and under unusual environmental and evolutionary conditions. And so it is tempting, given that these deposits have been so important in human history, to recognize this milestone as the formal beginning of a new geologic period. But the first production of black shales was not a global event; it did not leave a global stratigraphic horizon. And so the Devonian Period must end in another way.

The Devonian Period ends, as did the Cambrian and the Ordovician Periods, with a mass extinction, though the end-of-the-Devonian mass extinction has a complication. It was at least two, and may have been as many as seven, individual extinction events spread out over tens of millions of years. But it did end, as did the Cambrian and the Ordovician mass extinctions, with a cooling of the planet and with a glaciation. Sea level dropped and the shoreline regressed across the continent exposing more land and leading to more erosion.

When the planet warmed again and shallow tropical seas again covered Laurentia, new marine life and new plant life developed. Next would come the thick limestone deposits that cover the center of the continent. And the massive coal deposits that lie along the western edge of the Appalachian Basin. And the final growth of the Appalachian Mountains, which would be completed when a new supercontinent formed.

7

Fires, Forests, and Coal

Middle Paleozoic Era: 358.9 to 298.9 Million Years Ago
Carboniferous Period

I ts common name is sea lily. It has a flexible stalk with a broad base
that secures it to a rock or a submerged log or to a patch of hard sand
and keeps it from moving. At the top of the stalk is a feathery plume that
gives the appearance of a flower, hence the name. But it is not a flower.
Nor is it a plant of any kind. A *crinoid* is a marine animal. It has a mouth
and an anus, both located close to the base of a feathery plume. And the
plume itself is actually tentacles that filter food from the water and bring
that food to the mouth.

A crinoid is a type of echinoderm, closely related to starfish, sea urchins,
and sea cucumbers. Its first appearance in the fossil record was during the
Ordovician, which makes crinoids much older than dinosaurs. But, unlike
dinosaurs, they are still around.

Today crinoids live in deep water around the Bahamas and in the Straits of
Florida and along the outer fringes of tropical lagoons in the Chuuk Islands
and in Fiji. But if one wants to see crinoids that were alive when they were in
their heyday, when there were vast "meadows" of these "flowery" creatures
on the seafloor, their plumes swaying in rhythm with the waves, then one
should go to Burlington, Iowa, the self-proclaimed Capital of Crinoids.[1]

Fossil crinoids from Iowa (photo courtesy of Sanjay Acharya,
Smithsonian National Museum of Natural History).

After the end of the Devonian Period, after a brief retreat of the sea,
the sea advanced again. The shoreline moved across the continent from
southwest to northeast, the water deepening over what is now the Mississippi Valley. As a result, outcrops of crinoid-rich limestone are almost
continuous from Iowa to Alabama. They reach from western Illinois to
western Kansas. At Burlington, Iowa, a high bank along the Mississippi
River consists almost entirely of crinoid fossils. For that reason, this extraordinarily fossil-rich rock is known as the Burlington Limestone.

Nearly three hundred species of crinoids have been found in the Burlington Limestone. By comparison, fewer than a hundred species of crinoids
with long stalks are alive today. Those found in the Burlington Limestone
have been mostly fragments of what were once whole animals, an indication that these marine creatures had died quietly, their bodies slowly dis-
integrating after they died. That was not the case at another site, a quarry
more than a hundred miles away in Le Grand, Iowa, where the conditions
of the fossilized crinoids tell a different story.

Some of the most exquisite examples of fossilized crinoids found anywhere in the world have been found near Le Grand. Though they are fewer
in number than at Burlington, whole animals have been recovered, often in
clusters. The preservation of complete bodies indicates that these fossilized

animals are in a "death position"; that is, they died suddenly, probably when a large wave rolled over them, flattening an entire field of crinoids, ripping the individuals from their tight holds on the seafloor and killing them quickly, then covering them with a layer of limy mud.

The Burlington Limestone is one of several major limestone deposits found in the Mississippi Valley that formed during this geologic period—*sub*period, to be exact—that is known, quite conveniently, as the Mississippian.

The Burlington Limestone formed about 345 million years ago. The Salem Limestone, which is found in Missouri, Illinois, Indiana, and eastern Kentucky, is slightly younger. It has an age of 340 to 335 million years. It formed farther from shore and in deeper water than the Burlington Limestone, and so it has fewer fossils. Those that are most common are *foraminifera*, one-celled animals that made calcareous shells.

Because the Salem Limestone formed farther from land and in deeper water, it contains fewer layers of mud and silt that were washed in from the land than does the Burlington Limestone. That makes the Salem Limestone more homogenous. It also makes it more suitable as a building stone because it can be quarried into large blocks that would not be weakened or disrupted by the occasional thin layer of mud or silt. If one made an accounting, one would probably find that the Salem Limestone is the most common building stone used in the United States.

It has been used at more than thirty state capitol buildings. It is the interior walls and columns of the Lincoln Memorial in Washington, DC. It forms the border stones and steps of the Capitol Reflecting Pool, also in Washington, DC.[2] It is the stone that covers the exterior of the Empire State Building in New York. So much stone was needed in New York—and it had to be of a uniform color and texture—that it was quarried from a single large hole, the "Empire Hole," as it is known, located near the small town of Oolitic, Indiana.[3]

The celebrated group of Minerva, Hercules, and Mercury atop the facade of New York's Grand Central Terminal is carved from Salem Limestone. It was the stone used to build the National Cathedral in Washington, DC, and the complex of buildings known as The Crescent in Dallas, Texas. And it is the stone that covers thousands of residential buildings known as "greystones" in Chicago. And it was the stone used to reconstruct the Pentagon after the terrorist attacks of September 11, 2001.

Limestone deposited during the Mississippian subperiod is found across much of the central and western United States. It is the Redwall Limestone of the Grand Canyon. It is the Madison Limestone that runs from the Black Hills of South Dakota to Colorado. The dramatic cliffs along the Missouri River east of Helena, named the Gates of the Mountains by the Lewis and Clark Expedition, are Madison Limestone. So is Sleeping Indian Mountain east of Jackson Hole, Wyoming. The Escabrosa Limestone in Saguaro National Park in Arizona and the Leadville Limestone in the San Juan Mountains of Colorado have the same Mississippian age. All of these formed in the same shallow, clear, limy sea.[4]

◆

While great masses of limestone were accumulating on the seafloor, much was happening on land. In particular, after the Devonian extinction, forests again expanded and covered much of the land surface. That had the dual effect of pulling yet more carbon dioxide out of the atmosphere—the formation of the limestone deposits had done the same—and, through photosynthesis, injecting huge amounts of oxygen into the atmosphere. It was during this second expansion of forest, during the Mississippian subperiod and the immediate aftermath, that oxygen would rise to the highest level in the Earth's history.

The level of oxygen would eventually reach about 35 percent before it would go down again. (Today the oxygen level is 21 percent.) That increase would have a profound effect, most noticeably on the insects and other arthropods that were scampering around that were dependent on the diffusion of oxygen into their bodies for respiration.

Insects and other arthropods would grow to enormous sizes. It was a period of gigantism. A fossil dragonfly has been uncovered that has a wingspan of thirty inches. Its body was proportionally as large, making it the size and weight of today's rock pigeon or Mississippi kite. A spider with eighteen-inch legs then walked the planet. Millipedes more than eight feet long scurried through the forests, hiding and searching for prey and leaving behind tracks that look like the treads of miniature tractors. And there were scorpions, presumably deadly, that weighed as much as fifty pounds.

There was yet another major consequence of the high level of oxygen: For the first time in the planet's history, fires raged across the land.

The earliest evidence of fire on the planet is from the Silurian Period. It consists of what appears to be charred remains of small low-growing plants, though some scientists have suggested these were patches of fungi or algae, that lived 440 million years ago. The oxygen level was then about 13 percent, just at the threshold when fires could ignite and burn.

The evidence for fires during the Devonian Period is much clearer. The fossil site known as Red Hill in south-central Pennsylvania contains ample evidence of ancient fires in the form of charcoal, though it takes an experienced eye to find it. During the Devonian Period forests of *Archaeopteris* had spread quickly. And, yet, the fragmented fossils of *Archaeopteris* found at Red Hill do not show signs of having burned. The scattered pieces of charcoal are of shrubs and ferns, indicating that these were low-temperature fires that kept close to the ground and left the trees unscathed.

The earliest evidence anywhere of a major fire is in northwestern Ireland near Donegal. There a black, charcoal-rich layer overlies a sandstone that formed along the edge of a large estuary. The surrounding countryside was covered with dense forests when a wildfire swept through 325 million years ago. The fire burned across several thousand square miles.

Such huge fires were common during the Mississippian subperiod. The evidence is in the ample amount of soot that was left behind. Known as *fusain*, this black grimy material is found in many deposits that formed on land during this time.

It is especially prominent in Pennsylvania, Ohio, Indiana, and Illinois and is found as far south as Oklahoma and Alabama. This increase in fires was also caused by a shifting of the continents. Laurentia had already collided with Baltica and Avalonia to form a large continental mass, Euramerica. Because of its great size, much of Euramerica had a monsoonal climate that was subjected to alternating seasons of heavy rain and extreme dryness. It was, of course, during the dry season when fires were most common, though fires probably raged throughout the year. In fact, for tens of millions of years, much of the surface of the Earth was probably shrouded beneath a planet-wide haze of fire-produced soot.

How the soot-heavy air may have affected the evolution of animals is still debated. But it is almost certain that the giant dragonflies, the spiders with eighteen-inch legs, and the fifty-pound scorpions rarely, if ever, had a clear sunny day. Plants, however, did respond, especially, trees, which developed a new type of bark.

Archaeopteris and *Eospermatopteris* were both extinct. A new tree, *Lepidodendron*, now dominated the forests. It is often incorrectly described as a "giant club moss," but, if *Lepidodendron* has a living relative, it is more likely to be the often overlooked quillwort, a plant that is common to most swamps of North America and that few people can readily identify because it has a truly unremarkable appearance similar to that of a small sprouting onion. But in the distant past, *Lepidodendron* was a giant.

It grew to heights of nearly two hundred feet and stood on bases that were as much as six feet across. *Lepidodendron* consisted mainly of a central long trunk that was topped by a small crown of thin branches and tiny leaves. It was an arrangement that many have compared to that of a giant spear of asparagus.

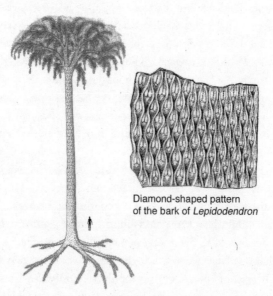

Diamond-shaped pattern
of the bark of *Lepidodendron*

Much of the early coal deposits formed from the bark of *Lepidodendron* trees. These trees could grow up to 160 feet (50 meters). Impressions of the diamond-shaped pattern of its bark have often been sold falsely as dinosaur skin (photo courtesy of Falconaumanni).

Fossils of this ancient tree are abundant, especially pieces of its trunk. The trunk left readily recognizable impressions in the muck and the mire of the swampy ground in which *Lepidodendron* lived. The impressions show a pattern of diamonds arranged in rows that spiral around the tree. On first look, it seems to be a pattern you might expect from an impression of, say, a dinosaur skin, and many fossil impressions of *Lepidodendron* have been sold as such by unscrupulous people who deal in fossils. But the spiraling diamond-shaped pattern is purely botanical.[5] Though *Lepidodendron* is often called the "scale tree," these are not animal scales. They are the scars left by leaves that dropped as the tree grew.

Lepidodendron was mostly soft tissue and contained little wood. Its structural integrity—the support for the long trunk—came from a ring of thick bark. The evolutionary development of this bark was probably an adaptation to the frequent fires of the period, a way to make the plant more resilient to the heat and flames. The bark grew to a thickness of a foot or so, which means much of the mass of this tree was in its bark.

Furthermore, *Lepidodendron* grew in dense stands of two hundred trees per acre or more. In comparison, modern trees that are comparable in size, like the Ponderosa pine and Douglas fir, usually number a few dozen individual trees per acre in today's old-growth forests.

It was this combination—thick massive bark and dense stands of swamp-land trees—that gave an unexpected turn to Earth's geologic history—and, as it turned out, to the course of civilization.

Storm waves and high winds were inevitable. Dense stands of *Lepidodendron* would topple into the swamps. And when they fell, they—and their massive rings of bark—would be cut off from the oxygen in the atmosphere and so be prevented from decaying. There they would lie, their high-energy carbon preserved in the swamp water.

Meanwhile, later forests would grow on top of them. Then they, too, would be toppled and add to the mass of accumulating energy-rich material. Sea level would rise and fall and rise again. With each rhythm, layers of sand and silt and mud would cover what was now a thick mat of organic material. Debris would wash down from mountains and bury the material further, compressing it and heating it and transforming the bark of *Lepidedondron* and the tree's other components, as well as the ferns and the other plants that grew nearby and the occasional long-legged spider or huge scorpion or giant dragonfly that got caught in the swampy muck, into a carbon-rich, civilization-altering combustible rock: coal.

◆

There are several reasons why coal became the most-used fossil fuel in the world. It burns slowly and at a steady rate. It can be used almost directly out of the ground with little or no processing or refining. It is easy to transport long distances and exists in vast quantities. Only a fraction of all of the coal that has ever formed has been taken out of the ground. There is enough coal still buried to supply the world with energy for hundreds of years.

The transformation of civilization by coal came in the eighteenth century. It was coal that powered the Industrial Revolution. It made it possible for industries to move from generating power by water wheels and windmills to producing it by more versatile and powerful steam engines, which derived their energy from the burning of coal. Factories could now be larger and more productive. Ships powered by coal-fired steam engines could cross oceans faster and more reliably than those that relied on wind and sails. Steam locomotives became commonplace, carrying people and goods between cities and across continents. And, in 1879, when Thomas Edison lit his first light bulb, he did so with electricity supplied by a coal-burning generator.

Today coal continues to run most of the world's economy. Almost half of electricity is produced by coal. The production of steel relies on blast furnaces that use coal. The high-temperature kilns that are used to melt limestone, silica, and iron oxide to produce cement, the world's most common building material, require coal.

Though it is used in great abundance, coal is not evenly distributed around the world. The vast majority of the deposits are found in just five countries. India, China, Australia, and Russia have significant amounts. But the king of coal, the country with the largest coal reserves and, until recently, the major user of coal, is the United States. And this is owed to the unusual geologic circumstances of Laurentia long ago.

Coal has formed in almost every geologic period. It formed before the Cambrian Period, which was before land plants. Such early coal deposits are few in number and amount and are probably owed to the accumulation of algal mats. Only during three distinct episodes has coal formed in vast quantities. And each time the southern section of Laurentia, what is now the central region of the contiguous United States, was perfectly poised to accumulate vast amounts of this energy-rich, easily recovered rock.

The most recent episode of coal formation occurred about fifty million years ago, when the global climate was extraordinarily hot and humid. This extreme state lasted a few million years. At that time, the world's largest swampy forests covered what is now northeastern Wyoming and southeastern Montana. It is from these forests that the largest coal seams in the world—some more than a hundred feet thick—developed, seen today around Gillette, Wyoming, in the Powder River Basin.

An earlier episode occurred about a hundred million years ago. These coal deposits are not as extensive as those of the other two episodes, though they are substantial when compared to what is found in most other countries. These are the coal deposits being mined today in Colorado, Utah, and New Mexico.

The great period of coal production—the one that is most often depicted in panoramic displays in museums and in textbooks—is the one that occurred immediately after the Mississippian subperiod when forests of *Lepidodendron* covered the swampy parts of the Earth. This is the coal of Appalachia and of the Illinois Basin. This is the coal that is being removed at strip mines in Illinois, Indiana, and Ohio and being dug out of tunnels in Kentucky, West Virginia, and Pennsylvania.

In a survey of the geology of North America, there is a noticeable change from the deposition of limestone during the Mississippian subperiod to the deposition of coal. For this reason, a second subperiod has been created, the Pennsylvanian. These are the only two subperiods in the Geologic Time Scale. And that has led to some confusion, especially among those who are first introduced to geology.

One must remember that it was British geologists who first devised the Geologic Time Scale. And they did so during the early nineteenth century, before much of the geology of North America had yet been surveyed. In devising the geologic time scale, British geologists noted that the coal that was then being mined in England and Wales and used to power the beginning of the Industrial Revolution came from a distinct period of geologic time, a period that followed the Devonian Period and that was labeled as the "Carboniferous." Then, near the end of the nineteenth century, after broad surveys of the geology of the United States had been completed, geologists in North America abandoned the Carboniferous terminology and introduced two new periods, the Mississippian and the Pennsylvanian. And that is how it stood for most of the next century. Finally, in 1999, members of the International Commission on Stratigraphy formally settled the matter and voted to recognize the Carboniferous as a geologic *period* and divided it into two *subperiods*, the Mississippian and the Pennsylvanian. And, yet, some British scientists continue to resist using the new designations, referring to the Mississippian and Pennsylvanian as the Lower Carboniferous and the Upper Carboniferous.

Nevertheless, if one wants to see and study great coal seams exposed in hillsides and road cuts, one should travel to North America, especially in Appalachia and the Illinois Basin. And that is what Harold Wanless of the University of Illinois did. For forty years, beginning in the 1920s, he sampled and described individual coal beds and the layers of sand and silt that lay between them. He consulted with others who were doing similar work. Maps were drawn. Individual beds were followed, some for hundreds of miles. Some beds were found that traced out broad river channels. But the most important discovery he made was that there was a natural rhythm to the deposition of coal. There was sandstone that had formed upon a surface that had been cut by sea waves. Then a layer of silt that had formed by the flow of freshwater. Then coal. Then sandstone, silt, and coal again. This rhythm was seen all across the landscape. And Wanless gave it a name: *cyclothem.*[6]

Cyclothems are common in the eastern coal regions of the United States. They are impossible to miss in the rock cuts along the highway that connects Pikeville and Sidney in eastern Kentucky. What is revealed in these road cuts are cutaway views of river deltas where the river channels shifted back and forth across valley floors and where plant material and, hence, coal accumulated. In Ohio there is a spectacular cyclothem exposed north of the town of Cambridge along I-77. Here the freshwater silt is at road level. Above it is the coal, then the sandstone.

Many cyclothems have been named. The Dennis Cyclothem is exposed a mile or so southwest of Dennis, Kansas. The Plattsburg Cyclothem can be found exposed along a road cut west of Altoona, Kansas. Of particular note is the Wabaunsee Cyclothem, which rises as a broad hump along the side of I-70 west of Topeka. This cyclothem has been traced for hundreds of miles from West Texas and across Kansas and Illinois and into the hills of Appalachia, an indication of how continuous swamps were during the Pennsylvanian subperiod. In Illinois, four cyclothems are exposed along the western bank of the Vermilion River near Lowell, Illinois, though some hiking is required to see these rhythmic cycles of sand, silt, and coal.

Wanless also proposed a cause for the rhythmic change. He suggested it was controlled by fluctuations in sea level caused by cycles of glaciation. When ice accumulated at one of the Earth's pole, water was drawn out of the oceans and sea level dropped, allowing a large swampy plain to form across what is now the eastern portion of the United States. Then, when the ice melted, sea level rose again, sending sand over the same area. It was a bold suggestion, one that was not immediately accepted because Wanless had no explanation as to why vast coal deposits had developed rhythmically during this particular period of time. But today we know why.

Wanless was working before the ideas of plate tectonics and shifting continents were widely accepted. And that is one of the keys. During the Pennsylvanian subperiod, Euramerica—the former Laurentia now enlarged by the additional of the continental bits of Baltica and Avalonia during the Devonian Period—straddled the equator. And it was the only continental mass that did so. All other continental material had gathered and formed the large landmass of Gondwana, which, during the Pennsylvanian subperiod, sat over the South Pole.

Global temperature had dropped since the Devonian Period. And an ice cap had formed across Gondwana. The ice cap now went through a series of advances and retreats. Each advance drew water out of the oceans and sea level fell. A retreat returned water to the oceans and sea level rose. But what had caused the series of advances and retreats? That, too, was unknown to Wanless, though, again, today it is clear.

One has to fast-forward to the most recent few million years of the Earth's history when cycles of ice ages were happening on the Earth. These, as will be discussed more fully later in this book, were caused by cyclic changes in the Earth's orbit around the Sun. Specifically, the shape of the Earth's orbit changes slightly from being circular to slightly elliptical and back again. That changes how heat from the Sun is distributed across the planet. And that leads to the advance, then retreat, of an ice cap.

That is what is on display at the hundred or so cyclothems that can be found in the eastern United States. The road cuts near Pikeville and the steep bank exposed along the Vermilion River tell of a time when Euramerica straddled the equator and the giant landmass of Gondwana was at the South Pole and was covered by a major ice cap. And the Earth was going through its normal oscillatory motions as it orbited the Sun.

But that is not the whole story of the Mississippian and the Pennsylvanian subperiods. There is one more critical element, one that involves the building of two mountain ranges and the final assemblage of the east coast of North America.

◆

Euramerica is the product of the collision of Laurentia and two much smaller continental landmasses, Baltica and Avalonia. Those two landmasses have since departed, separating from what is now North America when the most recent supercontinent broke apart. But it is important to realize that they and Laurentia were part of the same large landmass during the Carboniferous Period. And so as the limestones of the Mississippian subperiod and the coal beds of Pennsylvanian subperiod accumulated across the middle section of Laurentia, the same was happening on Baltica and Avalonia. And when they separated, Baltica became what is now northern Europe and Avalonia what is now much of the British Isles—which explains why there are extensive limestone and coal deposits in northeastern France, northern Germany, and Poland, as well as in England and Wales.

Besides the collision of Avalonia and Laurentia yielding an abundance of coal to England and Wales, it had another long-lasting effect: It created the Acadian Mountains, the second mountain mass to be added to the Appalachian Mountains.

The Acadian Mountains effectively replaced the Taconic, creating a massive range that was similar in location and in extent. They ran from Newfoundland to Georgia and Alabama. But when Avalonia pulled away from Laurentia—it had originally been part of Gondwana and had been rafted across an ocean by a seafloor tectonic plate—much of Avalonia remained intact. Only small pieces of it were left behind and are to be found on North America today.

The ones that are most easily seen are at the northern end. Avalonia is named of the Avalon Peninsula of southeastern Newfoundland, which is composed entirely of this ancient continent fragment and where one finds Ediacaran fauna at St. John's and Mistaken Point. It is the rocks of southern New Brunswick and of northern Nova Scotia. In Maine it is the slate and schist that forms the Schoodic Peninsula, a popular place where people gather to watch pounding surf during rough seas, and that forms Mount Desert Island. These two places are part of Acadian National Park. Here one can see dark veins of basalt wedged into light-colored granite. The wedging of those veins is a mark of the volcanic activity that occurred during the Acadian Orogeny.

The region around Boston, including Cape Ann, is part of Avalonia. So is all of Rhode Island and the southeastern section of Connecticut, including all of the coastal towns from New Haven to Stonington. The seam where Avalonia welded against Laurentia is well exposed at an outcrop of bedrock in Deep River, Connecticut. It is along a small ridge next to the entrance to Devitt Field, the town's baseball field. The rock on the west side is a dark schist of Laurentia; the rock on the east side is a lighter-colored gneiss with some laminations and is part of Avalonia. It is an example that the evidence of major geologic events can sometimes be found in unexceptional places.[7]

The collision between Avalonia and Laurentia produced the Acadian Mountains. The mountain building began about 380 million years ago and lasted for a few tens of millions of years. Afterward there was tectonic quiet across this section of the newly formed Euramerica. The mountains eroded, causing thick piles of sediments to accumulate across southern New York and central Pennsylvania, sediment that would provide the right amount of burial for the swampy forests to be turned into Pennsylvanian coal. After another sixty million years or so, the quiet was broken when Gondwana, which had been moving steadily northward, slammed into Euramerica. The result was the third and final episode in the formation of the Appalachian Mountains: the Alleghenian Orogeny.

◆

The aptly named Valley and Ridge Province is a series of closely spaced, long parallel valleys and ridges that run from New Jersey to Alabama. Here the rocks are of many different types and of many different ages. The harder sandstones and limestones form the tops of the ridges while the softer and more easily eroded shales comprise the floors of the valleys. This arrangement of long parallel ridges was an almost insurmountable barrier to early migrants who were traveling west to settle the Ohio Country and the Northwest Territory. Today it is still a headache for those who try to route and plan highways to connect distant communities. But to geologists, this regular arrangement of valleys and ridges is the classic example of a fold-and-thrust mountain.

The folds are best seen in Pennsylvania, where they culminate with such regularity and intensity that they are often known—with no vulgarity intended—as the Pennsylvanian climax. Though vegetation is often heavy across this part of Pennsylvania, there are places where a complete series of folds can be found. A notable one is along a new road cut near Macedonia, Pennsylvania, where giant waves in the rocks form the ridges.

To see the thrusts, it is best to go south. At Sharp's Gap in Knoxville, Tennessee, there is a cut through one of the more prominent ridges. Here one can see a Cambrian-age sandstone that has been thrust over an Ordovician-age limestone, a reversal of the usual stratigraphic order. This particular thrust is actually a pronounced feature in the Valley and Ridge Province, one that can be followed from Georgia almost all the way to New Jersey. It represents a fault, the Saltville Fault, where two large blocks have slid against each other. The maximum throw along the fault—that is, the total amount the two blocks have slid against each other—is more than two miles.

The inevitable conclusion to be reached from the folds and the thrusts is that a major section of the Earth has been compressed and shortened. We can also conclude that the Earth's crust has buckled and slid because of a great collision. And where there is a great collision and compression, there is heat and the melting of rock, rock that may find a way to rise and cool and solidify near the surface. That is the origin of the granite found between Atlanta and Richmond, including the granitic dome of Stone Mountain near Atlanta, where the largest bas-relief in the world has been carved, the Confederate Memorial Carving.[8]

The folding and thrusting is also the origin of the Shenandoah Valley in Virginia and the Great Valley in Tennessee, where an abundance of soft rocks has rendered the landscape smoother than the Valley and Ridge Province. This action is also the origin of the high hills of the Blue Ridge, including the Great Smoky Mountains, a massif that was pushed up by the collision of Eurameria and Gondwana.

The collision did not occur simultaneously along the entire stretch of the Alleghenian Orogeny. It began when what is now southern Morocco pushed against the New Jersey shoreline. Then it progressed south. Because Gondwana was so huge—South America was then attached to Africa—and because it was rotating slowly clockwise, what is now Colombia, Venezuela, and Guyana rammed up against Euramerica—the landmass of Cuba was caught up in the middle of this collision—and formed the Ouachita Mountains of western Arkansas and eastern Oklahoma, the Marathon Uplift of southwestern Texas and a major mass of mountains that reached across Chihuahua and Sonora in Mexico.

In all, a two-thousand-mile-long belt of mountains was formed. It stretched from New Jersey to Baja California. As described above, the eastern half is well exposed. The western half, the Ouachita-Marathon-Sonora segment, is mostly hidden beneath younger rocks. But there are a few places of easy access that a glimmer of the effects of this orogeny can be seen. Nearly vertical beds of sandstone are revealed in road cuts east of Marathon, Texas. Those who have the time to stop and study these beds in detail and identify the sedimentary structures—the cross-bedding, the graded bedding, and so forth—will learn that the stratigraphic top of each bed is to the east.

In the southern part of the Ouachita Mountains is the Broken Bow Uplift that runs from Little Rock, Arkansas, to Broken Bow, Oklahoma. Here there are parallel valleys and ridges that are reminiscent of the Appalachians. And, where exposed, the ridges can be seen to be composed of tilted and folded and fractured rock. But the most intriguing feature here that is linked to the Alleghenian Orogeny is not at a rocky outcrop. It is the scalding water at Hot Springs National Park.

The hot springs are on the southwestern slope of the Zig Zag Mountains, so named for the unusual chevron pattern in the valleys and the ridges when viewed from above. But there is no volcanic activity here, no geologic hotspot, that supplies heat to the water. Instead, the water is heated by the natural geothermal gradient—that is, by the normal increase in temperature as one descends into the Earth.

The source of water is rain that fell about four thousand years ago and has been slowly percolating through the mountains. The water has penetrated several thousand feet down into the Earth along cracks created by the tremendous pressures produced along this region of North America when Gondwana slammed into Euramerica. Then, because of the unusual chevron pattern of the valleys and ridges, the hot water is funneled into a small area. The flow of hot water is further confined—and pressurized—by impermeable layers of chert that guide the hot water to the surface. The result is a series of closely spaced, hot artesian wells.

Large bathhouses have been built over the wells, the original owners touting the therapeutic benefits that these miraculously hot waters could provide. For many years, especially during the first half of the twentieth century, thousands of visitors came each year to patronize the establishments in order to partake of the waters and to be pampered by bath attendants and massage therapists who could contribute to a visitor's sense of well-being. There was even a minor industry in the bottling and selling of the waters as a curative agent.

But, alas, there is no firm evidence of a medicinal quality to the waters at the hot springs of Arkansas, except from the general feeling of relaxation that one gets by soaking in a hot pool. The waters are safe to drink and lack the sulfurous odor of most hot springs, a characteristic that betrays the nonvolcanic source of the heat.

There is also a steady supply of hot water. Enough water comes surging out of the ground each day to fill an Olympic-sized swimming pool. And, regardless of the season, it comes out at a remarkably constant temperature of 143° Fahrenheit (62° Celsius), which is midway between the temperature of normal bathwater and of a hot cup of tea or coffee.

There are also public pools at three different temperatures. The one with the fewest number of patrons is filled with water that has come directly out of the ground. It is rare that anyone withstands this water for more than a few minutes. A second pool is mixed with a sufficient amount of cold water—which also emanates from the Zig Zag Mountains—to provide a pleasant soak. The third is of water that is cool enough for children to play in and where most people congregate. And beneath them, providing the pathways—the plumbing—for rainwater to percolate downward then rise to the surface, is a system of fractures produced by the Alleghenian Orogeny.

◆

That coal originated from plant material—and was not, as was once generally assumed, a precipitate of seawater—was first proposed in 1718 by French botanist Antoine de Jussieu. He had been working in the hills around Saint-Chamond near Lyon digging down through a bed of shale where he was finding impressions in the shale that were reminiscent of tropical ferns. He continued to dig. Eventually the shale passed down into a bed of coal where more fernlike impressions were found. From such slender evidence, he came to a grand conclusion. The strangeness of coal—the fact that it was a rock that could burn—came from its origin as plant material that had grown in a tropical climate. Nearly a century passed before his idea was widely accepted.

By then, coal was driving the Industrial Revolution. That meant there was great interest in understanding more about this remarkable rock. Industrialists wanted to know where it could be found and how it could be removed from the ground and transported efficiently. From that came the ideas of drawing geologic maps and of building networks of canals. Those who were promoting the fledging science of geology were wondering why coal had accumulated where it did and why it appeared in thick beds that extended over large distances. Among those early geologists who were keen to answer such questions was Charles Lyell.

In volume two of his immensely popular book *Principles of Geology,* first published in 1832, Lyell suggested that coal beds had originated from great masses of driftwood that had accumulated at the ends of major rivers where they reached the sea. As an example, he cited the Mackenzie River in Canada, where travelers had reported seeing such huge piles of driftwood that had been prevented from being swept out to sea by barrier islands and shoals. This suggestion by Lyell was the high point of the drift theory of coal accumulation. He soon changed his mind—and modified subsequent editions of his book—after he made his first trip to North America.

Lyell was on the continent of North America for twelve months beginning in July 1841, traveling through both the United States and Canada.

His first major stop was at Niagara Falls, where he noted that the slow erosional rate of the falls, indicated by historical records, and the existence of a long, narrow gorge downstream of the falls confirmed his notion that the Earth had a deep history. He traveled to Pennsylvania, where he examined coal mines and compared them to the ones in Britain. He spent time in the Appalachian Mountains and described the system of parallel ridges and valleys as being arranged like "so many wrinkles and furrows." He went to Savannah, Georgia, where he studied the recently discovered bones of a mastodon. And he made a special trip to a remote location in Nova Scotia, one that required weeks of travel, in order to see an ancient forest of fossilized trees exposed along the Joggins Cliffs.

The Joggins Cliffs run for miles along the southern shoreline of the Bay of Fundy. Here the tidal range is high and winter storms are frequent and harsh. The two combine to erode away the steep sea cliffs at a ferocious pace. Long sections of new cliff faces are exposed every few years. What is visible today might be gone next week. But then there will be a new collection of fossils to study.

The rocks exposed in the Joggins Cliffs are Pennsylvanian in age, a time when life was prolific and diverse. Fossilized trackways made by the giant millipede *Arthropleura* have been found here. So, too, are there imprints of the giant dragonfly *Megasecopterid*. There are fossilized cockroaches and amphibians, marine shells and snails. There are large casts of *Lepidodendron* trees still in upright positions. And there are multiple layers of coal sandwiched between layers of shale and sandstones.

It was the relationship between the upright trees and the layers of coal that interested Lyell and caused him to make a trip to Nova Scotia. Before his departure for North America, he had read a description of the Joggins Cliffs and its fossilized trees written by Abraham Gesner, a Halifax physician, who had been to the site many times and had studied it and who had effused in publication that it was a place "where the delicate herbage of a former world is transmuted in stone."

Gesner met Lyell in Nova Scotia and guided him to the site. They spent three days walking the shoreline and climbing as high as possible up the steep cliffs. The first of the upright trees that they saw had no part of the original plant preserved except the bark, which was now a tube of pure coal filled with sand. The bark was a quarter of an inch thick and was marked on its outer side with irregular longitudinal ridges and furrows. The diameter of the tree was sixteen inches at the base and fourteen inches at the top, five feet and eight inches above the base, where it had broken off. A second tree was nine feet in length and rested on a seam of coal, one foot thick. On the same foundation of coal stood two other large trees, each about two-and-a-half feet in diameter and fourteen feet long. And there were many other, similar examples. At each one the base of the tree rested on a bed of coal or shale, never sandstone. Moreover, there were multiple levels of upright trees. Lyell counted ten. (Today, after much research, more than sixty distinct levels of upright trees have been identified, indicating a long succession of forests.)

From these observations, one can conclude that coal had not formed by the accumulation of driftwood but by the repeated flooding and burial of dense swamp forests. Lyell made the correction in subsequent editions of his book. Today this all seems obvious, but there was a time when it was not so clear.

◆

There is much, much more to see and consider at the Joggins Cliffs. The strata, originally horizontal, have been tilted an immodest 20°, a consequence of the Acadian and Alleghenian orogenies. That has brought more than four million years of continuous geologic history within easy reach of anyone who strolls along the beach for a mile or two.

It is at Joggins Cliffs that the earliest land snail has been found. And the earliest known reptile, *Hylonomus lyelli*, the progenitor not only of dinosaurs but also of all birds and mammals. Hundreds of skeletons of early reptiles have now been found at the Joggins Cliffs. Most were discovered at the bottom of stumps of upright fossil trees where the unfortunate individuals were apparently trapped by floods or rising tides and died.

The fossils are so well preserved and in such abundance that there is a term for such a site: *Lagerstätte*. It is a German word that originally meant an area that contained mineable concentrations of ore. Paleontologists now use it to identify an area of exceptional fossil preservation. The Archean Gunflint Chert and the Cambrian Burgess Shale are Lagerstätten. So is Fossil Butte of the Green River Formation in Wyoming, where tens of thousands of fossilized fish have been found, and the Carnegie Quarry in the Morrison Formation in Utah, where more than fifteen hundred dinosaur bones have been uncovered.

In North America there are two Lagerstätten that represent the coal age of the Pennsylvanian subperiod. Joggins Cliffs is one. The other is near Mazon Creek near the town of Braidwood, Illinois, known both for its diversity and abundance of fossils and for the unusual way the fossils were discovered.

Until the 1970s, Pit 11 of the Peabody Energy Company of St. Louis was an active coal strip mine. For more than fifty years, coal was removed by first stripping away an upper layer of shale and piling it up into isolated pyramids. It is from these great heaps of spoils that hundreds of thousands of fossils have been found.

The fossils are inside iron concretions, hard iron-rich balls about the size of a fist that formed soon after the death of an organism. A careful tap of a rock hammer splits a concretion open. Inside might be a beautifully preserved fern leaf, a millipede, a scorpion, or what is locally known as a Tully monster, *Tullimonstrum*, a sluglike marine creature with a long proboscis with eight tiny teeth at the end. Except for the teeth, *Tullimonstrum* has no hard parts. Its existence is only known from the Mazon Creek fossil beds. Because its fossils are abundant, this creature, which grew up to fourteen inches in length, must have once swarmed as schools in seas that once cover what is now northern Illinois.

More than three hundred species of plants and four hundred species of animals have been found at Mazon Creek. These include fish hatchlings and juveniles with egg yolks still attached. There are jellyfish and shrimp and horseshoe crabs. Much of our knowledge of Pennsylvanian insects comes from the 150 species that have been found inside Mazon Creek concretions. Some concretions contain pieces of bark of the tree *Lepidodendron*.

Though diverse and abundant, the Mazon Creek fossils represent an end. Soon after their formation—309 million years ago—the production of coal stopped abruptly. And it did so because the Carboniferous rainforests collapsed.

◆

The simple explanation for the Carboniferous Rainforest Collapse is that the climate shifted suddenly from hot and humid to cold and dry. It happened as the collision of Euramerica and Gondwana was ending and the supercontinent Pangea had formed.

Having a single large continental mass meant that the distances were too great for monsoonal rains to reach most of the interior of the continent. There was also a greatly reduced length of coastline, which limited the amount of area that marine life could migrate to as the climate changed. And there was the huge rain shadow produced by the long range of mountains produced by the Alleghenian Orogeny and that now stretched across the middle of Pangea. This rain shadow also contributed to a drying of the land.

Lepidodendron became extinct, as did many other plants. The Carboniferous Rainforest Collapse was only one of two mass extinctions of plants on the planet. The other would come fifty million years later during the greatest extinction in the Earth's history, the Permian extinction.

Many amphibians also went extinct. They relied on water to deposit their eggs and on water for juveniles to develop into adults. And so this was the time when the amniotes, those four-legged animals that produced hard-shelled eggs that came packaged with their own watery environments, began to proliferate. And foremost among the amniotes were reptiles—*Hylonomus*, in particular—because they not only laid watery-filled, hard-shelled eggs but also had scaly skins that could protect them from the increasingly arid conditions.

And there is one final twist to the story of the abrupt end of coal production and the growth of Pangea. During the final stages in the assembly of Pangea, a giant shallow basin was formed in the region now known as West Texas. It was here that some of the world's largest reservoirs of oil would accumulate.

8

The Great Dying

Late Paleozoic Era: 298.900 to 251.902 Million Years Ago
Permian Period

Members of the Chamber of Commerce of Midland, Texas, have recently been promoting their city with a new slogan: "Midland: In the Middle of Somewhere." Gone is the image, so the members hope, of a small West Texas town where the ground is scorched by the Sun and where the air is filled with wind-driven dust. The new image is one of prosperity—great prosperity. And it is owed to what lies thousands of feet beneath Midland.

Midland sits at the center of a huge basin, the Permian Basin. The basin stretches from just south of Lubbock to nearly the Rio Grande River and from San Angelo westward to the southeastern part of New Mexico. It is filled with sedimentary rocks that were deposited mainly during the geologic period that came after the Carboniferous—that is, during the Permian Period (hence the name of the basin). In fact, the thickest section of Permian rocks anywhere in the world is in West Texas and New Mexico. In places the accumulation is nearly thirty thousand feet. And it is what is contained within those rocks that has made Midland both prosperous and the focus of national and international attention. Here, beneath Midland and a wide expanse of West Texas, is the world's largest reservoir of oil.

The last statement needs a qualification. The first commercial oil well in West Texas began production in 1921. By 1993, a respectable, though not a record-shattering, fifteen billion barrels of oil had been recovered from the region. An assessment of the nation's oil resources made two years later by the United States government concluded that there were only a few billion barrels of oil yet to be pumped in West Texas. An assessment made in 2018 estimated that there were at least twenty-seven billion barrels and perhaps as much as seventy-two billion barrels. If the last number proves true, then the amount of oil that lies in the Permian Basin rivals that of the Ghawar field in Saudi Arabia, currently the world's largest oil field. To put this in another context: If Midland and the other communities of the Permian Basin were part of OPEC (the Organization of Petroleum Exporting Countries), then the amount of oil in West Texas would make them the fourth largest member of OPEC, after Saudi Arabia, Iran, and Iraq.

But why was there a jump in the amount of oil that might lie beneath the Permian Basin? There were no major new discoveries of oil in West Texas. It was simply a matter of how oil is taken out of the ground.

The first estimate, made in 1995, was based on the amount of "conventional" oil: that is, the amount of oil that could be recovered by the means then available, by the drilling of a vertical well down into a rock layer where oil, under natural pressure, could flow into the well and be pumped up to the surface. Since then, major advancements have been made in oil-drilling technology. It is now possible to do directional drilling, which means it is possible to drill a well in any direction, even horizontally, and, as a well is being drilled, to change direction. It is also possible to recover "unconventional" or "tight" oil. This is liquid oil that is trapped between closely packed grains of shale. To allow the oil to flow, the shale must be fractured by injecting a fluid at high pressure into a well, a process known as fracking. It was two advancements—directional drilling and fracking—that greatly increased the estimate of how much oil could be recovered from the Permian Basin.

Today, thanks to these advancements, four million barrels of oil are pumped each day from wells in West Texas. That represents 40 percent of all of the oil currently being produced in the United States. The recovery of oil from the Permian Basin is one of the main reasons that, since 2018, the United States has been the world's largest producer of oil.

And that raises a question: Why did so much oil accumulate in such a small region of the continent?

◆

The source of most oil is microscopic plants and animals—phytoplankton and zooplankton, often cyanobacteria or foraminifera—that grew in inland seas or large bays where they multiplied into great numbers. When the phytoplankton and zooplankton died, they settled to the bottom of the inland sea or bay where they mixed with mud and silt that was being washed off a nearby land surface, forming an organic-rich ooze. If the ooze was then buried by more silt and mud, sealing it off from oxygen and decomposition by bacteria, the temperature of the ooze would rise, causing it to change chemically into a waxy substance known as kerogen. If burial continued, the kerogen could be changed into oil.

The amount of additional burial was the key. If too much material was added and the burial was too deep, the temperature would rise too high and only natural gas would be produced. If not enough material was added and the burial was not deep enough, then temperatures would not be high enough to convert the kerogen into oil molecules. That sweet spot in temperature, the "oil window," is between 140° and 250° Fahrenheit (60° and 120° Celsius), which corresponds to a burial range of five thousand to twenty thousand feet.

And then there is one final condition for a reservoir of oil to form. Because oil is a liquid and is lighter than the surrounding rock, it will seep upward and reach the surface unless it encounters a barrier that stops the flow of oil and causes it to accumulate.[1] Such barriers are often impermeable rock layers of salt or gypsum or similar material that has formed by the evaporation of seawater.

The Permian Basin met these conditions. And it met them multiple times.

Most oil-rich regions have one or two places in the rock sequence where oil has formed and accumulated. In the Permian Basin there are *eleven*! In Midland, one will hear conversations about the multiple "plays" in the Wolfcamp Shale or the Bone Springs Formation.[2] It is these multiple oil plays that gives West Texas the distinction of having one of the largest—if not *the* largest—reservoir of oil in the world.

And if one also asks the question: Why were these conditions met multiple times in the Permian Basin? The answer is: It was a consequence of the collision of Euramerica and Gondwana and the formation of the supercontinent Pangea.

At the western end of the line where the two great landmasses collided, a large inland sea formed, one that was connected to the greater waters of the ocean by a narrow seaway. Much is known about this inland sea because it is now the location of the Permian Basin. It is one of the most well-studied geologic regions of the world because of the oil. For example, it is known that the inland sea actually consisted of three smaller seas, each one contained within a separate basin. The smallest of the three was the Marfa Basin, which was located closest to the ocean. The Midland Basin was the largest and the farthest east. And between the two was the Delaware Basin, where much of the oil has been discovered, and which was connected to the Marfa Basin and to the ocean through what is known as the Hovey Channel.

The collision occurred close to the equator, and so the water was warm and life was prolific, fed mineral nutrients from material washed down from mountains that had grown during the Marathon-Ouachita segment of the Alleghenian Orogeny. The collision was also causing the surface to buckle and the three basins to deepen. Moreover, because the Hovey Channel was narrow, the three basins were occasionally cutoff from seawater. And the water within the basins evaporated, leaving behind layers of impermeable salt and gypsum.

This was repeated for almost fifty million years. There was a proliferation of both phytoplankton and zooplankton that left an organic-rich ooze. Then a drying of the basins and burial of the ooze beneath sediments and salt and gypsum. The burial heated the ooze and changed it to kerogen, then to oil. Seawater again covered the land and life again proliferated. Another layer of ooze, another drying of the basins, and another oil-rich layer formed. And the cycle continued.

Much of this story is told by the samples retrieved in the nearly two hundred thousand wells that have been drilled into the Permian Basin. Some are exploratory wells; many are now oil producers. But there is little that a classically trained geologist, roaming the countryside, examining outcrops, and taking samples, can tell about this region. When one flies over Midland, Texas, and the Permian Basin, one sees a flat plain covered by endless fields of oil wells, the existence of the wells the only proof that something of geological interest had happened here.

But to see and touch and climb over the evidence, one must go to the edge of the Permian Basin, specifically, to the edge of the Delaware Basin where a broad limestone reef formed. As fortune would have it, a much later mountain-building event—the Laramide Orogeny, which would form the

Rocky Mountains and uplift the Black Hills of South Dakota—pushed up
segments of this great reef so that great vertical sections of the rocks that
formed in a Permian sea are now exposed.

◆

The Capitan Reef is exposed in three places. The shortest segment comprises the Glass Mountains near Alpine, Texas, where about four miles
of the reef can be seen. A second is north of Van Horn, Texas, along the
Apache Mountains, where fifteen miles are exposed. The greatest section,
both in length and in vertical relief, is a forty-mile stretch that runs from
Whites City in New Mexico to Pine Springs, Texas. It is the prominent
escarpment that can be seen west of the local highway that links the two
towns. At the southern end are the spectacular cliffs of the Guadalupe
Mountains. At the northern end is the equally intriguing network of caves
known as the Carlsbad Caverns.

Such caverns are common in ancient limestone reefs because the carbonic acid that forms by the absorption of carbon dioxide into rainwater
works on the limestone to dissolve it. That is how the famous limestone
caves of Mammoth Caves in Kentucky and Wind Cave and Jewel Caves
in South Dakota formed. But the limestone caves of the Carlsbad Caverns
are slightly different—and more majestic—because these caves formed
close to a major oil reservoir.

The oil contains sulfur. And when sulfur reacts with water, sulfuric
acid forms. That is what happened along the Capitan Reef. These were
not huge ponds of acid, but enough sulfuric acid was formed to seep
through and to be more efficient at dissolving the limestone than the less
corrosive carbonic acid. But through this dissolution of limestone, more
than three hundred caves have formed, many with huge rooms. The
largest cave chamber in North America is the Big Room, nearly two thousand feet long, more than a thousand feet wide, and nearly three hundred
feet high, at Carlsbad Caverns.

But the true showcase of the Capitan Reef is the vertical cliff that tops
out at El Capitan, Guadalupe Peak, and Shumard Peak. Exposed here
is a full section of the reef, one that took several million years to grow.
Unlike modern reefs, which are composed primarily of colonial corals,

Capitan Reef is composed of the skeletal remains of calcite-secreting organisms—mostly sponges, but also crinoids, brachiopods, a few stromatolites, and an abundance of fusulinids, tiny animals that look like rice grains scattered across a surface. After death, the hard skeletons of these animals were cemented together by algae, on which their successors continued to grow. Thus, a giant reef grew upward and outward into the Delaware Sea.

There is no lack of places to see these fossils. There are steep trails that go from the base to the top of the reef.[3] Along the way, as one endures what is usually a hot and dusty day, it is worth taking the time to imagine what it must have been like 260 million years ago when one could *swim* to the top and poke into small crevices and find, among a wide variety of other types of marine animals, trilobites and ammonites—both now extinct.

The growth of the reef ended when the narrow seaway that connected the Delaware Sea to the ocean, which had been closing intermittently, closed permanently. That produced a third remarkable feature in this region of West Texas: a broad deposit of thinly layered salt, gypsum, and organic-rich limestone.

In cross section, it looks like the cutaway of a thinly layered torte. The white layers dominate and are separated by extremely thin layers of dark brown. The white layers are salt and gypsum and the dark ones are brown limestone. This is the Castile Formation, famous for its many repetitive laminations.

The alternating pattern of white and brown can be followed vertically through the exposed section of the Castile Formation for nearly fifteen hundred feet. Horizontally, some layers have been followed for more than seventy miles. There are at least 261,162 of these thin layers. And each one has been measured. The average thickness is 0.07 inches, less than two millimeters.

What is recorded in these many laminations is a seasonal drying, then replenishment of water by rain and groundwater of a broad shallow lake that once extended across much of the Delaware Basin. Except for obvious disruptions between some of the layers, it is an unbroken record of conditions along the western coast of equatorial Pangea. Climate specialists who have examined the record see a remarkable stability of the climate, one that lasted for more than a quarter-million years. Those who are concerned about celestial matters have uncovered a slight one-hundred-thousand-year oscillation that is almost certainly due to a slight periodic change in the Earth's orbit around the Sun. Many are attracted to these exposures for the unexpected artistic expression that nature has given. In particular, there are sets of tiny waves in the laminations that affect only some layers. On viewing them, they look like exquisite line paintings. Some geologists have suggested that the wavy laminations formed when earthquake-produced waves passed through what was a water-soaked rock during the Permian Period.

◆

The Permian Period is also on display at the Grand Canyon. To recap the geologic history displayed in the canyon walls: At the bottom of the canyon are the Zoroaster Granite and the Vishnu Schist, both remnants of the two supercontinents Columbia and Rodina. Above them, and best seen in the eastern part of the canyon, are the tilted beds of a series of rocks known as the Unkar Group, which are sediments that accumulated in basins as Rodina was breaking apart. Next is the Great Unconformity and, above it, are the massive horizontal beds.[4]

Beginning at the bottom, the Tapeats Sandstone, the Bright Angel Shale, and the Muav Limestone all formed during the Cambrian Period as the sea was deepening and the shoreline was moving eastward. Then there is a long hiatus, another unconformity. No rocks of the Ordovician or Silurian periods are preserved at the Grand Canyon, nor can rocks of those periods be found anywhere in northern Arizona or New Mexico. Either the rocks did not form, or, more likely, they have been completely eroded away.

Atop the Muav Limestone is the Temple Butte Formation, Devonian in age. These are sedimentary rocks that accumulated along a flat coastal plain for tens of millions of years. Above the Temple Butte Formation is the Redwall Limestone, one of the massive limestone deposits that was laid down during the Mississippian subperiod. Above the Redwall is a sequence of rocks known as the Supai Group. The lower part of the Supai Group is Pennsylvanian in age, contemporaneous with the formation of the vast coal deposits elsewhere on the continent and that formed in shallow seas and lagoons. From the upper part of the Supai Group to the top of the South Rim are rocks of Permian age.

With one exception, all of the Permian-age rocks at the Grand Canyon are sediments laid down by water. Above the Supai Group is the Hermit

Formation.[5] These are river sediments washed down from nearby mountains. By the time the rivers draining these mountains reached the region of the Grand Canyon, they were sluggish, meandering streams wandering across a vast plain, much as the modern Mississippi moves across low-lying portions of Mississippi and Louisiana. Above the Hermit Formation is the Coconino Sandstone. And above the Coconino Sandstone are the Toroweap Formation and the Kaibab Limestone.

Both the Toroweap and the Kaibab were deposited in shallow water along a broad tidal flat. In places, the flat was more than a hundred miles wide. Both formed in tropical water, but there are few fossils in the Toroweap, many more in the Kaibab. Fossils in the Kaibab Limestone include clam-like brachiopods and crinoids. When one is standing on the South Rim of the Grand Canyon, one is standing on the Kaibab Limestone.

The Coconino Sandstone is the exception. It is the white ring that runs through the canyon just below the canyon rim. It is composed of pure quartz sand, ranging in color, so the authorities say, from bleached ivory to cream. The striations are a type of cross-bedding that indicates this rock consists of petrified sand dunes that formed on dry land. Its presence in the walls of the Grand Canyon is a sign of the drying of the planet during the Permian Period.

Follow the Bright Angel Trail into the Grand Canyon, the most popular trail that leads to the bottom of the canyon. About a mile down the trail is the contact between the Coconino Sandstone and the underlying Hermit Formation. Within the Hermit Formation, beginning at the base of the Coconino Sandstone and going downward, are vertical gashes, some as much as a foot wide and up to twenty feet long and filled with the ivory-and cream-colored sandstone. These are cracks that formed in the brick-red Hermit Formation as it dried, further indication of aridification during the Permian Period.

But where did this vast amount of sediment that now forms the upper layers at the Grand Canyon originate? The chemistry and mineralogy of the rocks show that little, if any, of this material washed down from the Marathon-Ouachita segment of mountains formed during the Alleghenian Orogeny. Instead, from drainage patterns within Hermit Formation, the Toroweap Formation and the Kaibab Limestone show that the sediment came mainly from a range of mountains to the northeast known as the Ancestral Rocky Mountains, so named because they were located along a segment where the modern Southern Rocky Mountains are today. But there is a fundamental problem in understanding these ancient mountains: How did they form?

◆

The Ancestral Rocky Mountains ran from eastern Utah through southeastern Colorado and into northern New Mexico. That trend is almost perpendicular to the trend of the Marathon-Ouachita Mountains, and so it seems unlikely that the Ancestral Rocky Mountains were part of the Alleghenian Orogeny and the compression and folding and uplift produced by the collision of Euramerica and the South American portion of Gondwana. And so the question remains unanswered, in part; details of the Ancestral Rocky Mountains have been overprinted by the later Laramide Orogeny that formed the modern Rocky Mountains. But that the ancestral mountains existed is beyond question because so much debris was shed from those mountains and can be found across a major part of the western United States.

The Ancestral Rocky Mountains consisted of two large mountainous masses known in geology as the Uncompahgre Uplift and the Ancestral Front Range, located roughly in the current locations of the San Juan Mountains and the Front Range. They began their rise in the Pennsylvanian subperiod and were greatly eroded away by the end of the Permian Period.

Debris of the Ancestral Front Range is most famously found at the Garden of the Gods near Colorado Springs, Colorado. The entrance is through a narrow gateway between two gigantic gateposts, each one a slab of salmon-colored Permian sandstone that has been uplifted and tilted by the Laramide Orogeny that formed the modern Front Range. Almost every major feature is debris of the Ancestral Front Range, including Balanced Rock, Steamboat Rock, the Sleeping Giant, the massive Tower of Babel, and the spectacular towering pinnacles known as the Three Graces of Cathedral Spires. The rock formations at the Garden of the Gods—the Fountain Formation and the Lyons Formation—are among the thickest and most colorful rocks exposed in Colorado today.[6]

The Uncompahgre Uplift also shed a great volume of debris. Some of the debris moved southwestward into what is now the Paradox Basin, where alternating layers of salt and limestone formed and a significant amount of oil accumulated. The total depth of salt and limestone is more than fifteen thousand feet.[7]

And more debris from the Uncompahgre Uplift is found in the Permian-age layers at the Grand Canyon and at the equally spectacular Canyon De Chelly located in northeastern Arizona within the Navajo Nation. At Canyon De Chelly the Permian-age rock is primarily contained within one massive unit, the De Chelly Sandstone, which, like the Coconino Sandstone, is derived from windblown sand that accumulated into giant sand dunes. Those giant dunes have been cut away and are revealed in cross section in the high cliff faces that characterize the canyon. In those cross sections are long sloping striations—the striations are giant cross beds of sand—that indicate that the wind that formed the dunes came primarily from the north.[8]

The De Chelly Sandstone once covered much of the region known as the Four Corners, where the corners of four states meet—Utah, Colorado, New Mexico, and Arizona. This region was later raised, probably during the Laramide Orogeny, then eroded deeply. The deep erosion stripped away much of the De Chelly Sandstone, leaving one of the most spectacular landscapes on the planet in its stead.

Monument Valley is a red-sand desert located along the Arizona-Utah border and known for its many clusters of towering mesas and buttes. Major rivers—the San Juan and the Colorado and their tributaries—did the cutting, aided by an occasional ferocious wind. Much of the cutting and windblown erosion probably occurred during the last few million years. What remains today are rocky sentinels. West Mitten Butte, East Mitten Butte, and Merrick Butte each stand about a thousand feet above the desert floor. From Artist's Point, one can see the three buttes as well as many more distant peaks. John Ford's Point was named for the Hollywood director who used the panoramic views of Monument Valley in many of his films, including *Stagecoach*, *My Darling Clementine*, and *Fort Apache*.

Study the mesas and the buttes of Monument Valley—and the films of John Ford—and one will see the same four geologic units. At the base of each mesa or butte is an apron of debris, the Organ Rock Shale, which is contemporaneous with the Hermit Formation. Above the Organ Rock Shale are the massive vertical walls of the De Chelly Sandstone. And atop the De Chelly Sandstone are two easily eroded layers, the Moenkopi

Formation and the Chinle Formation. Both formed during the Triassic Period, the geologic period that followed the Permian.

Look again at the mesas and buttes of Monument Valley. What one is seeing is a transition in physiography and in climate. Both the Organ Rock Shale and the Moenkopi Formation formed on broad tidal mudflats next to an ocean on the edge of the continent. The intervening De Chelly Sandstone represents a dry period when large sand dunes accumulated on the continent and when the edge of the ocean had greatly retreated. And the Chinle Formation is a deposit that formed after a dramatic and geologically brief change in climate—one that lasted a few million years—when the world's climate was humid and when rain fell almost continuously across the planet.

One more Permian-age deposit needs to be explored. It, too, is composed of sediments from the Ancestral Rocky Mountains, a further indication of how massive those mountains were.

When crossing the plains of the Texas Panhandle south of Amarillo, it is easy to miss the great chasm of Palo Duro Canyon because it lies entirely below the level of the plain and because the Red River that runs through it is of little consequence, certainly not at the scale of the Colorado River. And yet, this is the second largest canyon in North America, more than a hundred miles in length, up to twenty miles wide, and as much as eight hundred feet deep.[9]

The rocks of four geologic periods are exposed in the canyon. The upper veneer of rocks formed in the last twenty million years, either from debris washed down from the modern Rocky Mountains or from sand blown into the area by wind. The bulk of the canyon walls are rocks of the Triassic Period. These are the more colorful and varied exposures in the canyon, ranging from lavender to pink to gray shales that are easily eroded to the orange-colored sandstones that form cliffs. Beneath this palette of earth tones are the equally striking red hues, ranging from carmine to vermilion, of the rocks of the Permian Period.

It is the contact between the Permian and the Triassic Periods at Palo Duro Canyon that is of particular interest. The transition is easy to follow along the canyon walls. It is especially striking at the two most prominent features in the canyon, the spire known as Capitol Peak and the end of a broad ridge known as Triassic Peak. At the base of each peak are the furrowed red beds of the Permian. Higher up are the shales and the sandstones of the Triassic Period. Where the two meet is a discernible line, one that marks a major event in the history of life. It is along this line between the Permian and the Triassic periods that the greatest mass extinction occurred on the planet.

◆

A decade ago, the Permian extinction was described as a time when life almost ended on the planet. It was then estimated that 90 to 95 percent of all of the different types of plants and animals, both marine and terrestrial, had become extinct. That extreme assessment has now been tempered—somewhat.

Improvements in the age-dating of rocks and in the ability to measure isotopic compositions of rocks, and hence to characterize past environments, have brought the Permian extinction into focus. It is still the greatest mass extinction, but its magnitude has been reduced. It is now thought that 60 to 80 percent of all of the different plants and animals suddenly disappeared. That is still a startlingly high percentage, but it is far from the nearly complete catastrophe that was once portrayed. And it is still adequate to describe the Permian extinction as "The Great Dying."

One major change in perspective came when it was realized that what was once thought to be a single extinction event was actually two distinct ones. The later one is still known as the Permian extinction and was the greater of the two. The earlier one—they are separated by a mere eight million years—had an extinction rate of about 60 percent, which put it in the ranks of being a mass extinction. And that high rate might have been a factor and had set the stage for the later and more catastrophic Permian extinction.

The earlier event, the Guadalupian extinction,[10] had a greater impact on terrestrial life than marine life. Nearly half of land plants became extinct. Almost all of four-legged tetrapods, which were numerous and scurrying across the landscape by that time, died out.

The Permian Period was when the first large-bodied animals appeared. Among the largest was a group known as *Dinocephalians*. They were a broad group of animals that ranged across the planet and that occupied a wide range of niches. They included herbivores, carnivores, and omnivores. The largest was *Tapinocephalus*, an herbivore with a stocky, barrel-shaped body and a massive bony skull. From the arrangement of the bones and the shape of the skull, this large animal may have engaged in head-butting either to establish territory or to challenge rivals and secure mates. *Tapinocephalus* grew to more than ten feet in length and weighed as much as two tons. It and the other *Dinocephalians* died out during the Guadalupian extinction. The tetrapods that did survive the Guadalupian extinction were much smaller and, from where their fossil remains have been found, lived in underground burrows.

The Permian extinction had an equally devastating effect on both terrestrial and marine life. It marked the end of trilobites and of eurypterids, the giant sea scorpions that had existed since the Ordovician Period. It was the end of the guillotine-jawed placoderms that were once the terror of the sea. It was also the end of the giant reefs, such as the Capitan Reef of West Texas, which were composed of skeletal sponges held together by the calcareous secretions of algae. The new reefs, which would not appear for another twenty million years or so, would be built of colonial corals and would build features such as the Great Barrier Reef. The Permian extinction was also the only mass extinction of insects. Those few that survived would give rise to some of today's most common and troublesome insects, including grasshoppers, locust, crickets, and cockroaches.

The Permian extinction occurred 252 million years ago and lasted an astonishingly short 100,000 years or less. A great deal of interest has been drawn to it because it was so devastating, but that interest has been hampered because there are so few sites in the world where sediment was being deposited across the Permian-Triassic boundary. That is, there are few sites where the rock record is continuous across the boundary.

The Permian-Triassic boundary evident in the walls of Palo Duro Canyon is actually an unconformity. There are more than twenty million years of sediment accumulation missing. And that is true across almost all of North America.[11] So investigators of the Permian extinction have had to work elsewhere. And they have found a number of good sites in southern China.

The sites in southern China tell a story of rapid climate change. The extinction began with a rapid cooling of the planet. That produced massive polar ice caps, which produced a sudden drop in sea level. The drop in sea level, in turn, exposed new land surfaces that were subject to erosion, which probably accounts for the lack of sites where there is a continuous rock record across the Permian-Triassic boundary.

After the initial rapid cooling, there was an equally rapid heating of the planet. That was accompanied by dramatic changes in the atmosphere. There was a sudden injection of carbon dioxide and of sulfur. The amount of metallic atoms, especially nickel, increased substantially. And, to add to the mystery, those who have examined the spores produced by trees and ferns of this time period have noted a large number of mutated ones, suggesting the Permian extinction was also a global mutagenic crisis.

What could account for so many dramatic changes happening simultaneously and over such a short time period of a hundred thousand years or so? In 1979, it was shown that a huge meteor impact had been one of the components that contributed to the extinction of the dinosaurs. Might an earlier impact have caused the Permian extinction?

Much effort has gone into searching for evidence of an impact-related cause for the Permian extinction. And no clear evidence has been found. There is no impact crater, no meteorite fragments or other indicators that a large meteorite hit the Earth. And so the search for a cause for the Permian extinction has required scientists to look elsewhere. And they have found it. Instead of an extraterrestrial cause for the extinction, it was triggered by an event that came from inside the planet.

One of the most voluminous volcanic eruptions of the past five hundred million years covers what is now Siberia. Known as the Siberian Traps, they are massive lava flows that are piled up like giant steps—hence, the name "trap," derived from the Swedish word *trappe*, which means "step." These lava flows, now greatly eroded, once covered an area comparable to the area of the contiguous United States and are as much as a mile thick. And they erupted 252 million years ago, coincident with the Permian extinction.

Great quantities of sulfur dioxide and carbon dioxide were emitted during these eruptions—gases that would have opposing effects on the climate. Sulfur dioxide easily dissolves in water vapor and forms sulfuric acid, which then adheres to tiny particles in the air—many of those particles were also products of the eruption, tiny fragments of lava that were spewed high into the air—to form clouds of aerosols. The volume of the eruptions was so great that clouds of aerosols blanketed the planet, shielding the surface from direct sunlight and causing the planet to cool. Hence, as indicated by the deposits in southern China, the initial phase of the Permian extinction was a period of cooling.

The sulfur-rich aerosols were eventually removed from the atmosphere by rain—acid rain that would have devastated plant life—and were absorbed into the land surface and added to the oceans. That left an atmosphere that was greatly enriched with carbon dioxide, a greenhouse gas, and so the global temperature started to rise.

But did the Siberian Traps emit enough carbon dioxide so that the global land surface temperature reached 113° Fahrenheit (45° Celsius), as indicated by a careful study of the deposits in southern China and a temperature at which most plants and animals would die? That, in the opinion of many, is unlikely. And so an additional means is needed to raise the planet's temperature.

During the Carboniferous Period—the Mississippian and Pennsylvanian subperiods—Siberia was an isolated continental mass, far from both Gondwana and Euramerica. It was also located close to the equator. And so, as happened on Euramerica, vast deposits of limestone formed.

When the eruption of the Siberian Traps occurred—Siberia was now attached to Pangea—molten lava shot up through the limestone and released huge amounts of carbon, which went into the air and reacted chemically with oxygen to form carbon dioxide, which contributed to the warming of the planet.[12]

But what about the mutated spores? Why would they be associated with a massive volcanic eruption?

Vast amounts of coal also formed in Siberia during the Carboniferous Period. When the molten lava shot through these deposits, it not only released additional amounts of carbon but also a wide range of other chemicals. In particular, there were chlorine and fluorine. The chlorine and the fluorine combined with oxygen and carbon dioxide already in the air to produce a whole range of complex chemicals, including long-lived chlorofluorocarbons, the type of chemicals that were being released recently by refrigerators and air conditioners and that were causing a decrease in the amount of ozone high in the atmosphere. The same happened during the Permian Period. The layer of ozone that had protected life on the planet for more than two billion years began to disappear. And, as it disappeared, ultraviolet radiation from the Sun, radiation that could produce mutations, reached the surface. And that may be the reason for the mutated spores.

And so what was the killing mechanism that caused the Permian extinction? The voluminous eruption of lava produced by the Siberian

Traps may have produced a domino effect that led to an unusual sequence of major events that disrupted the atmosphere and the ocean, each one a potential mechanism for widespread death. Or there may have been one or two mechanisms that were more lethal than the others. That is yet to be decided. What is certain is that both plants and animals, marine and terrestrial, suffered and were decimated.

The abrupt change in the fossil record brought on by the Permian extinction also had an influence on geologic thought. Here, so one nineteenth-century geologist concluded, was a global event that could be recognized around the world. For that reason, it was a natural break in the fossil record that could be used as a fundamental division of geologic time.

◆

In 1788, with the publication of his book *Theory of the Earth*, Hutton suggested that the history of the Earth could be read from the rock record, though he did not provide a means as to how the history might actually be read and deciphered or how the geologic history of the planet might be formalized. In 1815, English geologist and canal engineer William Smith took an understanding of the Earth a step further when he published the first geologic map of any significant part of the world. The map was of Wales and England. The key to producing the map, as Smith often acknowledged, came after he realized that nature had left a definite sequence of fossils that could be used to identify and distinguish different strata separated by many miles. Then in 1859, Darwin in *On the Origin of Species* provided a means through natural selection for the succession of fossils to occur. The final step came the next year when John Phillips, Smith's nephew and occasional helpmeet and, eventually professor of geology at the University of Oxford and president of the Geological Society of London, suggested that the *entire* history of the Earth could be deciphered and understood and events put into sequence by using the history of life as it was revealed in the fossil record.

In the mid-nineteenth century, Phillips had as wide and varied an experience with fossils as anyone in the world. He had described and illustrated fossils in a series of books. He also took note that there were fossil groups, such as trilobites and giant tree ferns, that were restricted to the oldest strata. And that later in the rock record, during a period of "middle life," the gigantic bones of the "terrible lizards," or dinosaurs, appeared. And, after the dinosaurs had vanished, another period that led up to the present time contained mammals, birds, and flowering plants.

Moreover, he recognized that between each of these periods of ancient, middle, and recent life, there was a dramatic decrease in the number of life-forms. As he stated in his 1860 book *Life on Earth*: "Periods of extraordinary *abundance* alternate in every great series of strata with other periods of comparative *scarcity*." And one of these period of comparative scarcity corresponded with what is known as the Permian extinction. (The other corresponded to the extinction of the dinosaurs at the end of the Cretaceous Period.)

Phillips used this to formalize the division of the history of life—and hence, the history of the Earth—into three geologic eras: Paleozoic, Mesozoic, and Kainozoic, the last one known today as the Cenozoic. He even suggested that these three eras of vastly different life-forms represented three separate creations of life, but he soon abandoned the idea.

It was also not lost on him that the three fundamental periods in the history of life—ancient, middle, and recent—had a parallel with the three divisions of human history—at least, human history as seen from the viewpoint of an English Protestant. There was the antiquity of ancient Greek and Roman civilizations that was followed by the Middle Ages that was dominated by corrupt Catholicism that was, in turn, followed by the glorious period of the Enlightenment that continues today, the era of modern history.

Phillips and his contemporaries—and many people still today—saw the tripartite stages of human history and of the history of life as representing, in each case, a progression toward a state of perfection. Others saw these stages as inevitable consequences of physical laws. Regardless of the perception, dividing each into distinct eras has the benefit of focusing discussions. Moreover, abrupt disruptions in the fossil record have been used to define and divide small intervals of geologic time.[13]

◆

And so the Permian extinction brings to a close the Paleozoic Era, the era of ancient life. As life-forms repopulate and new forms dominate the Earth, a new geologic era begins, one of "middle life," the Mesozoic.

9

A Grand Staircase

Early and Middle Mesozoic Era: 251.902 to 145.700 Million Years Ago
Triassic and Jurassic Periods

The Le Fevre Overlook and Rest Area is located along the side of Highway 89A south of Fredonia, Arizona. It is little more than a widening of the road and a small pavilion. I once sat here for a few hours studying the distant hills. During that time, a dozen or so cars drove past. One car did stop. The occupants got out, stretched their legs, spent a few minutes surveying the same scene I was studying, then got back into their car and drove away. They clearly did not understand the significance of what they had seen.

Sixty miles south of the overlook is the Colorado River and the Grand Canyon. Partway up the canyon wall is the Great Unconformity. Paleozoic rocks are found from that point up and out of the canyon and across the landscape to the Le Fevre Overlook and Rest Area. At the overlook is the familiar Kaibab Limestone.

Look north from the overlook. The horizon is about sixty miles away. Within those sixty miles are five prominent cliffs, each one running for scores to hundreds of miles. The four nearer cliffs contains rocks of the next geologic era, the Mesozoic. The farthest cliff, which can barely be seen from the overlook as it forms just part of the horizon, contains rocks of the Cenozoic, the geologic era in which we live.

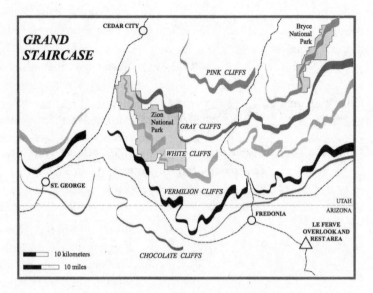

Steps of the Grand Staircase north of the Grand Canyon
in southern Utah and northern Arizona.

It was Clarence Dutton in 1882, in his publication *The Physical Geology of the Grand Cañon District*, who first called attention to the remarkable arrangement of the five cliffs as "the great stairway." The more grandiose name by which this feature is known today came later, in 1924, from Charles Keyes, also a government geologist and an expert on desert land-forms, who called it the "Grand Staircase."

It was Dutton and Powell who named the individual cliffs based on the color of each one. The Chocolate Cliffs are the ones farthest south and nearest to the Le Fevre Overlook and Rest Area. Beyond them are the Vermilion Cliffs, the White Cliffs, and the Gray Cliffs. The fifth and the farthest one north are the Pink Cliffs.

There is a broad stroke in the history of life revealed at the Grand Staircase. The oldest rocks are found in the Chocolate Cliffs, rocks that formed in a dinosaur-free world. The next three steps—the Vermilion Cliffs, the White Cliffs, and the Gray Cliffs—have rocks that formed when dinosaurs were everywhere. And by the time the rocks of the Pink Cliffs were forming, the dinosaurs had vanished.

There is also a record of dramatic climate change recorded at the Grand Staircase. Consider first the general outline of the Permian-age mesas and buttes of Monument Valley. The mesas and the buttes sit on broad bases composed of soft rock—to be precise, the Organ Rock Shale—that washed out of the Ancestral Rocky Mountains during a period of a wet climate, and that weathered into gentle slopes. Above the soft shale are the massive vertical walls of the De Chelly Sandstone that formed from dry windblown sand in an environment that was much like today's Sahara Desert. Then there was a return to a wet climate as the mudstones and the freshwater limestones of the rocks that cap the mesas and the buttes formed. This same wet-dry-wet cycle of climate is writ large at the Grand Staircase.

The same mudstones and freshwater limestones that cap the rocks of Monument Valley are those exposed along the lowest riser of the Grand Staircase, the Chocolate Cliffs. If one then moves up the stratigraphic ladder and climbs up the Grand Staircase, one encounters a long period of the accumulation of windblown sand during an extremely dry climate, one that was interrupted by brief periods of wet climates.

The long period of aridity is indicated by the three major cliffs of the Grand Staircase, all of accumulated, windblown sandstone, the Vermilion Cliffs, the White Cliffs, and the Gray Cliffs. And between each of these is a brief period of a wet climate when soft erodible layers formed on obviously water-soaked floodplains and mudflats.

It is quite easy to see the five steps up close. The Grand Staircase-Escalante National Monument is the natural choice, but there are few reliable roads. A better choice is Zion National Park in Utah, where modern rivers have cut deeply into a section of the Grand Staircase.

Within the national park, there is a small outcrop of Permian-age Kaibab Limestone. There are also exposures of the slightly older Triassic-age mudstones and freshwater limestones of the Chocolate Cliffs, the Moenkopi Formation and the Chinle Formation. And immediately above them are the hard red sandstones that form the Vermilion Cliffs. And the considerably younger Cretaceous-age Dakota Sandstone is found in the northern part of the park and forms the base of the Gray Cliffs. And between these—between the relatively small exposures of the Vermilion Cliffs and the Gray Cliffs—is the showcase of Zion National Park, the high vertical walls of sandstone that form the White Cliffs.[1]

In southern Arizona, Nevada, and California it is called the Aztec Sandstone. In Wyoming it is the Nugget Sandstone. But the greatest expanse of this geologic formation, covering eastern Utah, northern Arizona, and western Colorado, is the Navajo Sandstone.

Wind scour and giant cross-bedding of the Navajo Sandstone at an exposure known as The Wave in northern Arizona. (Photo courtesy of Diana Echert)

It is one of the largest ancient sand dune deposits on Earth. It covers an area of more than 135,000 square miles, equal to the area of the state of Montana. In places it is more than two thousand feet thick. And the thickest deposits are those exposed in the walls of Zion Canyon. The great diagonal patterns, such as those at Checkerboard Mesa in Zion National Park, are the cross beds that indicate the growth and migration of ancient sand dunes.[2]

The Navajo Sandstone and other great sandstones of the Mesozoic found in the American Southwest—the Moenave that forms the Vermilion Cliffs; the Wingate that forms the Canyonlands and the Entrada that forms the great arches of Arches National Park, both in Utah—indicate prolonged periods of dry climates and persistent strong winds. And the dry climates and strong winds came because all of the major continental masses had assembled into a third supercontinent, Pangea.

◆

Pangea, from the Greek for "entire world," was in the last stages of being assembled when the first part of the Mesozoic Era, the Triassic Period, began 252 million years ago. The supercontinent would begin to break apart in several large pieces, which would become the major continents of today, by the middle of the next geologic era, the Jurassic Period, about 180 million years ago.

During those intervening ninety million years or so, Pangea was the only major mass of dry land on the planet. It was surrounded by a single large ocean, Panthalassa. It stretched nearly from pole to pole and had the general shape of the letter C with the opening of the letter pointed to the east and occupied by an equatorial body of water called the Tethys Sea. That arrangement of a single large landmass surrounded by a single large ocean meant that weather conditions were extreme. It was a period of *megamonsoons*.

Though the word monsoon is often used to describe an individual rainstorm, technically a monsoon is the intense rainy phase of a seasonal weather pattern that is followed by a dry phase. It occurs because there is a significant difference in how much solar energy can be absorbed by land and how much by the ocean. The ocean can absorb a great deal of solar energy, and it does so with almost no change in temperature. By contrast, the land can absorb a small amount of solar energy and, when it does, the land surface heats up considerably.[3]

During summer months sunlight heats up both land and oceans, but the land temperatures rise more quickly. As the land surface warms, the air above it expands and an area of low pressure forms. Meanwhile, the ocean remains at a lower temperature and the air above it remains at a higher pressure. This difference sends ocean storms over the land, storms that carry moist air inland, which is dropped as rain. It also means that the higher the summertime temperatures were over the land, the more intense and prolonged the rainfall would be.

Summertime temperatures over the interior of Pangea often exceeded 140° Fahrenheit (about 45° Celsius). That produced unbelievably intense storm conditions—megamonsoons—with individual storms that were thousands of miles across and that lasted for months. The warm equatorial waters of the Tethys Sea added to the scale and severity of the storms by feeding energy to the storms. Because the Tethys Sea straddled the equator, such severe storms moved westward onto and across Pangea, drenching the eastern coastlines of the supercontinent. By the time a storm finally reached the western half of the supercontinent, it would have lost most of its water, and so those regions of Pangea were left in perpetual drought.

Amazingly, the effect of the megamonsoons can actually be seen within the great sandstone deposits of western North America. Recall that the presence of the Navajo Sandstone and the other deposits indicate prolonged periods of dry climates and persistent strong winds. Now examine individual sand grains.

Pick up a piece of Navajo Sandstone. Most of the sand grains are quartz. A few are calcite. A very small number are zircons. It is the age of those grains of zircon that is the key.

More than a thousand grains of zircon from the sandstones have been found and their ages determined. And the ages show that these zircons formed long before the three-hundred-million-year-old Ancestral Rocky Mountains. And so the source must be somewhere else.

The ages of the zircons cluster around five hundred million years and 1.1 billion years, the times of the Taconic Orogeny and of the Grenville Orogeny. And so those zircons found in the Navajo Sandstone and the other sandstones solidified from a rock melt those many years ago, during those mountain-building events. From there, they were transported westward.

To understand how they were transported, it must first be realized that the geography of North America during the Mesozoic Era was completely different from what it is today. During the Mesozoic Era, a long range of high mountains existed across the center of Pangea, mountains that had formed by the collision of Gondwana and Euramerica, and that included the Alleghenian section of the modern Appalachian Mountains. When the great Mesozoic storms of the megamonsoons swept across Pangea, they encountered those high mountains, dropping great quantities of rain, which weathered and eroded the surface, the water carrying the debris down a vast west-flowing river system, one that was comparable in size to the Mississippi River and its many tributaries today. But the rivers moved the sand grains only to the western region of the supercontinent. They now had to be moved by wind.

And so came the dry phases of the megamonsoons. Tremendous winds did come up and blew the sand grains southwestward, the direction indicated by the pattern of dunes in the Navajo Sandstone. And it is across what is now Utah and northern Arizona, which were near the equator during the Mesozoic Era, that many sand grains came to rest, where they accumulated into a series of high dunes seen today in the spectacular vertical cliffs of Zion National Park.

But if one goes back and looks again at the sandstone steps in the Grand Staircase, one will notice that there was a period, between the Chocolate Cliffs and the Vermilion Cliffs, when the intensity of the megamonsoons increased. With the increase came an extinction. In the aftermath of the extinction, new forms of life dominated the land. For the first time forests would be filled with what is now a very common type of tree, the conifer. And a new type of animal would roam the planet, its many variations occupying almost every available niche. These would be the dinosaurs.

◆

To a paleontologist, rocks of the early Triassic Period are boring. Fossils are few and far between because life had not yet recovered the abundance or the diversity it had before the Permian extinction. A few species of fossilized clams might be found. The occasional shell of a land snail or of a tightly coiled ammonite might be uncovered. When one does discover a fossilized specimen, it is almost certain to be identical to ones already found elsewhere in the world. Find a scallop-like *Claraia* in Utah and identical specimens have probably been uncovered in Italy or Iran or China.

There is also a monotony to the rocks. Look at the thinly layered rocks of the Virgin Formation in the Muddy Mountains east of Las Vegas, Nevada. These were deposited during the early Triassic Period under the sea. Since the Cambrian Period, such rocks have shown signs of disturbance—bioturbation—as marine animals crawled through the muck and mud searching for food. But the thin layers of the Virgin Formation are undisturbed. One can sense the lifelessness of the planet in these rocks.

And so it was for more than thirty million years. The intense rainstorms of the megamonsoons swept over the land. The future of life looked bleak.

Then, in a geologic instant, the climate changed—for the worse. But from its aftermath the first spasms of renewed life appeared.

A sudden intensification of planet-wide rain was first noticed by Michael Simms and Alastair Ruffell of Birmingham University in the United Kingdom. Simms was working on crinoids of the early Triassic Period. (A few species of crinoids had survived the Permian extinction.) He noticed that after a partial recovery, the number of crinoids had suddenly fallen. By coincidence, Ruffell, who shared an office with Simms, had been examining rocks near his family home in Somerset. Those rocks had formed at the same time as the crinoids' demise and showed a thin strip of a gray mudstone running through a massive red sandstone. The existence of the mudstone indicated a sudden shift of climate from dry to wet and back to dry. Seeing the correlation and wondering why the crinoids had died, Ruffell quipped, "Perhaps, the crinoids didn't like the rain."

It was a flippant remark. In 1989, Simms and Ruffell published their idea that an episode of intense rainfall had caused a local extinction. Their idea was ignored for almost twenty years.

Then in 2006, there was a surge of interest in their idea after other geologists found a similar correlation and at the same position in the stratigraphic record at rocky outcrops around the world. It was found in Austria and Hungary, in Israel and India, in Argentina and in Antarctica, and in Japan.

The most visually impressive site where this shift of climate has now been recognized is in Italy at Tofana di Rozes. Tofana di Rozes is one of the most spectacular outcrops of the Dolomites in the Southern Alps. The southern face of the prominence rises thousands of feet. The exposure is mostly white limestone. Across the base are four, closely spaced dark lines. Those lines are mudstones that record the same episode of heavy rainfall and the same extinction event as at Somerset.

The episode of heavy rainfall lasted about a million years. It occurred 232 million years ago during the Carnian Age of the Triassic Era. And so it is known as the Carnian Pluvial Event.

The realization that such a sudden shift of climate had occurred did have an impact on the study of North American geology because it provided a solution to a long-standing puzzle: Why are there petrified remains of a large forest scattered across the landscape of eastern Arizona?

◆

The Painted Desert is a land of colorful badlands. The easily eroded siltstone, mudstone, shale, and a considerable amount of volcanic ash were deposited there during the Triassic Period. They are part of the Chinle Formation. At the time these rocks were deposited, eastern Arizona was a marshy floodplain cut through by sinuous river channels and dotted with innumerable ponds. And the surrounding distant hillsides were covered with dense forests.

Today the remains of some of these trees lie scattered on the ground, on dramatic display at Petrified Forest National Park, which is within the Painted Desert. These trees are an extinct conifer, *Arucarioxylon*, which grew as high as two hundred feet and often had bases more than six feet in diameter. In appearance, it probably resembled, and may be related to, the modern Norfolk Island pine.

It was probably the increased rain of the Carnian Pluvial Event that allowed a dense forest of *Arucarioxylon* to grow where a parched sandy desert had been before. And it was probably an especially heavy fall of rain and a strong wind that uprooted the trees. The accompanied floodwater then carried them as rafts down onto the floodplain, the transport stripping away needles and limbs and most of the bark. The trees were left in jumbles stranded on sandbars or trapped in logjams. Soon after, possibly carried by the same floodwater, a thick layer of silt settled on the trees, covering them and sealing out oxygen and preventing decay. Then, as the years passed, nearby volcanoes belched out silica-rich ash that settled over the region in thick blankets. The heavy rain of the Carnian Pluvial Event continued. Some of the rainwater dissolved the volcano-produced silica and carried it down into the ground through the jumble of tightly sealed dead trees. Over a long time, the carbon atoms of the trees were replaced by silicon atoms in the volcanic ash and the trees were mineralized into brightly colored stones. Here there are agates and jasper and amethyst, all forms of quartz. And the result, from a disruption of the planet's climate more than two hundred million years ago, was the world's largest concentration of petrified trees.

And that is only half of the story of the Carnian Pluvial Event told by the rocks at Petrified Forest National Park. The other half is less visible, but it is also found in the same colorful rocks of the Chinle Formation in the Painted Desert.

◆

The earliest evidence of dinosaurs in North America is found in the Chinle Formation. The dinosaur in question is *Coelophysis*, a carnivore that, though it began small, at later times would grow into individuals that were up to ten feet in length. It had hollow bones (the name means "hollow form") and was bipedal. Its long and slender neck and tail probably gave it extraordinary balance, and so it was probably a fast and agile runner. The head was narrow and had forward-looking eyes that gave it stereoscopic vision, the mark of a hunter.

An extraordinary discovery was made in 1947 when a mass grave of more than a thousand individual *Coelophysis* was found in northwestern New Mexico at a site known as Ghost Ranch. The bones of fully grown adults, male and female, and of juveniles were uncovered, and so the full range of growth development is displayed. *Coelophysis* was apparently cannibalistic because tiny baby *Coelophysis* bones have been found in the rib cage of a large *Coelophysis*. (Cannibalism is common among reptiles.) The skeletons of a few hundred individuals are complete and show no signs of scavenging, and so the animals were buried quickly, possibly by a flood that swept a herd of *Coelophysis* down a river channel.

Coelophysis were typical of early dinosaurs in their modest body size, certainly much smaller than the giants that would come later. They were also much fewer in number compared to the amphibians and other reptiles that lived at the same time. But the expansion of dinosaurs into many different forms and into many different niches began during the Carnian Pluvial Event. By the end, all three of the major groups had emerged; the ornithischians, which later included *Stegosaurus* and *Triceratops*; the saurischians, which gave rise to huge *Brachiosaurus*; and theropods, such as *Tyrannosaurus rex* and the birds.

But what caused the dramatic change in climate that led to the period of intense rainstorms known as the Carnian Pluvial Event? Megamonsoons did reach their climax during the final assembly of Pangea during the Triassic Period. But the Carnian Pluvial Event came on suddenly and lasted only a million years or so. What might have triggered it?

The culprit seems to be the same type of event that triggered the Permian extinction: a major volcanic eruption. And this one happened in North America.

The Wrangell Mountains of northwestern North America are composed of lavas that were erupted as huge volumes in only a few hundreds of thousands of years at the time of the Carnian Pluvial Event. The eruptions initially spread lava over the seafloor, but the lava became so voluminous that volcanoes emerged above the sea as lava continued to pour out.[4]

It is these voluminous eruptions that might have released enough carbon dioxide to warm the planet. And that might have increased rainfall by increasing evaporation rate over the oceans.

Maybe.

Exactly how a voluminous eruption leads to a major extinction is still debated. A host of kill mechanisms was proposed in the previous chapter to account for the pairing of the eruption of the Siberian Traps and the Permian extinction. A similar pairing can be made between the eruption of the Wrangellia Flood Basalts in northwestern North America and the Carnian Pluvial Event, which did have a major extinction at its start. And there are other such correlations yet to come.[5]

In fact, another pairing came just thirty million years after the Carnian Pluvial Event ended. This time the voluminous eruption of lava was not only coincident in time with a mass extinction, but it marked the beginning of the tearing apart of Pangea—and the creation of the Atlantic Ocean.

◆

When driving west on the upper deck of the George Washington Bridge that links Manhattan and New Jersey, look northward. Dark gray cliffs line the edge of the New Jersey side of the Hudson River. These cliffs are the Palisades, the rock that ended the Triassic Period.

The Palisades are the edge of a giant volcanic *sill* that has been broken and eroded away. A sill is a horizontal sheet of rock that was once molten that solidified inside the Earth. Here the sheet of rock, when molten, forces its way between layers of sandstone and shale. The Palisades Sill was the result. In places, it is more than a thousand feet thick. It can be followed for more than fifty miles from Staten Island northward to the town of Haverstraw in the appropriately named Rockland County of New York.

The Palisades Sill is the first rock seen when emerging from the western end of the Lincoln Tunnel that connects Manhattan and New Jersey and that runs beneath the Hudson River. The western abutments of both the George Washington Bridge and the Tappan Zee Bridge are anchored on this rock. It is the "granite" of Graniteville Quarry Park on Staten Island. The hard rock found in the park is a diabase, identical in chemistry and mineral content to basalt, but denser and harder because it cooled slower than basalt because it remained inside the Earth.

Some of the molten rock that rose up and fed and formed the Palisades Sill also erupted onto the surface. The lava is exposed as three long parallel ridges that cross through northern New Jersey and that are known as the First, Second, and Third Watchungs, the name derived from the original inhabitants, the Lenape, who referred to the three parallel ridges as *Wach Unks*, or "high hills."[6]

The volcanic activity represented by the Palisades Sill and the First, Second, and Third Watchungs was part of a massive outpouring of lava that occurred in four phases and lasted about six hundred thousand years. The formation of the Gettysburg Sill in southern Pennsylvania was part of this activity. Here the hard rock is exposed as a line of ridges known as Round Top, Little Round Top, and Cemetery Ridge, names familiar to Civil War historians. The volcanic activity is also responsible for Mount Pony in Virginia, where the Federal Reserve Board operates a radiation-hardened facility—surrounded, in part, by the hard volcanic rock—that serves as the node for all American banking and electronic transfers of funds. It is the volcanic rock exposed at the Orenaug Hills in Connecticut and the Pocumtuck Hills in Massachusetts and along the North Mountain Ridge that forms the southern shoreline of the Bay of Fundy.

And the eruption of lava was much larger than this. The same series of eruptions also formed the so-called Mosquito Basalt in the Maranhão Basin of Brazil in South America and large volcanic sills in Guinea and in Mauritania, both in Africa. A thick pile of lava flows in the High Atlas Mountains of Morocco also dates from this event.

In total, nearly a million cubic miles of lava was erupted. It covered an area of more than four million square miles. It is known as the Central Atlantic Magmatic Province. It is one of the most voluminous eruption of lava on continental land in the planet's history. It is certainly comparable in volume and in area covered to the eruption of the Siberian Traps. And it is much larger in volume and in area than the eruption of the Wrangellia flood basalts. Like the Siberian Traps, the eruption of lava that formed the Central Magmatic Province originated beneath Pangea, though unlike the Siberian Traps, that activity initiated the breakup of the supercontinent.

The breakup of Pangea did not begin with the volcanic eruptions. Instead, there was a prolonged period when the supercontinent was being stretched and pulled apart along a line that closely followed the line where the edge of Gondwana had pushed up against the edge of Euramerica. As the stretching and the pulling apart continued, a series of elongated basins formed along that line.[7] Eventually, a point was reached when several of the larger basins coalesced and from that a continuous long and deep rift formed between the two parts of Pangea that were moving away from each other. The floor of the rift was deep enough that seawater flowed in and flooded the rift, forming a long narrow seaway, somewhat similar in appearance to how the Red Sea between Africa and the Arabian Peninsula looks today.

And the spreading continued—as it does today. From it, the northern half of the Atlantic Ocean has formed. (The southern half began to form later from a similar rifting between what is now South America and Africa.) And if there is any lingering doubt that the eastern coast of North America and the northwestern coast of Africa were once adjacent, one needs only to study the geometry of the coastlines and the local geology. The rocks and fossils around the city of Dakar, Senegal, match those of southern Georgia. The coastline of Western Sahara matches the contours of the coastline and the geology between Chesapeake Bay and Boston. And the sequence of rocks found around the Bay of Fundy is the same sequence found in the Argana Basin in the High Atlas Mountains of Morocco.

But the opening of the Atlantic Ocean comes after the Triassic Period. The end of that geologic period is marked by the great volumes of lava that formed the Central Atlantic Magmatic Province and the great amounts of carbon dioxide and sulfur that were emitted and, in some manner, changed the climate. And the consequences this activity had on life were great because another mass extinction occurs.

This extinction saw an end to the dominance of amphibians, including a variety of large crocodile-like animals that played the roles of large herbivores and of large carnivores. They would be replaced by what, until now, had been an inconsequential group of small reptiles that would come to dominate the land in both number and size.

◆

It was inevitable that prehistoric people would have chanced upon the bones of dinosaurs, such was the abundance of fossilized relics in some parts of the world. They would have known from the general shape and from detailed features that these were the mineralized remains of animals that had once been alive, often very large animals.

An example of such knowledge was demonstrated in 1860 when the Jesuit priest Jean-Baptiste l'Heureux followed members of an Algonquian-speaking people of the North American Great Plains known as the Piegans down into a steep canyon in Alberta, Canada. They led him to a place where the fossilized bones of large animals—some individual bones measured nearly two feet in diameter—were tumbling out of the steep walls of the canyon. L'Heureux noticed that there were offerings of cloth and tobacco scattered around, an indication of the reverence that the Piegans held for the site. Later visitors collected some of the bones and studied them and decided that they were the bones of dinosaurs, mostly of two giant herbivores, the duck-billed *Hadrosaur* and the two-horned *Certaopsian*. Today the site is part of Dinosaur Provincial Park in Alberta, Canada.

The first reported discovery in North America of what was probably a dinosaur bone occurred in 1787. The discovery was reported at a meeting of the American Philosophical Society in Philadelphia, the meeting led by the society's president, Benjamin Franklin. The report was given by Caspar Wistar, a local physician, who described the discovery as "an exceedingly large Thigh-bone of some unknown species of animal lately found near Woodbury Creek in Gloucester County, New Jersey." Members of the society recommended that Wistar return to the site and search for the remaining skeletal parts of the animal. Whether he did is unrecorded. The bone that he described has since been lost, but because the bones of *Hadrosaur* were later found at the site, Wistar's discovery was almost certainly a bone of that animal.

The next notable discovery of dinosaur bones in North America came in 1818. Solomon Ellsworth, a local farmer, was using gunpowder to dig a new water well in the red sandstone in East Windsor, Connecticut. (This red sandstone is part of the Portland Formation in the Hartford Basin.) As shards of rock were being blasted out, Ellsworth noticed that several contained small bone fragments. He took them to Nathan Smith at the Yale School of Medicine. Smith thought they might be human bones. Not until 1915 were the few fragments of bone recovered by Ellsworth recognized as those of a dinosaur, a small herbivore, *Anchisaurus.*

A few more discoveries of fossilized bones that were later identified as belonging to dinosaurs were made in North America during the first half of the nineteenth century, but they were of little consequence. The study of these ancient bones had moved across the Atlantic, where important discoveries were being made in the Jurassic-age rocks of southern England, discoveries that would spur on discussions about what type of animals these giant fossilized bones represented.[8]

One of the key figures in the discovery of fossilized bones in southern England was a young Dorset woman, Mary Anning. In 1811, at age twelve, Anning found and unearthed an almost complete skeleton of *Ichthyosaurus,* an ancient marine reptile similar in size and appearance to a modern dolphin. Later she would find, and be renowned for the discovery of, several skeletons of plesiosaurs, another marine reptile, this one with large-lobed fins, and for several pterodactyls, the flying reptiles one often sees in dinosaur-themed movies. She was also sought out by university professors and amateur naturalists who wanted her assistance in finding the best specimens for their collections.

And there were other discoverers of fossilized bones. In 1822, Mary Ann Mantell traveled to Surrey with her husband, Gideon Mantell, a physician who was visiting a patient, when she chanced upon several fragments of bone and teeth in the gravel of a road that had recently been paved. She showed these to Gideon, who was so intrigued that he spent years searching local rock quarries for more specimens. The teeth were of special interest because the whole specimens that he found were huge—several inches long—and because they were chisel-shaped and fluted on one side. These were clearly the teeth of a giant plant eater. Mantell named the creature *Iguanodon* because of the resemblance of the bones to those of the modern iguana.

In 1824, William Buckland of Oxford University found bones in the nearby village of Stonesfield of a reptile with dagger-shaped teeth, obviously a meat eater. From the size of the bones, Buckland estimated that the creature must have been at least forty feet long. He named it *Megalosaurus*. And in keeping with the conventional wisdom of the times, Buckland maintained that the animal that he had exhumed was nothing more than a lizard greatly enlarged.

Not until 1842 was it realized that these discoveries represented the remains of a new class of animal. Though only nine different types of giant reptiles were yet known, in that year Richard Owen announced that these ancient reptiles were sufficiently different from any living ones in the shape of their bones and in the arrangement of their teeth that a new name was required. He called them *dinosaur*, from the Greek meaning "terrible lizard."

The story now returns to North America. It was 1858, and Joseph Leidy, chairman of curators at the Academy of Natural Sciences in Philadelphia, was called by a member of the academy to New Jersey to see a new discovery. A complete set of large dinosaur bones had been found near Haddonfield. Leidy collected the bones and returned to Philadelphia, where he studied them for ten years.

Until then it had been assumed that all dinosaurs were quadrupeds—that is, that they moved around on four legs. But after his ten years of study Leidy concluded that the specimen he had removed from Haddonfield was not only surprisingly large—it must have been at least twenty-five feet long—but, based on its small forelimbs and long hind limbs, it must have stood in a semi-upright position; that is, this particular type of dinosaur must have been bipedal.

And that is how he positioned the animal when he assembled the bones and put them on display at the academy's museum in 1868. It was the first time that a dinosaur skeleton had ever been exhibited anywhere in the world. And it was an immediate sensation. Attendance at the museum doubled the next year. It tripled the following year. Copies of each of the bones found at Haddonfield were made and additional models of Leidy's upright dinosaur were exhibited around the United States and in England and in Europe. It was the beginning of the public's fascination with dinosaurs.

◆

Much can be learned from the skeletal remains of a dinosaur. Obviously, the size of an individual can be determined as well as whether it walked or swam or flew. It might also be possible to determine whether an individual was a male or a female and whether, at the time of death, it was an infant, a juvenile, or an adult. The shape and arrangement of the teeth tell whether it ate flesh or stripped leaves off a plant. But bones and teeth do not reveal much about the behavior of a dinosaur. For that one must seek out something that individuals left behind when they were alive, some type of mark that they left in the rocks. Such markings have been found. They are the tracks that dinosaurs left after slogging their way along muddy riverbanks and across coastal floodplains.

Such tracks have been found in large numbers. Several hundred tracks have been found in an area of less than half an acre at Clayton Lake State Park in New Mexico. More than two thousand were uncovered in 1968 during excavation of a new state building in Rose Hill, Connecticut. More than three thousand were revealed when, in 2000, a small hill was being leveled on a small farm in St. George, Utah. The leveling overturned slabs of sandstones that were later determined to be not only the tracks of dinosaurs, but also skin impressions where dinosaurs had fallen into mud, claw marks where they had climbed out of shallow holes, and, for the first time anywhere in the world, impressions made by dinosaurs as they swam in shallow water.

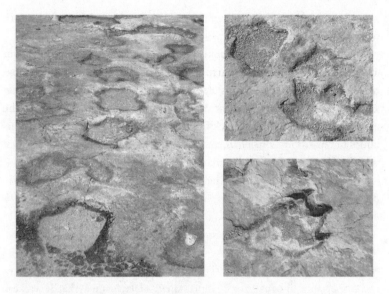

Dinosaur stomping grounds at Clayton Lake State Park in New Mexico.
There are hundreds of individual tracks at this site. The most common ones
are broad, three-toed tracks made by large herbivores, probably *Iguanodon*.
The smallest track at this site was made by a small dinosaur that stood about
a foot in height. The largest were made by thirty-foot-high individuals.

And names have been attached to indicate the abundance of such tracks.
In northeastern British Columbia there is the "dinosaur highway." Utah has
the "dinosaur stomping grounds." And a succession of areas where tracks
are especially dense that runs from near Boulder, Colorado, to Tucumcari,
New Mexico, is known as the "dinosaur freeway."

At one site along the dinosaur freeway near La Junta, Colorado, there
are more than a hundred trackways made up of more than fifteen hundred
individual footprints made by both bipedal and quadruped dinosaurs. Many
of the trackways are parallel and nearly equally spaced and are tracks made
by giant plant eaters, possibly *Apatosaurus* (formerly called *Brontosaurus*),
that were walking as a group with large and small animals intermixed.
There are also random trackways of flesh-eating dinosaurs, possibly *Allosaurus*, similar in appearance to the better known *Tyrannosaurus*. In one

case, the trackway of a flesh eater runs parallel to the group of plant eaters. Was this individual hunting the herd?

A more dramatic pair of trackways was found along the Paluxy River in Texas in 1944. One is a narrow trackway with huge footprints, possibly those of the herbivorous *Apatosaurus*, deeply impressed into what could only have been a mudflat. Each successive step is six feet from the next. Superposed on those tracks are the footprints of a large carnivorous dinosaur, possibly *Allosaurus*, moving in the same direction. Again it seems likely a predator was pursuing prey.

At Copper Ridge, Utah, there is the track of what may be an *Allosaurus* with alternating pace lengths indicating that this dinosaur might have been limping. At Mosquero Creek, New Mexico, there is another limping dinosaur, this one a plant eater that was traveling as part of a herd. From the pattern of tracks, it is obvious that this dinosaur had injured its left foot.[9]

At the Beneski Museum of Natural History at Amherst College there is a slab of sandstone on display that shows the near-perfect impression of the hind end of a small dinosaur that sat down on a muddy lakeshore about 160 million years ago, perhaps to take a drink. At Clayton Lake there is a spot that indicates that a large quadruped slipped in the mud and rolled partway on its right side. It then stretched its forelegs out in front to push itself up and used its tail to right itself. At St. George a biped that had been squatting in the mud put its small forearms down, claws curled inward, then shuffled forward before getting up and walking away.[10]

These are records of dinosaurs moving. They are migrating and they are hunting. They are drinking and swimming and resting and, occasionally, slipping and falling and injuring themselves. But there was a time when these markings were as mysterious to people as the giant fossilized bones they were finding. The man who first understood what these markings represented was Edward Hitchcock of Amherst College.

In 1835, the citizens of Greenfield, Massachusetts, were having the streets paved with slabs of sandstone cut from nearby rock outcrops. A local physician, James Deane, noticed that some of the slabs had impressions in them that looked like the three-toed markings left after the passage of a large bird. Deane sent drawings and casts of the supposed "turkey tracks" to Hitchcock, who had recently completed *A Report on the Geology, Mineralogy,*

Botany & Zoology of Massachusetts and asked his opinion about them. Hitchcock came to Greenfield, saw the impressions and the outcrops, and began a lifelong obsession collecting, studying, and trying to understand how these markings had been made.

Local people had known of such markings for decades, referring to them as tracks made by Noah's raven, a reference compatible with the idea that all rock markings attributed to animals must have been made during the biblical Noachian flood. But Hitchcock and many of his contemporaries were of an enlightened generation. He set out and searched for rocks with similar markings.

In all he would collect more than twenty thousand fossilized footprints in the Connecticut River Valley. The first specimen was collected from what is now the northern side of Holyoke, Massachusetts, on the eastern side of the Connecticut River. He coined the term *ornitho-ichnology*, the study of "stony bird tracks," for what he was doing, later shortening it to *ichnology*. And that reflects his deep belief, until the day he died, that giant birds had made these tracks, birds that may have stood as high as fifteen feet and that had moved as large flocks, walking over the muddy surface, followed by many others of analogous character, but of smaller size.[11]

Others soon doubted Hitchcock's conclusion. For many, the discovery of *Archaeopteryx* in 1861 settled the matter. *Archaeopteryx* had lived much later than the ancient animals that had made the footprints in the Connecticut River Valley. Because *Archaeopteryx* had the general appearance of a dinosaur but had feathers, birds as they are known today must have first appeared after *Archaeopteryx*; that is, birds must have appeared much later than the age of the rocks that Hitchcock had studied.[12]

The fossilized footprints in the sandstone beds of Massachusetts and Connecticut are those of dinosaurs. Some of the three-toed footprints were probably made by animals similar to *Coelophysis*, an early dinosaur whose skeletal remains have been found in the Chinle Formation in New Mexico. Others may be diminutive relations of *Megalosaurus*, first described by Buckland. But many, even to this day, have still not been clearly associated with the skeletal remains of known dinosaurs.[13]

◆

When dinosaurs were roaming the muddy sands of what is now the Connecticut River Valley and leaving their footprints, the colossal storms of the megamonsoons were at their most intense. Some were hemispherical in extent. But as Pangea began to break apart, the pattern of ocean currents changed. And that changed how heat was distributed in the oceans, which in turn lessened the intensity of the storms.

When the Triassic Period began, the storms were at their height. During the Jurassic Period, they began to lessen. The Jurassic Period began with a world in which much of the central and western portions of Pangea were extremely arid. That was the time when great seas of sand in the form of giant dunes covered much of what is now the southwestern part of the United States. That was when the Navajo Sandstone, on massive display at Zion National Park, was formed. But as the Jurassic Period continued and the narrow seaways that marked the breakup of Pangea continued to widen into what would eventually become the North Atlantic Ocean, separating North America from Europe and Africa, the climate became milder and the weather extremes became less.

Dense forests, mostly of conifers, grew close to the poles. Other vegetation, especially palms and ferns, grew as lush forests in a tropical band that extended far from the equator. The amount of carbon dioxide in the atmosphere was at least four times what it is today. The average air temperature may have been as much as 30° Fahrenheit (about 10° Celsius) warmer than today. Even the deep oceans warmed.

From this came more evaporation and more rainfall, but not the steady torrential rains of the Carnian Pluvial Event. These were milder and did not extend across all of Pangea. Large areas of the western half of the supercontinent still experienced major droughts.

And during this period of lush forests and milder climate there was a great abundance of animal life. Dinosaurs had succeeded in filling almost every ecological niche. And some of them had become very large. This can be seen at what was the real Jurassic Park along Dinosaur Ridge in Utah.

On August 17, 1909, Earl Douglass of the Carnegie Museum of Natural History in Pittsburgh, Pennsylvania, was climbing a knobby ledge scouting for bones near Split Mountain, a canyon cut by the Green River in Utah, when a peculiar sequence of shapes in the sandstone caught his eye. As he later wrote about it in his diary: "At last in the top of the ledge where the softer overlying beds form a divide—a kind of saddle, I saw eight of the tail bones of a Brontosaurus [Apatosaurus] in exact position." He continued.

> It was a beautiful sight. Part of the ledge had weathered away and several of the vertebrae had weathered out and the beautifully petrified centra lay on the ground. It is by far the best looking Dinosaur prospect I have ever found.

And it was. It was something that both amateur and professional paleontologists dream about: a nearly complete skeleton of a large dinosaur.

Douglass proceeded to dig. The original eight large bones soon became hundreds. And the bones of this individual were commingled with those of other huge dinosaurs. This was a bonanza, more than a "one strike," as Douglass would write in a letter to the museum's benefactor, Andrew Carnegie, who had been hoping that something "as big as a barn" would be discovered for the museum. Digging proceeded, Douglass now supported by a half-dozen other men. New specimens appeared: a small plant-eater known as *Dryosaurus*, a heavily armored *Stegosaurus*, and another large *Apatosaurus*.

It took five years to prepare the original skeleton of the *Apatosaurus* for display at the museum. It measures seventy-one feet long and stands fifteen feet tall at the arch of the back. The species is identified as *Apatosaurus louisae*, named after Louise Carnegie, wife of Andrew Carnegie. It is on display in the Hall of Dinosaurs at the museum.

And this was just the beginning. Douglass and his many successors have continued to dig along Dinosaur Ridge in Utah. In 1915, the area was declared Dinosaur National Monument. The place where thousands of dinosaur bones have been found, representing more than a dozen large individuals, is called Carnegie Quarry.[14]

Despite dramatic artistic illustrations of large plant-eating dinosaurs being attacked and consumed by carnivorous ones, the dinosaur remains in the Carnegie Quarry tell a different story. The western interior of North America was then mostly dry with a few rivers slowly flowing through the area, much like the braided system of channels that forms the Platte River today. The site of the Carnegie Quarry was along one of the those rivers. There was a drought and these large animals came both to drink and to

wallow in the cool water. These animals died individually from malnutri-
tion and dehydration and possibly from disease or parasites. When rain
finally returned, flash floods washed the carcasses into a small area where
they are found today at the quarry.

The rocks of the Carnegie Quarry are part of a large geologic unit known
as the Morrison Formation that reaches from Montana and North Dakota
to Arizona and New Mexico. The Morrison Formation is either exposed
or underlies all of Wyoming and Colorado. The fossil-rich nature of this
formation was known long before Douglass made his important discovery.
In 1806, William Clark of the famed Lewis and Clark Expedition that
crossed the continent had dug out what he thought was the rib of a large
fish more than three feet in length. Because he recorded the discovery
along the southern bank of the Yellowstone River below the present town
of Billings, Montana, it was almost certainly the bone of a dinosaur.

But the real fame of the Morrison Formation came in 1877 when
Arthur Lakes, a Denver teacher and seeker of dinosaur bones—his
intention was to sell them for profit—found several giant bones along the
Front Range of the Rocky Mountains.[15] He contacted two paleontologists on the East Coast about his discovery, Edward Drinker Cope of
the Academy of Natural Sciences in Philadelphia and Othniel Charles
Marsh of the Peabody Museum of Natural History at Yale College. Cope
and Marsh were already in a treacherous rivalry to obtain dinosaur bones.
Lakes's offer to sell his remarkable find and lead an expedition to uncover
more intensified the rivalry and led to what newspapers were reporting
as the Bone Wars.

The story of the Bone Wars is a long one that ended with both Cope
and Marsh in financial ruin and with the first discovery of many dinosaur
species that are familiar today, including the first *Apatosaurs*, *Stegosaurs*, and
Allosaurs. It also led to the intentional destruction of an untold number of
dinosaur bones as each man tried to inhibit the work of the other.

◆

Soon after the last grains of sand and the last slurries of silt and mud
settled and were consolidated into what is now the Morrison Formation,
the Jurassic Period came to an end. But there is a problem here: What

exactly marks the end of the Jurassic and the beginning of the next geologic period, the Cretaceous?

Some geologists maintain that there was a mass extinction that can be used to mark a division between the Jurassic and Cretaceous periods. Others suggest that if an extinction did occur—and some have questioned it—the event was limited geographically and was not global in extent, and so cannot be used to distinguish between two geologic periods. In short, not everyone agrees that a sudden shift occurred in the planet's history between these two periods.

Some of the problem seems to be the fault of the history of geology. The Cretaceous Period, one of the first geological time periods to be named, was first formally defined in 1822 by d'Omalius d'Halloy for a thick strata of white chalk cropping out in the Paris Basin; whereas, the Jurassic Period was first introduced in 1829 by Alexandre Brongniart for an unusual set of marine fossils found in the Jura Mountains that run along the France-Switzerland border. Neither man was concerned with defining where the boundary between the two time periods should be located within the rock record. And so two centuries have passed as generations of geologists have struggled with the problem.

Currently members of the International Commission on Stratigraphy have taken a radical step and have abandoned the use of fossil succession to define the boundary—which is how all other geologic periods of the Phanerozoic Era are defined—and, instead, are using a change in the Earth's magnetic field. (One could almost feel the hearts of professional geologists being broken when the decision was made. Geologists are a conservative bunch, especially when it comes to defining their holiest scepter, the Geologic Time Scale.)

So, for now, the boundary between the Jurassic and the next geologic period, the Cretaceous, is defined as the top of the normal magnetic polarity stripe known as Chron M19n. It occurred 145.7 million years ago and is coincident with an "explosion" in the number of small, globular microorganisms known as *Calpionella alpine* that can be found in Mexico and across Central and Eastern Europe—and so there is reason for the conservatives in the world of geology to hope that the Jurassic/Cretaceous boundary will, eventually, be defined by fossil succession. Thus, as this illustrates, a considerable effort goes into precisely defining the boundaries of the Geologic Time Scale.

But for those who choose to look at the history of the Earth through a wider lens, it is sufficient to note that there is a significant change in vegetation between the Jurassic and the Cretaceous periods—the first flowers and the first broadleaf trees appeared during the Cretaceous—and that there is a significant change in the types of animals, including in the types of dinosaurs. And, as d'Halloy noted, there was a surprisingly large amount of chalk deposited during the Cretaceous Period.

10

Western Interior Seaway

Late Mesozoic Era: 145.700 to 66.043 Million Years Ago
Cretaceous Period

G reat walls of white chalk, the ruins of a former world, rise high above the grassy prairies of western Kansas. These walls, known collectively as Monument Rocks, were once used by travelers as a way to measure their progress along a poorly marked trail that led westward to newly discovered gold fields in Colorado. That was more than a hundred years ago. Today these walls are used to tell a different story: that the center of North America was once pulled down below sea level, then sprang back.

The evidence, of course, is in the rocks themselves. Chalk is a soft, fine-textured limestone made up of tiny marine organisms that lived near the sea surface and dropped down onto the seafloor after they died. As their tiny limestone shells accumulated, an ooze formed, perfect for engulfing and preserving the remains of other animals that fell to the bottom of the sea.

And so it is undeniable that Monument Rocks formed beneath the sea. These rocks are the same age and formed in the same manner as the White Cliffs of Dover in southern England. But those rocks are at almost the same vertical position with respect to modern sea level as when they formed. The white chalky rocks of Kansas are not.

Search around Kansas. The fossilized remains of ancient oysters and giant clams can be found where chalk is exposed in modern stream beds.

Widen the search to include Nebraska and Iowa. A variety of marine fossils can be found, including large turtles and large bony fish, the latter as long as fifteen feet. Shark teeth are plentiful. A nearly complete skeleton of a long-necked plesiosaur that once swam in the ocean—the same giant marine animal that Mary Anning found in the chalk deposits of southern England—was found a few dozen miles west of Monument Rocks. All of this indicates that a long arm of the ocean once covered the central part of North America for a very long time.[1]

This arm of the ocean, known formally as the Western Interior Seaway, extended from the Arctic Ocean to the Caribbean. The opposing sides of the shoreline can be followed today by seeking out a hard sandstone that formed from debris that was washed down from surrounding hillsides. Along the former western shoreline, such sandstone forms the spectacular rimrocks along the Yellowstone River Valley between Columbus and Billings in Montana. It is the rock that comprises Dinosaur Ridge west of Denver, where dinosaur footprints are numerous.[2] It is this same hard sandstone that forms the cap rock that protects the ancient dwellings at Hovenweep National Monument in Utah and at Mesa Verde National Park in Colorado. And it is the same rock that the ancient Anasazi used to make tools to grind corn.

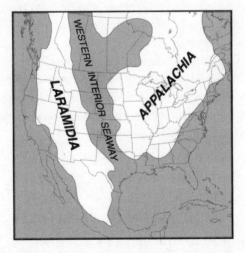

Seventy-five million years ago the dry land of North America consisted of two large masses, Laramidia and Appalachia, separated by the Western Interior Seaway.

The eastern shoreline is a bit more challenging to find. It is best displayed in the bluffs around Sioux City, Iowa. In particular, the sandstone is exposed at Sergeant Bluff, named for Army Sergeant Charles Floyd of the Lewis and Clark Expedition who, in 1804, three months into the expedition, became violently ill and died from what is believed to have been a ruptured appendix. He was the only member to die during the expedition. The 100-foot obelisk that now marks his final resting place was carved from the hard sandstone that formed along the eastern shoreline of the Western Interior Seaway.

At its greatest extent the Western Interior Seaway stretched from Montana to Minnesota, from eastern Arizona to central Oklahoma. At Monument Rocks in Kansas the water depth was at least six hundred feet. Monument Rocks now stand about three thousand feet above sea level. What allowed the central part of North America to be flooded with seawater?

In part, it was a global rise in sea level. The breakup of Pangea and the continued movement of North America away from Europe and Africa were accomplished by seafloor spreading. The mid-ocean ridge, where new ocean crust was being formed, was very active and a new and long segment of the ridge formed between North America and the other two continental masses. The mid-ocean ridge is hotter and shallower than normal seafloor. Consequently, the average depth of the oceans had decreased, which meant sea level rose, and seawater flooded the land.

During the Cretaceous, sea level was almost a thousand feet higher than it is today. But that is not enough to flood the central part of North America as the continent is today and form the Western Interior Seaway. Something else happened. And that something else lies deep beneath our feet.

It is time to introduce the Farallon Plate.

◆

When, during the Triassic Period, Pangea began to break apart and North America separated from Europe and Africa and began its slow movement westward—a movement that continues today—the western edge of North America began its encounter with the vast seafloor of the Panthalassa Ocean. Because the rocks of the seafloor, mostly basalt, are denser than those of a continent, which are granitelike in composition, the seafloor started to slide, or subduct, beneath North America and into the Earth. It did so as a slab tens of miles thick, a slab that represented the tectonic plate that lay under the Panthalassa Ocean, a tectonic plate known as the Farallon Plate.

After nearly two hundred million years almost all of the Farallon Plate has slid into the Earth and disappeared. Only two sections of this plate are still on the seafloor: the Juan de Fuca Plate that lies off the coasts of British Columbia, Washington, and Oregon, and the Cocos Plate that lies south of Mexico and Central America. Both the Juan de Fuca Plate and the Cocos Plate continue to slide beneath North America. But what has become of the bulk of the Farallon Plate? Where does it reside? And what is its configuration? For decades after the theory of plate tectonics was first proposed, scientists offered many ideas about the fate of the Farallon Plate, ideas that were little more than vague connections because there were no observations to support them. Then in the 1990s, a revolution began. A technique was developed that allowed scientists to see inside the Earth with remarkable clarity, a technique known as tomography.

Tomography was first used by the medical profession to image the inside of the human body by the use of X-rays. X-rays are absorbed by the different tissues of the human body by different amounts. And so, by sending X-rays through a human body at different angles, and after appropriate computations are made by a computer, a three-dimensional image, or CAT scan of a human body can be constructed. Seismic tomography works basically the same way with seismic waves.

In seismic tomography, a network of seismometers records the arrival times of seismic waves that have passed through the Earth from a distant earthquake. Because seismic waves travel faster through rock that is cold and dense, it is possible, after the waves of many earthquakes from different parts of the Earth have been recorded, to construct three-dimensional images of the hot and cold regions of the Earth's interior.

In 2003, a major project was begun to produce such images of what lies far beneath North America, from the surface to the core-mantle boundary, a distance of nearly two thousand miles. Billed as a "telescope" that looked inside the Earth, EarthScope consisted of vast arrays of hundreds of seismometers, some in permanent locations, others moved periodically to maximize the ability to produce three-dimensional images. The project lasted fifteen years. It amassed nearly seventy terabytes of data, more than four times the amount of data contained in all of the books in the Library of Congress. Work continues on processing and analyzing so much data. But, early on, one of the major results of EarthScope was an image of the Farallon Plate as it exists today inside the Earth.

The Farallon Plate is no longer a single coherent flat slab of ancient seafloor. Its descent into the Earth has caused it to be contorted and torn. It exists today as several large fragments that are arranged in a simple manner: Beneath the western extreme of North America the Farallon Plate dips down into the Earth at a steep angle until it reaches a depth of about four hundred miles. At that depth it encounters a barrier, an abrupt increase in density of the mantle. Below that depth the pressure is so great that atoms are pushed close together and denser minerals are formed. Because the downward movement of the plate is controlled by the pull of gravity, the Farallon Plate is unable to penetrate the density barrier and descend farther. Instead, it sits atop the barrier and lies beneath North America at a constant depth. But, near the eastern edge of North America, the slab of ancient seafloor has been able to overcome the barrier and plunges down more than a thousand miles to the core-mantle boundary.

Now go back in time to the Cretaceous. The front edge of the Farallon Plate has not yet reached a point beneath the mid-continent. Its continued movement is like a rigid plate pushing its way through a hot viscous fluid. As the plate continues to move, the drag on the plate pulls down the surface of the fluid. Eventually, the viscous fluid responds to the depression and the surface returns to its original level.

That is how the Western Interior Seaway formed. The Farallon Plate was being pulled by gravity down through a hot viscous mantle. The plate was moving to the east, and so a broad north-south trough formed across North America. The surface was pulled down almost a mile, temporarily. It was that temporary depression across the center of North America, and the global rise of sea level during the Cretaceous Period, that led to the water-filled Western Interior Seaway.[3]

More will be said later about the Farallon Plate: how it produced the modern Rocky Mountains, how it helps to direct heat to Yellowstone, and how it is responsible for destructive earthquakes that occur beneath the center of the continent. But, for now, there is a more immediate interest: how this once giant seafloor tectonic plate, most of it now lying deep beneath our feet, added the last great landmasses to the edge of North America and, thus, gave final form to the continent as it is known today.

◆

There is a kaleidoscopic feel to the rocky outcrops that one sees when traveling across the northern half of Washington State. Here there are no great layered-cake displays of geologic formations that are so impressively revealed at the Grand Canyon. Nor are there any vast geologic formations like the Sioux Quartzite, the Marcellus Shale, or the Navajo Sandstone that cover hundreds of thousands of square miles. Instead, in the northwestern corner of the United States—as well as along the entire western margin of North America—there is a jumble of giant crustal blocks that are known as terranes.[4]

A *terrane* is a crustal block that is geologically distinct from its neighbors, meaning it has had a different history, may have a different age, and contains a different sequence of rocks, including a different succession of fossils, than adjacent blocks. How such blocks became jumbled was first suggested in 1971 by Jim Monger of the Geological Survey of Canada.

Monger had been working in the mountains of British Columbia trying to decipher the history of this geologically complex and high rugged region of the continent. As he moved across successive mountain ranges, he noticed that there were distinct boundaries where there were sudden changes in the types of microscopic marine fossils. Others had noticed the same and suggested that these boundaries, which could run for hundreds of miles, some keeping a nearly constant distance from the modern coastline, represented narrow ancient seaways. But Monger suggested a radically different origin.

The theory of plate tectonics had just recently been proposed. Monger applied it, reasoning that the different fossil successions indicated that these crustal blocks had originated in different places, at different times. These blocks had been carried from their original locations to their present ones by shifting tectonic plates that pushed the blocks up against the edge of North America, thus building the edge of the continent outward. A few years later, Peter Coney of the University of Arizona, who had been studying such blocks elsewhere along the western margin of North America, added the adjective "exotic" to indicate that the rock formations contained within these blocks had not originated on North America.[5]

It is such exotic terranes that one sees when traveling across northern Washington. Begin in the eastern part of the state. Imagine a north-south line drawn through the city of Spokane. This line approximates the position of the shoreline of Laurentia. The rocks to the east are those of the ancient continent, while those to the west are a jumble of exotic terranes.[6] Now travel westward across those exotic terranes.

The rocks of the Okanogan Highlands are the southern extent of one of the first exotic terranes to push up against North America. Named for a mining town in Canada, the Quesnel Terrane runs from the highlands northward as a narrow slice through British Columbia and into the Yukon. Farther west is the Methow Terrane that has few exposures until one starts to climb up into the mountains of the North Cascades. Near the eastern entrance to North Cascades National Park one can see Crater Mountain, which is part of the Hozomeen Terrane. At the entrance one crosses from the Little Jack to the Skagit Terrane. On the west side of the North Cascades are the Shusan, the Chilliwack, the Roaring Creek, the Nason, and the Nooksack terranes. Now jump to Vancouver Island, where one stands mostly on the Wrangell Terrane, which runs northward into Alaska and includes its namesake, the Wrangell Mountains.

This collection of names is to illustrate how quickly one can move through a succession of terranes. Many are of oceanic origin, consisting of seafloor crust or a single volcano or a set of islands from a volcanic arc. Some include fragments of other continents. The top of Crater Mountain is rock that erupted on the seafloor during the Mississippian subperiod, when great sequences of coal were accumulating in Appalachia. Rocks of the Quesnel Terrane are much older. They include limestones and muddy sandstones that were laid down on the seafloor during the Devonian Period, when sharks first appeared in the ocean. Those ancient limestones and sandstones can be found today in the cliffs and caves north of Omak, Washington, at a place known as Cave Mountain.

One does not have to climb a mountain or descend into a canyon to march across several terranes. In places, a day-long walk will do. And one of those places is San Francisco.

Three terranes run through the city. The northeastern section of San Francisco where the steepest hills are found is part of the Alcatraz Terrane, named for the nearby island that houses the notorious prison; the same suite of rocks that underlies the prison also lies beneath the mansions on Nob Hill. To the southwest is the San Bruno Mountain Terrane that runs from the Cliff House on the Pacific Coast to San Bruno State Park and South San Francisco. In between is the Marin Headlands Terrane.

The Marin Headlands Terrane is the largest and most colorful of the three. It extends from Coyote Hills near Fremont to the city of San Francisco, where it forms Twin Peaks and lies beneath the eastern half of Golden Gate Park. From there, it is found on the northern side of the Golden Gate Bridge, where it is easily recognized as the highly contorted, red-ribboned chert. The formation of this terrane is told in the chert.

It was in a deep tropical sea thousands of miles away and more than a hundred million years ago that the chert formed. There was a steady slow rain of dead microscopic animals, radiolarians, that fell through the ocean water and onto the ocean floor. The calcium carbonate in the shelly skeletons dissolved under the pressure and the cold temperature of the deep ocean, but the silica part did not. Instead, the small particles of silica accumulated, like a fine snow. The result was a thick red radiolarian ooze. A tenth of an inch of the ooze accumulated every thousand years. The total thickness of the red ribbon chert at the Marin Headlands is several hundred feet. And so tens of millions of years must have passed for this single formation to have accumulated; a very long time, indeed.

From the place of origin somewhere in a distant tropical sea, the thick deposit of radiolarian ooze was carried atop a moving seafloor tectonic plate, in this case, the Farallon Plate, much like the way a series of items is moved atop a conveyor belt. As the Farallon Plate slid beneath North America, most of the tectonic plate and the debris that lay atop it, including most of the ooze, descended into the Earth. But a small part was scraped off and is left today as the ribbon chert on the Marin Headlands. And that is what is seen all along the western margin of North America: Oceanic tectonic plates, usually the Farallon Plate, slid beneath the continent and the upper reaches of the plate were scraped off.

And so, in this manner, exotic terranes, numbering more than a hundred, were accreted to the western margin of North America from Alaska to

Mexico. Alaska, in fact, is composed almost entirely of exotic terranes. But that is far from the end of the story. As the various terranes were pushed up against North America many were also sheared apart and shoved thousands of miles northward. The Quesnel and Wrangell Terranes are a series of slices of formerly single large terranes.

And some of the terranes assembled into continental masses of substantial sizes, the so-called *superterranes*. The Quesnel combined with the Cache Creek and the Stikine, both also large terranes, and with many smaller ones into what is known as the Intermontane Superterrane. It is the rocks of the Intermontane Superterrane that make up most of the Canadian Coastal Range.

Most of this assemblage and subsequent collisions with North America occurred during the Cretaceous. A significant one occurred before, during the Triassic Period, when the Sonomia Superterrane collided with the edge of Laurentia and caused the Sonoma Orogeny. The rocks of the Sonomia Superterrane are those that lie beneath most of western Nevada.

And there was a superterrane, the Siletzia, that collided with North America about ten million years after the end of the Cretaceous Period. These seafloor volcanic rocks are exposed as the Olympic Mountains in Washington. Drill a deep hole beneath Portland, Oregon, or Seattle, Washington, and you will find rocks of Siletzia.

Thus the western margin of North America, as it is seen today, was essentially completed fifty million years ago. Next would be a series of final adjustments as the Farallon Plate continued to slide beneath the continent. But, before those final adjustments were completed, while terranes were still stacking up and building the continent westward, Gondwana was continuing to pull away from North America. And the eastern and southern coastlines of the continent would soon have their current forms.

◆

The splitting of Pangea occurred close to the lines that followed the outlines of the various continental masses that had come together and formed the supercontinent. In the North Atlantic, the splitting occurred along the western edge of Avalonia, the continental mass that broke away from Gondwana during the Cambrian, was rafted across an ocean, then rammed against the edge of Laurentia to form the ancestral Acadian Mountains, the second phase in the building of the Appalachian Mountains. Today all that remains of Avalonia in North America are the eastern portion of New England, parts of Nova Scotia, Prince Edward Island, and the Avalon Peninsula of Newfoundland. The bulk of Avalonia is now on the other side of the Atlantic Ocean. It comprises essentially all of England and Wales and much of Belgium and extends as far east as Poland.

Farther south along the East Coast of North America, Pangea broke apart close to the line where Gondwana had run up against what is now North America. Then, as the two large continental masses slowly drifted away from each other, stretching the Earth's crust, a series of basins formed, including the Hartford, Newark, Gettysburg, and Deep River Basins. The largest basin that would form was the South Georgia Rift Basin, four hundred miles long and nearly a hundred miles wide. If one had been watching over the millions of years that it took for North America and Africa to finally free themselves from each other, one might have expected them to separate along the axis of this basin. But that is not what happened.

As the North Atlantic was continuing to widen and the South Atlantic was beginning to spread open, Africa was moving away from Europe and South America was moving away from North America. The place where these four huge continents were pulling away became very complex. And a major piece of Gondwana—that is, Africa—broke off and remained attached to North America. That piece is peninsular Florida.

Drill a deep hole beneath Montgomery, Alabama, or Augusta, Georgia. The fossils you will find will be identical to those that lie beneath Louisiana or the Carolinas. Drill a deep hole anywhere in Florida and those fossils will match those near modern Dakar, Senegal, in West Africa.

This block of continental crust is the Suwannee Terrane. Its northern limit runs from Mobile, Alabama, to Savannah, Georgia. The basement rock is Cambrian in age, about five hundred million years old. Most of it has remained below sea level since then. And so it is covered by thick seafloor deposits, mostly limestone.[7]

The Suwannee Terrane was not the only piece of Gondwana to break off. As South America moved away from North America and as the Gulf of Mexico and the Caribbean Sea formed between them and continued to widen—in a manner similar to how the North Atlantic was forming, by seafloor spreading—pieces of South America broke off. The region of Coahuila in northern Mexico near the Big Bend along the Rio Grande was one of those pieces. So was the Yucatán Peninsula, today identified by geologists as the Maya Terrane. There was also a splintering of landmass from what is now the coast of Venezuela. Those splinters of continental crust now form the string of large islands known as the Greater Antilles, which includes Cuba, Jamaica, Hispaniola, and Puerto Rico.

The splitting apart of North America and South America was roughly concurrent with a splitting apart of Europe and Africa. The result had a major impact on the subsequent history of the planet. For the first time in hundreds of millions of years an open seaway, the Tethys Sea, ran close to the equator. This seaway was mostly shallow water with little circulation of seawater. As a result, the seawater was stagnant and contained little oxygen. When marine microorganisms, which grew in abundance in the warm Cretaceous seas, died and fell to the seafloor, they did not decompose. Instead, great thicknesses of carbon-rich mud and silt accumulated.

Where the seawater was very shallow, such as along the Western Interior Seaway, the carbon-rich mud and silt became coal. Seams of coal in Utah, Colorado, and New Mexico date from this period. Where the seawater was deeper along the edge of the shoreline of the Tethys Sea, the great thicknesses of carbon-rich mud and silt were transformed into oil.[8]

◆

There is a simple reason why Texas is so rich in oil: Twice in the history of the Earth when the planet was sweltering beneath a humid hothouse climate and when marine microorganisms were growing in great profusion, most of what is Texas today was covered by a shallow and stagnant sea.

The first time this occurred was during the Permian Period when, as recounted in an earlier chapter, a large reef, represented in part by Guadalupe Peak, encircled the large inland Delaware Sea. From this was formed a great reservoir of oil with multiple tiers that lies beneath the Permian Basin of West Texas.

Much more recently, when sea level was again high and the climate was hot and humid, marine microorganisms again proliferated in great numbers and died and fell onto the floor of a shallow sea and were turned into oil that now lies beneath the broad coastal regions of Texas and Louisiana. It was the discovery and initial recovery of this oil that led directly to the worldwide boom in oil production that continues to this day.

A specific date can be given to this beginning: It was Thursday, January 10, 1901. For two years workers had been drilling into the earth at a wooden derrick and drilling rig positioned atop a low rise of ground known as Spindletop Dome near Beaumont, Texas. On that day the drilling had reached a depth of about a thousand feet when mud started to boil up out of the well. Soon after, the entire drilling pipe started to rise—something the workers had never seen before. It moved up and started going through the top of the derrick. The workers moved back. There was a brief deafening roar as gas was released, then silence.

After a few minutes, one worker went back to inspect the damage and to look down the well. He could hear a kind of bubbling deep in the ground. Soon oil started to come out of the wellhead. It came out in pulses, each successive pulse higher than the previous one. It seemed as if the well was breathing oil. Finally the momentum was so great that oil shot through the top of the derrick. It spurted skyward as a stream more than a hundred feet high.

It took nine days to cap the gusher of oil using heavy timbers and giant clamps. When the well was finally put into production, it yielded as much as seventeen million barrels of oil a year, ten times the amount of oil being produced within the United States before Spindletop.

A surge followed in the exploration and discovery of oil. By 1910, the United States was producing two hundred million barrels a year. Ten years later the number was nearly five hundred million barrels a year. By then, three-quarters of all of the oil being produced in the world was being produced within the United States.

This great increase in oil production led railroad companies to convert their locomotives from coal-burning engines to ones that used diesel fuel. The same conversion happened in steamship companies. The fledging automobile industry directed its efforts toward the development of gasoline-powered vehicles. The aviation industry has always been dominated by oil-derived fuels. And that is the situation today: More than 90 percent of the energy used in transportation comes from oil.

The increase in consumption of oil that was spurred on by the discovery at Spindletop led some people to predict that the amount of oil in reservoirs would soon run out. A prediction made at the end of the First World War saw the end of oil in the next several years. Instead, major discoveries of oil were made in East Texas and in the Middle East around the edge of the Persian Gulf, both former shorelines of the Tethys Sea. Another dire prediction of the end of oil came in the late 1940s. Again, it was upended by major oil discoveries along the coasts of Venezuela and Libya, also Cretaceous-age oil that had formed along the Tethys Sea. Then, in 1956, a new prediction was made, this one quite specific about when the peak in oil production in the United States would occur. This was a prediction made by Marion King Hubbert, a geologist working for Shell Oil in Houston, Texas.

Hubbert took a deeper look at the future of oil, noticing, unlike earlier forecasters, that there had been a decline in the rate of discovery of new oil reserves within the contiguous United States. Furthermore, he knew that one of the generalizations that can be made about an oilfield is that, initially, production rose rapidly as the most accessible oil was recovered, then, as time progressed, production reached a peak and, finally, dropped and followed an irreversible decline. He applied that generalization to all of the oilfields within the contiguous United States and predicted that a peak in oil production would happen in 1969 or 1970.

There was an outcry from geologists who worked in the oil industry, including his supervisors at Shell Oil, and from those who worked in government, in particular, at the United States Geological Survey. These detractors collectively cited the failed warnings of earlier predictions. They argued, as many had in the centuries before them, that the United States was a land of limitless natural resources, that only ingenuity was required to discover more oil.

As the early years of the 1970s passed, a peak in oil production in the United States did occur in 1970 at the time predicted by Hubbert. By then, the vast oil reservoirs beneath the North Slope of Alaska, including at Prudhoe Bay, had been discovered and were producing oil. Hubbert incorporated these oilfields into a new prediction: A secondary peak in oil production for the entire United States would occur in 1985. And it did.

A frenzy in oil exploration around the world followed. The number of new oilfields discovered increased, but the total volume of newly discovered oil decreased. That is, almost all of the giant oilfields that exist—which represent more than three-quarters of all of the oil on the planet—have already been discovered. And so, following Hubbert's insight and his methods, a prediction was made for the timing of a peak in worldwide oil production. The prediction was 2005. And a peak in production did occur.

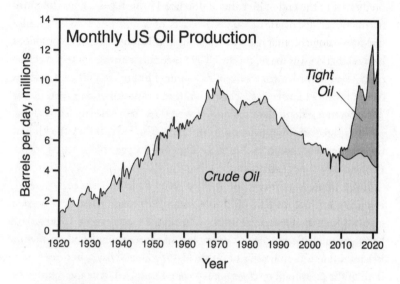

Monthly oil production in the United States. Crude oil, also known as conventional oil, is extracted from permeable rock and flows readily to the surface. Tight oil, or unconventional oil, requires additional effort, usually fracking, to be extracted. Since 2018, most of the oil produced in the United States has come from tight oil. (Data from US Energy Information Administration)

And so, today, we should be living in a world in which the production of oil is decreasing and oil is becoming a scarce commodity. But we are not. Since 2016, the United States has been producing more oil than it did in either 1970 or 1985. In 2019, the rate of oil production in the United States exceeded that of Saudi Arabia, where the world's largest reserves of crude oil are located. What happened?

Fracking.

The price of oil reached a low of $18 a barrel in November 1987 then started a rapid climb, peaking at $164 a barrel in June 2008. That gave an economic incentive to use a more expensive method, hydraulic fracturing, or fracking, to extract oil from rocks where the conventional method of drilling a vertical well and pumping out the oil would not work. It is fracking that has dramatically increased oil production in the United States, and that catapulted the country into the top spot in global oil production.

More than half of the oil produced in the United States today is produced by fracking. And of the oil produced by fracking, more than half of it comes from Texas. It mostly comes from the Permian Basin. A significant amount comes from the Eagle-Ford Formation that lies south of San Antonio. If Texas was an independent country, it would be the sixth-largest producer of oil in the world. Large amounts of oil are also produced by fracking of the Bakken Formation in North Dakota and in the shales found in the Anadarko Basin in Oklahoma.[9]

Just as there is a finite amount of crude oil on the planet, so, too, there is a finite amount of shale oil. If one totals how much remains of both, a quantity known as *proven* oil reserves, one finds that there are 1,780 billion barrels of oil worldwide. The current rate of consumption of oil worldwide is a hundred million barrels of oil per day. Simple division indicates that, if all of the proven oil reserves are recovered and used, it would take about fifty years to exhaust the world's total supply of crude oil and shale oil. But there is another *potential* source of oil.

In northeastern Alberta there is the largest known deposit of bitumen, a semisolid rock-like form of crude oil that is often referred to as "tar sands." This is material that formed during the Cretaceous Period at the edge of the Western Interior Seaway. It is equivalent to about two hundred million barrels of oil, about the same as all of the oil in Saudi Arabia. But, because it is solid, it is harder to recover and must be processed to be turned into oil. Nevertheless, the amount of potential oil could supply the entire world with oil for about five years.

The largest potential source of oil is in the form of oil shale.[10] About two-thirds of all of the oil shale in the world is in the United States: at the Green River Formation that is found across much of Utah, Wyoming, and Colorado. It is estimated that there is the equivalence of about four thousand billion barrels of oil in the form of oil shale, an amount that would extend the supply of oil by about a hundred years.

From this discussion of oil two things should be apparent. First, the age of oil, which began about a hundred years ago, will last only a few more generations. Even if the method of hydraulic fracturing is applied at a wide scale—so far, its use has been limited mostly to the United States— there is simply not enough crude oil or shale oil in the world to continue to supply the needs of modern society, which consumes oil mostly for transportation. Yes, there are a few giant oilfields yet to be discovered, but even ones as large as the Ghawar oilfield in Saudi Arabia or the Permian Basin in West Texas, the two largest reservoirs of oil in the world, will not supply enough oil to extend the age of oil by more than a few years.

And, second, the amount of oil that does exist on the planet was produced during the Cretaceous Period. About two-thirds of it formed between 110 and 90 million years ago. And the bulk of the oil, in the form of both crude oil and shale oil, is in North America. And this great abundance of North American oil is related to the formation of the Western Interior Seaway and the splitting of North America and South America. That is, the abundance of oil in North America is a direct result of the breakup of Pangea.

◆

Imagine standing at the edge of the Western Interior Seaway a hundred million years ago. Streams and rivers course down from nearby hills carrying great loads of sand and silt that are deposited in the many swamps and lagoons and along the edges of estuaries that line the coastline. Along the edge of the coastline the same loads of sand and silt are causing a long and wide beach to form. After countless years this distinct accumulation of sand and silt is buried by the addition of more river-borne sediments. It is compressed and consolidated and, eventually, sections of what is now a hard sandstone are exhumed. Exposures are seen today across a wide section of the middle of North America as the Dakota Sandstone.

Along the western side of the former seaway, the Dakota Sandstone has the form of a broad sheet, most of it still buried, that reaches from eastern Colorado to western Utah and into northern New Mexico. A small section of the sandstone can be found exposed at Arches National Park. Several more outline the edge of the San Juan Basin in New Mexico. Near Golden, Colorado, the Dakota Sandstone is the upended slab of rock known as

Dinosaur Ridge where hundreds of dinosaur footprints can be viewed. In southern Utah, it is the rock that lies along the base of the Gray Cliffs of the Grand Staircase.

Exposures of the Dakota Sandstone on what was the eastern side of the Western Interior Seaway are few and less spectacular. They are found mainly in small rock quarries or along the edges of stream beds or along low-lying rock cuts. They line up along a narrow corridor that runs from central Kansas through Nebraska and Iowa and into southern Minnesota. And here interest in the Dakota Sandstone might wane, nothing of particular importance to distinguish it from the many other layers of sandstone from elsewhere in the central part of North America. But there is something of importance here: Fossilized leaves of a new type of forest were found, a forest composed mainly of broadleaf trees. And the fossilized leaves were found in abundance.

The discovery was made in the summer of 1853 by two men who were making their first trip into the western half of the continent. Ferdinand Hayden was the junior member. Years later, he would attain lasting fame as the first geologist to enter the Yellowstone region and describe the many wondrous scenes. But in 1853, at age twenty-four, he had just graduated from the Albany Medical College in New York and was looking for something adventurous to do before starting his medical career. The senior member was thirty-six-year-old Fielding Bradford Meek, who was working at a natural history museum in Albany arranging and drawing fossils. Anxious to increase the number of fossils in the collection, the museum's director raised money to send Meek and Hayden on one of the first expeditions to survey the geology and collect fossils west of the Mississippi River.

Meek and Hayden left St. Louis by steamboat on May 21, 1853. They traveled along the Missouri River, reaching Fort Pierre, modern-day Pierre, South Dakota, twenty-nine days later, a distance of sixteen hundred miles. Because the current was swift and the river possessed many possible dangers, the steamboat ran only during daylight hours, which meant the two men had to wait for nightfall to search for fossils.

Once the steamboat was tied up and secured for the night, Meek and Hayden would set out, using candles or a lantern to light their way and carrying a pick or a shovel, and scale the nearby riverside cliffs.[11] They found mostly fragments of seashells, evidence that this region of the continent had once been covered by a sea. But when they scaled a low cliff that lined the river near present-day Dakota City, Nebraska, they discovered fossilized leaves.

Since then, many others have come to the same place along the Missouri River and have searched the nearby prairie for more fossilized leaves. And they have been richly rewarded. Thousands of specimens have been recovered.

One notable site is near Rose Creek near the Nebraska-Kansas border. More than twenty species of trees have been identified based solely on the appearance of the leaves. These represent a wide variety of species, including those that resemble modern-day willows, magnolias, and holly. Some display the distinctive three-lobed shape of the leaves of a sassafras tree. These leaves range in size from barely an inch long to about four inches in length and width. Others look identical to the simple leaves of the common household ficus.[12]

All in all, what is indicated here is the sudden appearance of a new type of forest, one that was different in its composition of trees from the great forests that preceded it. And it was not just the composition of forests that had changed. A new type of plant now dominated the planet—the angiosperm.

Unlike conifers and other gymnosperms that produce seeds that lack hard or thick coverings, angiosperms produce seeds that have protective coverings. They include trees that produce nuts, such as acorns, almonds, and walnuts. Plants that bear fruit or berries are angiosperms. And the production of a hard or thick covering—to produce a nut or a berry—required a radically new way of sexual reproduction, *double* fertilization—that is, the fertilization of a single female gametophyte by *two* male gametes. This is the most complicated means of sexual reproduction on the planet, more complex than the relatively simple meeting of a single male sex cell and a single female sex cell in animals. And this complication led to a radical innovation, a characteristic of all angiosperms: the flower.

All flowering plants—this includes trees, shrubs, and grasses that have small, barely visible flowers that usually escape notice—are angiosperms. No flowering plants existed 150 million years ago at the beginning of the Cretaceous Period. By a hundred million years ago, the middle of the Cretaceous Period, angiosperms dominated the planet, much as they do today.[13]

The great forests of conifers that had covered the land surface since the Triassic Period were now in retreat, pushed out of the tropics by the faster reproducing and more rapidly growing angiosperms. From this point forward in the planet's history, the stately groves of conifers—the redwoods, the pines, the fir, and the hemlock—would be found thriving only in the less hospitable regions of the Earth, at high latitudes or at high elevations. But this sudden and widespread occurrence of angiosperms in the fossil record presented a problem. Through his theory of evolution, Darwin envisioned the history of life as proceeding by slow, gradual changes, an idea that he encapsulated several times in his book *On the Origin of Species* by the Latin phrase: *nature non facit saltum*, "nature does not make a leap."

He had dealt with the sudden appearance of animals during the Cambrian Period by suggesting that too little of the fossil record for that period had yet been examined. A more thorough search for fossils of Cambrian age would probably reveal a gradual transition. But the angiosperms posed a more serious problem. Fossilized leaves of mid-Cretaceous age were not only being found in the central part of North America, but also in Greenland and in Norway and would soon be uncovered in Southeast and East Asia, South America, and Africa.

In 1879, in what is now a memorable letter written to the director of the Royal Botanical Gardens Joseph Hooker, Darwin addressed the problem succinctly: "The rapid development as far as we can judge of all the higher plants within recent geological times is an abominable mystery." The phrase "an abominable mystery" still echoes among those pursuing the question of the origin and evolution of angiosperms.

Little progress was made in solving Darwin's abominable mystery until the 1960s, when a remarkable sequence of fossilized leaves was uncovered on the Atlantic Coastal Plain of eastern North America in a rock unit known collectively as the Potomac Group. The Potomac Group, composed mostly of sand and gravel, runs northward from Richmond, Virginia, broadens between Washington, DC, and Baltimore, Maryland, then peters out near the northern part of Chesapeake Bay.

What is remarkable about the fossilized leaves within the Potomac Group is that they clearly show an increase in complexity and variety of form through time. The early leaves have simple, smooth outlines and their veins branch in irregular patterns. Later leaves include varieties with multiple lobes and with veins that follow more regular patterns. The lobes make the leaves less prone to being torn by strong wind and the more regular pattern of veins allows a more efficient transport of fluids. Both are decidedly evolutionary advantages.

And so Darwin's abominable mystery is allayed, somewhat, by a prolonged evolution of fossilized leaves found in the Potomac Group. But there is still the question of the origin and success of flowers.

That question, too, is answered, in part, by the Potomac Group. In the 1990s, tiny, fossilized flowers were found in thin layers of charcoal within the Potomac Group. Some of these early flowers are in excellent condition, the charcoal preserving the delicate three-dimensional forms. Though they lack petals and are very small—most measure less than a tenth of an inch in size—they clearly show the reproductive structures of flowers, such as carpels and stamens.[14]

The origin of angiosperms is, of course, tied to their genetics. Plants have a set of genes that are analogous to the Hox genes of animals. They are called KNOX genes (for knotted homeobox genes) and generate the entire above-ground body of all ferns, gymnosperms, and angiosperms. Exactly how KNOX genes differ in ferns, gymnosperms, and angiosperms is still being worked out.[15]

The question as to why angiosperms dispersed widely about a hundred million years ago is also still open. The only certainty is that no geological catastrophe was involved—or at least, none has been identified that would be analogous to the wide dispersal of conifers after the Permian extinction.

Flowers require pollinators, and so, perhaps, the dispersal of angiosperms was tied to the simultaneous increasing diversity of pollinators, namely, insects. But early flowers were small and plain and probably did not attract pollinators. By the time insects had diversified, many flowers were already large and complex.

Perhaps it was the double fertilization that is unique to flower-producing angiosperms that was the primary factor. One fertilization produces the seed and the other a food store that is rich in carbohydrates, proteins, and often fat that surrounds the seed, giving it a higher probability to survive.

Or, perhaps the rapid dispersal of angiosperms was driven by the more rapid life cycle of angiosperms compared to gymnosperms. That would explain why the dispersal is so evident as an abundance of fossilized leaves along the former shoreline of the Western Interior Seaway. River banks and coastal plains would have been disrupted frequently by floods and storm surges. That would have shifted the sand and mud and would have washed away plants. The faster-growing ones, meaning the angiosperms, would have recovered quicker and would have been reestablished sooner.

In short, there is still much to Darwin's abominable mystery of the origin and rapid dispersal of angiosperms and of the flower. Angiosperms do display an incredible amount of variation and their origin from the first small-flowered plants to planet-wide dispersal by the middle of the Cretaceous Period did occur over a few tens of millions of years. And, yet, a century and a half after Darwin and his contemporaries first formulated the question, the origin of the angiosperm flower remains among the most difficult and most important unresolved topics in evolutionary biology.[16]

◆

The story of the Cretaceous Period is not yet at an end. One more tale needs to be told. It is a tale that begins with the discovery of what is probably the world's most famous dinosaur, the ferocious carnivorous giant known as *Tyrannosaurus rex*.

Edward Cope died in 1897 and Othniel Marsh died two years later. Thus ended the so-called Bone Wars, the frantic pursuit, and often willful destruction, of large dinosaur bones in the western United States. But, as the nineteenth century was closing, a second, less intense competition was already underway.

In 1891, Henry Fairfield Osborn, who had studied geology at Princeton and who was mentored by Cope, founded the Department of Vertebrate Paleontology at the American Museum of Natural History in New York. Five years later, the steel industrialist and wealthiest man in the country Andrew Carnegie founded the Carnegie Museum of Natural History in Pittsburgh. Each man was fascinated by the skeletal models of giant dinosaurs that had been constructed from the discoveries made by Cope and Marsh. And each man was determined, using his family's fortune, to finance the discovery of more spectacular bones and make them the centerpiece of his museum.

The bones of the odd-looking *Triceratops*—an elephant-sized dinosaur with three horns sprouting from its head, two from the brow and a third, smaller one from the snout, and with an immense rounded frill masking the back of its neck—were first discovered in 1888 near Denver when a cowboy on horseback saw what appeared to be a large bone sticking out of the ground. Unable to reach it, he lassoed it, breaking off what looked to him to be part of a large horn and a skull. He sent the specimen to Marsh at Yale College. At first, Marsh thought it was the horn and the skull of a large prehistoric bison. But similar horns and skulls began to arrive at the university, and Marsh decided that they represented a new group of dinosaurs, one that he named *Triceratops* for the three prominent horns.

By 1892, at least fifty skulls of *Triceratops* had been sent to Marsh, but no significant part of the skeletal body had yet been recovered. That would be the goal of Osborn and of Carnegie. Each man would send a team of collectors to search for the skeletal remains of *Triceratops*. One of the men sent by Osborn was Barnum Brown, a young man who, for his ability to discover and recover large skeletons, would soon be known to colleagues and friends as "Mr. Bones."

Born in central Kansas in 1873 and named for the great showman P. T. Barnum, who was touring his circus through Kansas at the time of Brown's birth—his parents were apparently influenced by the colorful advertising posters plastered on barns, trees, and town buildings—this future collector of large dinosaur bones was fascinated by fossils from an early age. By the time he enrolled at the University of Kansas—his original intention was to study engineering—he had amassed a vast collection of fossilized bones and shells that he had found exposed in the walls of stream beds, relics of ancient animals that had once lived near the shoreline of the Western Interior Seaway. When he learned that paleontology could be a profession, he dropped engineering and focused on the collection and the study of fossils.

After graduation, Brown was hired by Osborn to work at the museum as a curator of dinosaur bones and sent him on several trips to collect additional fossils. Twice he was sent to search for the bones of *Triceratops*, once in Wyoming, then in South Dakota. He returned with an abundance of fossils—fossilized leaves, turtle shells, an assortment of animal bones—but only a few fragments of bone that were *Triceratops*.

Then in May 1902, one of Osborn's close friends, a Dr. William Hor-
naday, arrived at the museum with photographs he had recently taken in
Montana that showed what appeared to be the skull of an elephant. Osborn
and Brown recognized it as a *Triceratops*. Brown set out the next month
to investigate.

He arrived by train in Miles City, Montana, recording in his notebook
that dinosaur bones were on display in the lobby of a local bank. From there,
he and a small team of assistants, a guide, and a cook rode for five days on
horseback over a boundless prairie that he described as having a "striking
beauty" and arrived at a place where there were canyons hundreds of feet
deep with steep sides and where there was a somberness to the landscape
relieved only by brightly colored bands of clay that could be traced for miles.
This was Hell Creek, a tributary of the Missouri River.

He found the exact spot where the photographs had been taken. Exposed
on the surface was a skeleton and large skull, six feet long, of *Triceratops*. Then,
even before the cook's call for dinner, as he continued to examine the area, he
saw something quite different. It was the bones of an unknown creature.

The next day Brown and his crew began to dig. At first, the ground
was soft, but as they excavated, the rock was firmly cemented, so hard that
a pick could make but a slight impression. And so they resorted to dynamite, scraping away the debris with a horse-drawn plow. By then Brown
understood that something momentous had been found. It was Saturday,
July 19, and he wrote of it in a letter to Osborn, informing him "the femur,
pubes [*sic*], humerus, three vertebrae and two underdetermined bones of a
large Carnivorous Dinosaur" had been uncovered. Then he added: "I have
never seen anything like it from the Cretaceous."

This was the discovery of *Tyrannosaurus rex*. A total of thirty-four bones
were recovered from the site. Six years later, in 1908, Brown found another
specimen, this along Big Dry Creek, also in Montana, where a total of
143 bones were recovered, representing about half of the skeleton, including
a complete skull. This was the first look at one of the largest land carnivores
to ever exist on the planet.

From these two discoveries, the first full skeleton was constructed and,
in 1915, it was put on display at the American Museum of Natural History.
This is the same *Tyrannosaurus rex* that is on display at the museum today,
though its pose is decidedly different from the original one.

For a long time all dinosaurs, especially the large ones like *Tyrannosaurs rex*, were thought to be slow, lumbering beasts, somewhat akin to modern lizards.[17] And so when *Tyrannosaurus rex* was first put on display in 1915, it was positioned in an upright Godzilla-like pose—that is, the skeletal model stood nearly erect with its tail dragging on the ground behind it and with its two legs splayed out away from the body. But the view of *Tyrannosaurus rex* has changed. Instead, their closest living relatives are not lizards but birds.

The idea that dinosaurs and birds are closely related goes back to the 1830s when Hitchcock described the dinosaur tracks in the Connecticut River Valley as "stony bird tracks." He envisioned the animals that made those three-toed tracks as giant birds. But few of his contemporaries were convinced because he offered no supportive evidence. In 1868, British biologist Thomas Henry Huxley addressed members of the Royal Institution of Great Britain and suggested a dinosaurian origin for birds, much of his argument based on the recent discovery of the feather-covered, reptilian-like *Archaeopteryx* in Germany, an animal that was clearly a *transition* from a reptilian dinosaur and to a bird. And so the debate continued for another hundred years without much new information and without a resolution.

Then, at last, progress was made. In 1964, Yale paleontologist John Ostrom unearthed the astonishingly birdlike *Deinonychus* in southern Montana. It had long arms that looked almost like wings and a lithe body that indicated an extremely active and agile animal. It was much smaller in size than *Tyrannosaurus*—*Deinonychus* stood no higher than about six feet—and from the appearance of the bones and the teeth, it was a carnivorous dinosaur.

But did it have feathers? That was something that Ostrom could not determine from the site in Montana. More than thirty years passed. Then, in 1996, an announcement came from Liaoning Province in northeastern China. The remains of a dinosaur similar to *Deinonychus* had been found surrounded by halo of feathers. The connection was now clear: Birds did not evolve from dinosaurs; birds *are* dinosaurs. Their skeletons show that birds can be traced back to the same group that includes ferocious carnivores as *Tyrannosaurus rex*. That realization caused paleontologists to pause: Exactly what type of animal was *Tyrannosaurus rex*? How should this ancient tyrant of the "lizards" be depicted in museums?

More than twenty species of feathered dinosaurs have been found, most in China. They range in size from thirty-foot-long cousins of

Tyrannosaurus rex covered with featherlike fuzz to crow-sized dinosaurs with full-on wings.

And many additional specimens of *Tyrannosaurus rex* have been discovered. The most complete skeleton discovered to date was found in South Dakota in 1990. Named Sue for the discoverer, Sue Hendrickson, more than two hundred bones were recovered, making the skeleton about three-quarters complete. It is on display at the Field Museum of Natural History in Chicago, where it is positioned in a birdlike stance with its head and body leaning forward and its long tail outward and horizontal. The largest *Tyrannosaurus rex* yet discovered—about 60 percent of the bones were recovered—was found in Saskatchewan in 1991. Nicknamed Scotty, this carnivorous monster weighed about twenty tons and measured about forty feet from nose to tail.[18] It is also the longest-lived *Tyranno-saurus* yet found. Fossilized dinosaur bones have growth rings, like trees, and the ones for Scotty indicate it died in its early thirties. It can be seen at the Royal Saskatchewan Museum in Regina, Saskatchewan. It, too, is positioned in a birdlike stance.

And so is the original skeletal model of *Tyrannosaurus rex*, the one based on the discoveries of Barnum Brown. In 1993, it was remounted in what is now regarded as a more accurate and natural, horizontal pose. It is on display on the fourth floor of the American Museum.[19]

The discovery made by Ostrom set off a renaissance in dinosaur research. There are now more new species of dinosaurs—fifty or more a year—being discovered than at any time in the past. These were active, warm-blooded animals. Some migrated in groups. Some cared for their young. They evolved into many diverse forms and occupied—it might be more accurate to say dominated—almost every ecological niche. They grazed, they flew, and they hunted. Their behavior was complex. They were a very successful group of animals.[20] And no dinosaur has been more intensely studied and written about—and incorporated into popular culture in films and drawings and cartoons—than *Tyrannosaurus rex*.

Tyrannosaurus rex existed for less than two million years at the end of the Cretaceous Period. It lived exclusively in North America on the western side of the Western Interior Seaway on what was then a large island continent, separated by wide spans of water from all other large landmasses. Its bones and its tracks are found in the Hell Creek Formation, a geologic formation first defined and described by Barnum Brown. Its bones have been found mostly around the Hell Creek region of Montana. A substantial number have also been found in northwestern South Dakota and northeastern Wyoming. It has been recovered at three sites in Saskatchewan and at one site in Utah and one site in Colorado. How much farther it ranged is still to be determined.

Tyrannosaurus rex had the largest brain of all of the dinosaurs. Detailed examinations of the several skulls that have been recovered indicate that its ability to smell was extraordinary. Its eyes faced forward and looked over a narrow snout. Because of the wide spacing between its eyes, the visual acuity of an adult *Tyrannosaurus rex* was ten times that of a human. What a human can visually separate at a distance of one mile, a *Tyrannosaurus rex* could do at four miles.

And the end came quickly. The Western Interior Seaway was making its last retreat—as a new range of mountains, the Rockies, was beginning to grow. The two large island continents would soon be one and North America would be the single large landmass that it is today. But before the island continents unified, there was a cataclysm, a mass extinction of almost all animal forms on the planet, including the non-avian dinosaurs. *Tyrannosaurus rex* and *Triceratops* were extinguished in a geologic blink of the eye.

The Cretaceous Period and the Mesozoic Era ended abruptly. But what was the nature of the cataclysm? Was it a force that arose from within the Earth or one that came from outside?

11

A Calamitous Event

Chicxulub Meteor Impact: 66.043 Million Years Ago

Eugene Shoemaker was always talking about the Moon. One Saturday afternoon, many years ago, another student and I were enlisted by him to assemble some rather large equipment in his laboratory. Afterward he invited us to his home for dinner. As we dined, he asked each of us about ourselves. That satisfied, he shifted the conversation to the Moon. Conveniently, a globe of our nearest celestial neighbor was behind him and we spent the remainder of the evening talking about the origin of various lunar features.

There was another time when I and other students were camped with him out in the Mojave Desert. It was shortly after sundown and slightly more than a quarter Moon shone in the night sky. Through the clear desert air, it was possible to see a few individual features on the lunar surface. The crater Copernicus was conspicuous, the edge of long shadows cutting across the crater. Gene spoke about what it must be like to stand within that large crater. It was also possible to see the large, dark circular area that marked the thick basaltic lava flows that filled the Serenitatis Basin. With some imagination, one could identify where along the edge of the basin the sixth and final lunar landing had been. That mission had been made by Apollo 17 and Gene should have been on that flight, but, as he freely acknowledged, his adrenal glands had failed him. He had Addison's disease and, though he was widely regarded as the world's authority on lunar geology and history, NASA had judged him medically unfit for spaceflight.[1]

And then there was the time that I went to Meteor Crater with Eugene Shoemaker. We were in northern Arizona and there were several hours to pass before he had to catch a flight in Phoenix. So the day was spent at the crater that he had studied for many years, his work changing the way geologists looked at impact cratering, emphasizing the important role it has played in our planet's history.

Meteor Crater is best described as a giant bowl surrounded by a raised rim. It is three-quarters of a mile in diameter and the center of the bowl is six hundred feet below the level of the surrounding plain. It was the simplicity of form, and how the crater suddenly disrupted a flat desert landscape, that puzzled people. Furthering this confusion, tons of meteoric iron, mostly in the form of small fragments, had been found lying on the plain immediately surrounding the crater. From that, it was generally thought that this feature, known originally as Coon Butte for the profile of the raised rim, was the result of a meteor impact.

But that needed to be proven. And so, in 1891, Grove Karl Gilbert, chief geologist of the United States Geological Survey, visited the crater, intending to establish its origin.

He spent several weeks at the crater. With a survey crew, he made a detailed topographic map, showing that the volume of the prominent rim was approximately the same as the volume of the bowl. To him, that was evidence *against* an impact origin because he assumed that a large impacting body, if it existed, would still be lying beneath the crater, and so the volume of the crater bowl should be substantially *less* than the volume of the rim. He also took magnetic measurements at the crater, again deciding that no large body consisting of magnetic-disturbing, meteoric iron was buried beneath the crater. From this, Gilbert rejected an impact origin of the crater. In its place, he suggested that Meteor Crater had formed by an unusual type of steam explosion. According to him, the fact that fragments of meteoric iron were found around the crater was just a coincidence.

And that became the gospel, repeated by the next few generations of geologists. But a mining engineer, Daniel Barringer, who had made a fortune from a silver mine near Pearce, Arizona, was not so sure.

A barrel of a man who stood at five-foot-nine and weighed upward of two hundred pounds, a lover of whiskey and cigars, Barringer would later recall first hearing about Coon Butte during a "casual conversation"

with a Mr. Samuel J. Holsinger while the two men were on the veranda
of the San Xavier Hotel in Tucson, Arizona. It was October 1902, and
Holsinger had said that he had never seen the crater, but that he had
heard that tons of iron had been found lying on the surface and had been
shipped from the area.

Barringer was intrigued, so he did a quick calculation. If an impact
had formed Meteor Crater, then the fragments of iron found scattered
around the crater must represent just a tiny fraction of the total amount
of iron involved in the impact. Considering the size of the crater, Barringer estimated that tens of millions of tons of raw iron must be buried
just beneath the crater floor. By mining it and selling it at the current price
of about a hundred dollars a ton, he would become the world's first bil-
lionaire. But he feared that if he appeared interested someone might "jump
the claim." And so he worked secretly—and quickly. Arizona was then
a territory under the jurisdiction of the federal government. He used his
political connections, which were considerable, and by March 1903 his claim
was approved by the federal government, the paper signed by President
Theodore Roosevelt.

The digging of mining shafts and the drilling of holes started immediately. Seven major shafts were dug into the crater floor. Additional shafts
and several deep trenches were dug around the perimeter of the crater.
Dozens of holes were drilled, the deepest one reaching more than a thousand feet below the crater floor. Through it all, a slight concentration of
meteoric iron was found only once, near the bottom of the deepest hole.
Elsewhere, the digging and the drilling revealed nothing more than highly
fractured rock and the occasional pocket of loose sand.

Barringer died in 1929, and the effort ended immediately. By then, nearly
a million dollars had been invested. It seemed that the impact theory for
the origin of Meteor Crater was dead. But then, in the 1950s, Shoemaker
arrived and, as he would often say, he "saw the crater with fresh eyes."

He examined the crater in great detail, mapping the location of every
small fault, noting how various geologic units had been disrupted and
displaced. He collected rock samples and had them analyzed for mineral
content, discovering that a high-pressure form of quartz, coesite, was
present. Until then, coesite had been produced only in laboratories in rock
samples that had been subjected to extremely high pressure. It had never

been found in a volcanic rock. But here it was in nature. And there was only one process in nature that could have generated the extremely high pressures that were necessary to produce this mineral: the high-speed impact of a meteorite.

When I went to Meteor Crater, the debate about the origin of the crater had long been settled, resting, in part, on the discovery of coesite. And yet, I wondered. A high level of expertise was needed to produce coesite under laboratory conditions and to analyze a rock and determine whether coesite was present as one of the minerals. Might something else be present that Gilbert or Barringer could have seen—and that I could see now—that would have convinced them that Meteor Crater was the result of a meteorite impact? I asked Shoemaker. He let out a loud laugh—which is how he always responded when I asked him a question—and said there was. He would show me the "weenie-in-the-bun."

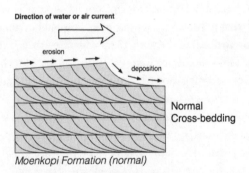

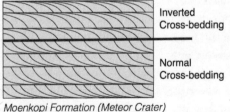

The upper beds of the Moenkopi Formation at Meteor Crater are upside-down, indicating the rocks of this formation have been folded back on themselves creating the "weenie-in-the-bun."

When the meteorite hit, it passed through a series of familiar rocks. The bottom-most was the Coconino Sandstone, a white sandstone formed of windblown sand. Above it was the Kaibab Limestone, a pale yellowish limestone that had formed at the bottom of a shallow sea. And covering the surface was a veneer of dark reddish-brown rocks, the Moenkopi Formation, a mudstone that had been deposited by the repeated washing of mud and silt across a coastal plain. Each surge of mud and silt left a thin layer. If one looks closely, one sees that the accumulation of so many thin layers has left a distinct sweeping pattern known as "cross-bedding." One advantage of recognizing cross-bedding in a rock is that, from the direction of the sweep, one can determine whether a particular rock is in its original position or whether it has been turned upside-down.

We scrambled along the crater rim and found a place where torrential rain had washed out a deep gully. We examined the walls of the gully. Along the bottom was the pale yellowish rock of the Kaibab Formation. Above it was the reddish-brown rock of the Moenkopi Formation with cross-bedding that indicated that it was in its original horizontal position. But partway up the wall of the gully there was a line that could be drawn with a finger. The rocks above were still those of the Moenkopi Formation, but the cross-bedding showed they were *upside-down*! And above that, again, was the Kaibab Formation.

When the meteorite hit and penetrated into the ground, the impact sent pressure waves outward, somewhat like an underground explosion. That had caused the surface layers to rise and to be peeled back. As they were being peeled back, the layers landed upside-down, accounting for the abrupt inversion of stratigraphy within the gully walls. Because the Moenkopi was a reddish-brown and the Kaibab was a pale yellow, as we stood there, it looked very much like a pale yellowish bun had folded back over a red hotdog—the "weenie-in-the-bun."

It required a careful examination to see this. But once seen, it was clear what had happened. And either Gilbert or Barringer could have made the same observation—if they had come to the crater with fresh eyes.

This abrupt peeling back and folding over and inversion of stratigraphy cannot be produced by volcanic activity. But it is a natural result of a meteor impact. As Shoemaker often said: "The physics is entirely different. A volcanic eruption is about expansion as gases escape, while a meteor impact is about a single, sudden compression. The features they produce are entirely different. But to distinguish them requires a careful look."

And so this is how Meteor Crater formed. About sixty thousand years ago, a small, iron-rich asteroid, about 150 feet across, slammed into the Earth. The impact sent out a strong compressive wave that pulverized the rock to a depth of a thousand feet or so. The rocky surface layers were peeled back and pushed upward and outward to form the crater rim. So much energy was released by the impact that almost the entire asteroid was vaporized, which explains why Barringer was unable to find a large body of meteoric iron. Gilbert was wrong in assuming that most of the body landed intact and was of a large size. All that remains of the asteroid are the hundreds of thousands of iron fragments that have already been collected and that are still scattered around Meteor Crater today.

◆

The idea that stones could fall from the sky was largely dismissed by even the most learned of individuals until undeniable evidence was presented early in the nineteenth century. As an example, in 1795, newspapers in London reported that a ploughman working in a field near the village of Wold Newton in Yorkshire had seen a rock fall from the sky and land near him. He retrieved the rock—according to him, it had made an inconsequential crater—and gave it to the landowner. The landowner, seeing an opportunity to make money, publicized the ploughman's story and put the rock on display in Piccadilly across from the famous Gloucester Coffee House and charged a fee of one shilling to anyone who wanted to see it. Among those who came was Sir Joseph Banks, the president of the Royal Society. Banks would eventually entertain the idea that meteorites might have an extraterrestrial origin, but it was never clear whether he ever believed the ploughman's story.

The turning point in understanding the origin of meteorites came years later. On the afternoon of April 26, 1803, a shower of thousands of meteorites rained down on the town of L'Aigle in Normandy, France. Thousands of people witnessed the event. The young scientist Jean-Baptiste Biot was sent by the French Academy of Sciences in Paris to investigate. He collected some of the stones and he interviewed many of the witnesses, both clergymen and laymen. In July, he issued a report in which he stated that the nature of the stones that had fallen was sufficiently different from any stones found in the rocky outcrops around L'Aigle, and that he had received consistent stories from a wide range of credible people, so that he had to conclude that the event had occurred and that the falling stones must have originated somewhere beyond the Earth. If anyone was skeptical of Biot's report, and, at the time, there was a great deal of skepticism between British and French scientists over the work their respective institutions were conducting, the feeling left when another shower of meteorites, this time over North America, happened a few years later and settled the matter.

At dawn on December 14, 1807, Nathan Wheeler, treasurer, town clerk, and justice of the peace of Weston, Connecticut, happened to be standing outside his house when, as he described it, he watched as "a globe of fire" moved across the sky. It traveled from north to south, passing almost overhead. He heard three loud and distinct explosions, similar to those made by a four-pounder cannon, followed by a continued rumbling, like that of a cannonball rolling over a floor. He did not see any stones fall nor did he collect any, but many of his neighbors did.

Edwin Burr was standing in the road in front of his house when the stone fell. The noise it made when it hit was loud, and Burr, who was within fifty feet, immediately searched for it. The morning was still dark, and another half hour passed before he found it. Most had been reduced to powder, but there were a few fragments. The largest one he collected was about the size of a goose egg.

Elijah Seeley was asleep and awakened by the noise. Later that morning when he went to tend to his cattle grazing in a field, he noticed the ground was torn up "as if by some violence" and, looking closer, found "a strange looking stone" at the same place. Ephraim Porter later said that he distinctly heard a loud sound and saw dust rising from a nearby hill, but, because he had never heard that stones could fall from the sky, assumed that lightning had struck the ground. Hearing that others had found stones in his vicinity, he was induced to search and came up with a mass of stone at the place where he supposed lightning had struck. It, too, had broken into fragments; the original weight had probably been about twenty pounds.

Benjamin Silliman, the newly appointed professor of chemistry at nearby Yale College, quickly learned of the event. He enlisted the aid of a colleague, James Kingsley, a professor of classical languages, and the two went to Weston to interview witnesses and to collect samples of the recently fallen meteorite. Much to their alarm, they saw people hacking the stones to pieces, dissolving them in crucibles, heating them on forges and pounding them on anvils in hopes of finding gold or silver. After considerable effort—and a great deal of persuasion—Silliman and Kingsley managed to purchase a few pieces. Years passed, and the largest unbroken piece, a thirty-pound specimen, was eventually found and purchased by Yale and is now on display at the Peabody Museum.

Silliman also wrote a report, concluding, as Biot had, that meteorites did originate somewhere from beyond the Earth, though where was still debated. Many would suggest they were rocks that had been thrown out during eruptions of lunar volcanoes. (They are not.) Silliman happened to favor the idea that they were pieces of rock shed by comets. The view today is that, though a few might have originated on comets, the vast majority originated in the asteroid belt, the collection of small rocky objects that orbit the Sun between Mars and Jupiter, and that a few have certainly come from the surface of the Moon and a very few from Mars.

And so, by the early nineteenth century, the extraterrestrial origin of meteorites was settled. But impact cratering was still not considered an important geologic process. And that is how the situation remained until the middle of the twentieth century when a man, working quite alone and who is seldom remembered by anyone except those specialists who study impact cratering, managed to change that impression. And he did so not by investigating possible impact features on the Earth, as Shoemaker would soon do, but by looking at the Moon.

Ralph Baldwin had graduated from the University of Michigan in 1937 after completing a careful study of the star Nova Cygni III, a star that showed an irregular pattern of brightening and dimming. That should have ensured for him a career as an astronomer, but the 1930s were not a time to begin a career in any field. The Great Depression was still crippling the nation. He did work as an instructor of astronomy at a few universities and as a presenter at the Adler Planetarium in Chicago. But none of this was steady work. And so he returned home to Grand Rapids, Michigan, where he began working for the Oliver Machinery Company, where he designed woodworking and metal-sawing machinery and eventually became president and chairman of the board. But his passion for astronomy never waned.

While still working at Adler Planetarium, he had seen some remarkably detailed photographs of the Moon. They showed a multitude of craters, from large lunar basins to the smallest features one could see on the photographs. The accepted explanation for these craters was that they had been produced by volcanic activity. It was an explanation that Galileo had given early in the seventeenth century when he turned what was the world's first telescope to look at the Moon. That was the only process that Galileo knew produced craters, and so that is what they must be. But Baldwin saw something different. Lunar craters varied so much in size and in their individual features, such as the structure of rims or the shapes of the floors, that he thought impact cratering must be the dominant process.

For nearly a decade he did his research. He showed his work to astronomers, but they were not interested because to them the Moon was more of a nuisance, shining brightly in a night sky and making it difficult to study faint objects. Geologists were not interested because they thought nothing important could be added to their science by looking through a telescope. Baldwin wrote papers about his work and submitted them to scientific journals. But they were rejected and labeled as unimportant. Then he decided to publish his observations and his reasonings as a book. *The Face of the Moon* was published in 1949. The sales were a disappointment. No one seemed interested. Then Eugene Shoemaker read the book.

As Shoemaker would say: "Finally, after centuries, someone with a keen eye was trying to understand the geology of the Moon." Impact cratering was important on the Moon, and it must be important in the history of the Earth. And so a new science began: astrogeology. And it would have developed at a slow pace as more geologists became interested. But a floodgate opened, one that was not owed to scientific interest but to geopolitical concerns. In 1961, the United States decided to race the Soviet Union to the Moon. To be successful, much more needed to be known about the lunar surface: how features had formed and what to expect when people actually landed and walked around and collected rocks on the lunar surface. And that meant more impact craters had to be discovered and studied on the Earth. And much of that effort was led by Eugene Shoemaker.[2]

◆

Meteor Crater in Arizona was the first of many features to be identified as impact craters. More than two hundred are now known. And among them I do have a favorite: Upheaval Dome in southeastern Utah.

Located in the Island of the Sky District of Canyonlands National Park, close to the seasonal tourist town of Moab, the impact crater known as Upheaval Dome was once considerably larger than Meteor Crater, but erosion has removed more than a mile of rock layers, leaving only a remnant of what once existed. But there is an advantage to the deep erosion: The deep structural features formed by the impact are clearly revealed.

During the first minute or so after the impact, the crater at Upheaval Dome did have the simple bowl shape of Meteor Crater, but on a much grander scale. This transient crater was several miles across and more than a mile deep. But the steep walls could not stand up to gravity, and so they slid down toward the center of the crater, causing the crater to widen and to shallow. The result was a *complex* impact crater, one with a mountainous central peak and a flat floor and a jumbled crater rim. In basic form, it probably resembled how the crater Copernicus on the Moon looks today. But then erosion took its toll.

Standing inside Upheaval Dome today, one sees the erosional remnant of the central peak and the slumping of the rim. At the center are the grayish-green rocks of the Moenkopi Formation. This was the foundation of the central peak. Moving outward and higher in the stratigraphic order is a ring of rocks of the slightly younger Chinle Formation. Above the Chinle Formation and lining what looks to be an inner crater are the rocks of the Wingate Sandstone. Farther out is a circular moat of the Kayenta Formation. And farthest from the center, forming a circular belt of inward-facing cliffs, is the Navajo Sandstone. This is the same sequence of Mesozoic rocks exposed as the Grand Staircase, but here those of the Moenkopi have been pushed up several hundred feet while those of the Navajo Sandstone have slid down. It is one of the finest examples of a complex crater to be found anywhere on the planet. And there is another intriguing feature, located about thirty miles away, that probably formed at the time of the impact.

In Arches National Park one finds the familiar sequence of rock layers running from the Moenkopi to the Navajo. All of these are undisturbed.

Then well above the Navajo is the Morrison Formation, which also lies undisturbed. But between the Navajo and the Morrison is a curious layer of rocks that seems to have waves in it. This is the Carmel Formation. These are rocks that formed in a tidal mudflat. They are found over wide areas of Utah and Colorado. They normally lie flat, like the Moenkopi and the others. But here they are contorted and folded. The folds vary greatly in size, suggesting a chaotic origin. What might have produced them? The shock waves that radiated outward from the Upheaval Dome impact.

The water-soaked mud that would later become the Carmel Formation was then lying on the surface. As the shock waves came rippling through, slabs of mud were squeezed and upended and put into their current chaotic positions. And so what is seen today is the passage of those strong shock waves frozen in time in the rocks of Arches National Park.

A similar disruption from another meteor impact is found along the banks of the Missouri River in Nebraska. This impact formed the Manson Crater, now completely buried by later sediment, that lies a few hundred feet beneath the town of Manson, Iowa. That something unusual must lie beneath Manson was known early in the town's history because the groundwater consisted of anomalously soft water, meaning that the water lacked ions such as calcium or magnesium. Today it is known that this anomaly is a direct result of the meteor impact: The impact instantaneously vaporized a surface layer of limestone, which would have supplied the calcium and magnesium, and left behind highly pulverized granite and gneiss.

The Manson impact occurred seventy-four million years ago, near the end of the Cretaceous Period when the Western Interior Seaway still covered this region of the continent. Though in this case it was not the passage of strong shock waves through the rock that caused a permanent disturbance, but a succession of tsunamis that were produced when the giant meteorite hit the planet. The effects of these giant and devastating waves can be seen in Niobrara State Park along the Nebraska side of the Missouri River about two hundred miles west of Manson. In the cliffs along the edge of the river is a cream-colored layer of rock, up to ten feet in thickness, that is composed of a chaotic mixture of limestone, gravel-sized quartz, fragments of sandstone and siltstone, and large, irregular pieces of dark shale. If the quartz grains are examined under a microscope, a feature known as *shock lamella* is revealed, which simply means that these mineral grains have been subjected to a shock of high pressure, additional evidence of an impact crater.

These are just a few of the impact features that have been found in North America. In total, twenty-eight have been found in the United States, thirty-one in Canada, and one—a very large one—in Mexico. A complex impact crater that is about the same size as Upheaval Dome and with a major part of the rim still intact sits along the eastern side of the Coosa River near Wetumpka, Alabama. A remnant of a central peak can still be discerned, as well as slump blocks along the crater rim and slabs of rocks exposed along the river bank that were lifted and tilted by the impact.

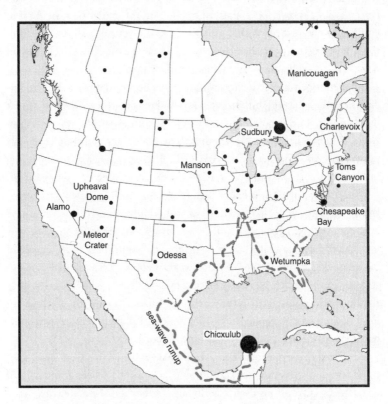

Impact craters in North America. The dashed line indicates how far the giant sea waves produced by the Chicxulub impact ran up onto dry land.

The most spectacular central peak produced by an impact in North America is Mont des Èboulements in Quebec. This peak stands at the center of a thirty-four-mile-wide complex impact crater. Only the northern and the western parts of the crater are exposed; the southern and eastern parts are covered by the St. Lawrence River. Because of this impact, the Charlevoix region is inhabitable today. Otherwise, the area would not have been eroded down and would have remained a continuation of the high plateau of the Laurentian Mountains.

The Alamo Crater in Nevada left giant piles of highly fractured rock that form many of the mountain ranges north of Las Vegas.[3] The Chesapeake Crater is responsible for the wide entrance to Chesapeake Bay. That crater was almost sixty miles in diameter, making it similar in size to the lunar crater Tycho, known for its pattern of long, bright, radiating rays. As if to emphasize the vast reach that a single large impact can have, debris from the Chesapeake impact has been found as far west as Brazos County in Texas and as far south as the island nation of Barbados.[4]

As mentioned earlier, a barrage of meteors occurred during the Ordovician Period about 460 million years ago and may be responsible for a great burst in animal diversity.[5] Given the current limits on the age-dating of rocks, it is impossible to tell whether these craters formed simultaneously or over a period of a few million years. But it is certain from the impacting bodies that formed these craters that the same type of meteorite, an L-chondrite formed these craters. Moreover, meteoric fragments recovered from each site show that each one was suddenly exposed to cosmic rays at the same time, about 468 million years ago. And there is only one way for that to have happened. A catastrophic collision must have happened between two asteroids 468 million years ago, a collision that sent out streams of rocky debris. And a stream of that debris, which was orbiting the Sun, rained down upon the Earth. It occurred during the Ordovician Period. And so this geologically brief period of increased meteorite bombardment is known as the Ordovician Meteor Event.[6]

Realizing that a stream of debris could have rained down from space onto the Earth and that craters of substantial size could have been formed, each one causing disruptions that would reach far beyond the crater rim, what might be the effect of a truly huge impact? What would such a punctuation in the geologic record look like? And what might be the consequences?

◆

I began my search for the most calamitous moment in the Earth's recent history—here "recent" means the last several hundred million years—by driving south from Denver on I-25. Along the way I passed a half-dozen semi-trailer trucks that had been blown over the previous night by strong wind gusts. A brush fire had also swept across a section of I-25 during the night, the flames fanned by the same wind. Each one was a reminder of the immediate destruction that can be caused by a strong wind.

I left I-25 at Trinidad, Colorado, Exit 13B, and drove several miles west along the Highway of Legends Scenic Drive. I crossed the Purgatoire River, which, given the nature of the event I was seeking, is an ominous name. I then took a few turns along a gravel road, arrived at a small parking lot, walked a quarter mile along a dirt path and found myself faced with evidence of an ancient, planet-wide catastrophe.

Here it is worthwhile to pause. There is no grandeur or drama to the landscape here. This is not the Grand Canyon. Nor is it Meteor Crater. What is here is a thin layer, no more than an inch thick, of what seems to be ordinary clay exposed in an unspectacular section of rock. But the story of catastrophe is writ large here. To see it, one must climb a steep slope of loose talus and wedge oneself beneath an overhang of brown massive sandstone.

Continuing for several feet beneath the sandstone is a dark mudstone interleaved by paper-thin layers of black coal. This is the Raton Formation. It is named for the Raton Basin, a broad topography depression, a depression that formed when the Rocky Mountains were starting to rise. During the Cretaceous Period, this was a low-lying plain crossed by slow-moving rivers that dropped mud and silt. It was then a landscape that resembled the Louisiana bayous of today. That is how the dark mudstone and interleaved layers of coal formed. But as these beds were forming, as the sediment was accumulating in what geologists refer to as a low-energy environment there was a disruption. The disruption is indicated by the thin layer of light gray clay.

The clay consists almost entirely of *kaolinite*, a common product of chemical weathering. Its presence would not be noteworthy except that this is a highly unusual type of kaolinite. It did not form by the weathering of individual mineral grains, but by the weathering of tiny glass beads. It is these glass beads that are the key. They can be seen with a magnifying hand lens. They are most readily recognized in small patches. And they can form in only one way: as tiny droplets of molten material that then rained down from the sky.

Now realize that this thin layer of gray clay with tiny glass beads once blanketed the entire planet. It has been found on all continents, including Antarctica. It has been recovered in deep-sea drill holes. It is a record of a global event, one that occurred in a geologic instant, when the entire planet was engulfed by a fireball.

The idea that the entire planet was once briefly in such a fiery state was first realized by geologists only a few decades ago. Walter Alvarez of the University of California, Berkeley, had been studying the transition from the Mesozoic to the next geologic era, the Cenozoic, in a series of limestone deposits near Gubbio, Italy. This also marked the transition from the Cretaceous Period to the Tertiary Period, and so the transition is often referred to simply as the K-T boundary.[7] And there is one more important aspect to this transition: The end of the Cretaceous Period—and so, the end of the Mesozoic Era—marks the disappearance of the dinosaurs.

As Alvarez studied the rocks, he noticed a thin layer of clay disrupting the sequence of limestone exactly along the location of the K-T boundary. If he could determine how long it had taken the clay layer to accumulate, he reasoned, he could decide whether the dinosaurs had disappeared suddenly or if their disappearance had been gradual. He asked his father, Luis Alvarez, a Nobel prize–winning physicist, for advice. The elder Alvarez suggested measuring the amount of iridium in the limestone and clay layers. Iridium is rare in the Earth's crust, but is always raining down upon the Earth at a steady rate as micrometeorites. And so if the amount of iridium was small, the clay had been deposited quickly. If it was large, a considerable amount of time had passed during its accumulation.

The measurements were made. The amount of iridium in the limestone indicated a reasonable rate of accumulation. But the amount of iridium in the clay was remarkably high, many times higher than what was found in the Earth's crust. If the iridium in the clay had accumulated at the same rate as in the limestone, then the thin layer of clay that was so pervasive along the K-T boundary around the world must have taken many tens of millions of years to form. But that was inconsistent with the fossil evidence and with many other indicators. And so there was only one alternative: A high concentration of iridium must have been added to the surface of the Earth at the time the K-T boundary formed. But how?

Luis and Walter Alvarez published their work in 1980. They noted that a mass extinction had also occurred at the time of the K-T boundary. And so they proposed that the extinction and the iridium anomaly were related. After considering several possibilities—that the extinction and the iridium anomaly had both been produced by the explosion of a nearby supernova or by the Earth and the solar system passing through a galactic cloud—they decided that only one mechanism was consistent with both the biological and the physical evidence: A large asteroid had struck the Earth, formed an impact crater, and some of the dust-sized material ejected from the crater reached the stratosphere and was spread around the globe. The dust had prevented sunlight from reaching the surface for a long time, stopping photosynthesis and causing the food chains to collapse, leading to the mass extinction. The iridium anomaly had come from the asteroid itself. Meteorites have an iridium concentration hundreds of times higher than in rocks of the Earth's crust. And so when the asteroid hit the Earth and threw material high into the air, the material that rained down and formed the clay layer along the K-T boundary was enriched in iridium.

The Alvarezes' claim caught the public's attention immediately. It was widely reported in newspaper and magazine articles. Multiple documentaries were made about the father-and-son team who had solved one of the great mysteries of science: the demise of the dinosaurs.

Many theories had been proposed for the sudden disappearance of the dinosaurs. They had grown too large and required too much food to survive. Their brain sizes were too small and they were too slow to compete with the smarter and more agile mammals. Those same mammals were feeding on their eggs. A global plague had killed the dinosaurs. The climate had become too hot (or too cold) for dinosaurs, which were then thought to be cold-blooded animals, to survive.

But the Alvarezes ended the speculation. Their theory was a simple one. The end of the dinosaurs did not come from any prolonged changed conditions on the Earth. It had come suddenly from beyond the planet. A giant asteroid—from the concentration of iridium in the clay layer, the Alvarezes had estimated it was about six miles in diameter, about the size of the island of Manhattan—had slammed into the Earth, sending up huge clouds of debris that encircled the planet and that had caused the mass extinction and had ended the reign of the dinosaurs.

There was a great flurry of activity to find additional evidence to support or refute their claim. Other clay layers found along the K-T boundary were analyzed. And each one showed a high concentration of iridium, including the one near Trinidad, Colorado. Fragments of mineral grains with shock lamellae were now searched for and were found in the same clay layers. It was now possible to explain the worldwide occurrence of glass beads: They were part of the fiery debris that had been produced by the impact and that had rained down on the planet.

But there was an important component missing in their theory: Where was the lethal crater that had caused global destruction?

◆

As it happened, in the 1970s, geologists working for the Mexican state-owned oil company Pemex had conducted magnetic and gravity surveys near a Yucatán town called Chicxulub hoping to find oil. The surveys revealed a large circular feature with concentric rings hidden beneath the surface. Several exploratory wells were drilled. The cores from those wells were examined, but none gave any indication that a substantial amount of oil might lie beneath the surface. And so the geologists moved on.

Several years after the Alvarezes made their famous proposal that a huge impact had caused the extinction of the dinosaurs and ended the Cretaceous Period, Alan Hildebrand, then a student at the University of Arizona, began to compile additional evidence in support of the impact. From that compilation, he convinced himself that the great crater must lie somewhere in the Caribbean. Then, in 1990, while attending a scientific conference in Houston, he first learned of the magnetic and gravity surveys conducted near Chicxulub and how they revealed a large buried circular feature. At the time, any large circular feature was thought to have had a volcanic origin. But Hildebrand saw it with different eyes. The feature was more than a hundred miles across—much larger than any known volcanic crater. He obtained samples of the drill cores and examined them. Within those samples he found the telltale signs of an impact crater: Quartz grains were revealed to have the microscopic parallel markings known as shock lamellae that are produced only by the passage of high-pressure, compressive waves. And there were glass beads in abundance, similar to those found worldwide at the K-T boundary.

Much work has now been done at this immense crater, confirming its impact origin many times over. Additional magnetic and gravity surveys have been conducted, revealing more about its form and structure, even though half of it lies beneath the water of the Gulf of Mexico and all of it lies buried beneath a half-mile-thick layer of limestone deposited long after the crater had formed. In 2016, a deep drill hole was dug just off the coast of the Yucatán Peninsula. It penetrated nearly a mile into the seafloor, passing through the limestone and into a thick section of intensely fractured rock and down into hard granitic rock that existed before the crater formed.

And so it is now firmly established: A complex impact crater with a central peak and two concentric rings does lie buried beneath the northern coastline of the Yucatán Peninsula. The outer ring is the rim of the crater, which has a diameter of 120 miles, making this the third-largest impact crater known on the planet.[8] The city of Mérida, the capital city of Yucatán, lies entirely within the boundaries of the crater. Detailed studies show that the closest piece of land to the impact point is along the coast at Chicxulub Puerto, a modest seaside village of a few thousand people and, because of its fortuitous location, the namesake of the crater. And a remarkably precise age has been determined for the crater: It formed 66.043 million years ago, plus or minus eleven thousand years.

But there is no geologic intrigue here for anyone who wants to ramble over an impact crater. The only immediate reminder of the significance of the site is found in the small town square at Chicxulub Puerto where local artists sell ghoulish paintings of dead dinosaurs. And yet, there is a feature that can be seen and visited, and that contains individual scenes of exquisite beauty, that formed as an indirect consequence of the impact.

A *cenote* is a sinkhole, usually cylindrical in shape, that formed where acidic rainwater has seeped into the underlying limestone, undermining the rock and causing it to collapse and exposing a pool of groundwater underneath. In the eastern part of the Yucatán Peninsula there are several thousand cenotes arranged in a scattershot pattern. In the northern part of the peninsula there are about two hundred arranged along a long arc.

It is the arc, as magnetic and gravity surveys have shown, that runs along the outer ring of Chicxulub Crater.

That there is a correspondence between the outer crater ring and the overlying arc of cenotes is easy to explain: A system of arcuate fractures created by the impact, which are responsible for the location of the ring, have allowed rainwater to seep to deep levels, making it more likely for sinkholes to form. And those that did form created a local environment of enchanting beauty. And each one owes its existence, ultimately, to a deadly and disastrous asteroid impact that happened long ago.[9]

◆

The intense scientific scrutiny that has been applied to Chicxulub Crater—after Meteor Crater, Arizona, it is probably the most intensely studied impact crater in the world, which is all the more remarkable when one considers that, because it is completely buried, no one has actually ever *seen* the crater—led to detailed computer simulations of the impact as the best way to understand how the crater formed and what the immediate consequences were.

Each simulation begins, of course, with a large rocky body hitting the Earth at high speed. The speed of impact was several miles per second, the orbital speed the body would have had as it traveled around the Sun. At that speed, the body would have passed through the bulk of the atmosphere in a few seconds. After another twenty seconds, it would have reached its deepest penetration into the Earth, a depth of about twenty miles. A huge, cup-shaped crater, about a hundred miles in diameter, would have formed as material was being excavated and tossed into ballistic trajectories and as other material was being compressed by the impact and pushed outward. Then gravity would have taken over.

This huge, cup-shaped crater would have been a transient feature, the walls too high and too steep to stand up to the pull of gravity. And so the walls of the crater would have started to collapse inward, sliding toward the center of the crater and forming a central peak. This transient crater was so large that two separate collapses would have occurred, an initial one that formed not only a central peak but also an inner ring, and a later one that produced an outer ring that became the rim of the final crater. In all, several minutes would have passed for the impact and the collapse to be completed and for the crater to take the general form it has today. And then, even though the crater had formed, the destruction beyond the rim was just beginning.

The high speed and high energy of the impact would have momentarily heated the atmosphere. From that, a blast of super-heated air radiated outward from the impact point at speeds in excess of hurricane-force winds. A firestorm grew. The intense heat and strong winds would have scoured the landscape and flattened and burned any forests that then existed. The heat and strong winds would also have incinerated any plants or animals that were within a few hundred miles from the impact.

As the firestorm spread outward, behind it, also surging away from the impact point, came a massive wall of dense, roiling dark clouds. Contained within those clouds was much of the rock that had been shattered and pulverized by the impact. Much of it shot out from the newly formed crater at low angles, behaving much like a liquid, flowing and swirling and, finally, after it settled, forming a blanket of continuous ejecta, looking much like the blankets of ejecta seen around the rims of large lunar craters.

A small fraction of the material ejected from the crater was actually thrown out into space. And a very small fraction of that material was given sufficient velocity to escape the Earth's gravity and remain in space and orbit the Sun. But the majority, in the form of sand-sized particles, reentered the Earth's atmosphere. During reentry, atmospheric friction heated the particles and caused them to melt and to form into tiny droplets that quickly cooled and solidified and fell back onto the Earth. As these particles rained down, they accumulated into a thin layer that once covered the entire planet. During the tens of millions of years since the impact, erosion has removed much of this layer. But there are places where it can still be found.

These droplets that were ejected into the air by an asteroid impact, flew ballistically a short distance in space, then reentered and were heated by the atmosphere and fell to Earth are the glass beads found in the Raton Basin near Trinidad, Colorado.[10] The thin silvery layer that contains these beads, sandwiched between two coal seams and capped by a massive sandstone, is one of the best exposures of ejecta from the Chicxulub impact to be found anywhere in the world.[11]

It is the wide occurrence of this layer that signals that this was truly a global event. An alien being observing the Earth at the time would have seen the initial brilliance produced by the impact itself followed by more than a day of the entire planet glowing a dull red as the droplets reentered the atmosphere. To a life-form on the planet, say, a *Triceratops* or a *Tyrannosaurus*, if it was not near enough to the impact to die instantly, then it would have seen the entire sky glowing red, the intensity of the red increasing steadily as more and more droplets reentered the atmosphere. And with the glowing sky came the heat.

Anything still alive would have suffered from the broiling heat. Only those animals that were able to take immediate refuge underground or underwater had a reasonable chance of survival. For the others, if they were not charred and burned and dead in a second round of firestorms set globally by the increasing heat, then they would almost certainly have been seared by the increasing heat and died slow, agonizing deaths.

And then there were the giant sea waves. Global sea level was still high at the time of the Chicxulub impact, and Yucatán was covered by several hundred feet of seawater. The large asteroid would have made a respectable splash, one that would have generated a ripple of large waves that spread out across the Gulf of Mexico and into the Atlantic and Pacific oceans, but these ripples were of little import compared to what came next.

The truly giant waves came from the formation of the crater. The crater was excavated in a matter of minutes, but it took almost an hour for seawater to pour in and fill the crater. And as the seawater poured in, the inertia behind the new currents overfilled the crater, and so there was a large wave of water that pushed outward. If one was standing along the southern coast of North America at the time, one would have seen the shoreline rapidly retreat, maybe more than a hundred miles. Then, within the next few hours, a high wall of water returned, surging far inland for hundreds of miles, flooding essentially all of what is today the southeastern part of the United States. Where the modern cities of Augusta, Georgia, and Memphis, Tennessee, are located was swept over and flooded by hundreds of feet of water.

In several places this surge of water left a telltale signature. One such place is along a short stretch of the Brazos River south of Waco, Texas. This region was, at the time of the impact, under about fifty feet of seawater. For more than a million years before the event, a thick layer of mudstone accumulated, laid down on what for those many years was the bottom of a quiet sea. But when one looks at this mudstone, one sees an abrupt change. Within it is a single layer of thick sandstone with jumbled rocks, an indicator of geologic violence. It is the only layer of sandstone within the sequence of mudstone. And so something must have happened abruptly, then ended with the same abruptness. Examine this sandstone closely. Among the jumbled rocks are glass beads, the same beads that fell from the sky across most of the planet.

This single layer of sandstone represents the passage of a set of giant sea waves—the largest wave was probably three hundred feet in height—each one ripping up rocks from the seafloor and carrying them far inland. Then, during the backwash, the returning waves pulled rocks from the dry land down and into the sea and mixed them with those already scraped off the seafloor. All the while, a heavy rain of glass beads fell from the sky. And after the waves had settled and after calm had returned to both sea and land, the deposition of mud and the creation of mudstone resumed at the bottom of a now quiet sea.

And then there was the severe ground shaking caused by the impact. More energy was released in the brief moment of the impact than all of the energy released by all of the earthquakes that have rocked the planet during the last several hundred years.

Across much of North America, the ground would have swayed back and forth by several feet. Massive landslides would have tumbled down from major mountain ranges. The new debris would have clogged rivers and caused some to change course. New lakes would have been created. Coastal plains and floors of shallow seas were especially vulnerable because these regions are usually covered by loose sediments that would have "liquidified"—that is, they would have momentarily flowed freely—during the period of intense shaking. And that is clear in undersea explorations that show that the shallow waters along the Gulf of Mexico and along the Atlantic coast as far north as New York are littered with great piles of debris loosened by the impact.

And there is other evidence of the incredible amount of shaking. Severe ground shaking causes water in lakes and narrow inlets to slosh back and forth, forming standing waves, or *seiches*, that can run up high on a shore, then pull back. In 1899, an earthquake in Alaska caused a seiche to run up the narrow inlet to Yakutat Bay, where the water rose as high as sixty feet onto the shoreline. Needless to say, this great wave caused all kinds of havoc to sea life and animal life that lived around the bay.

And the same thing happened during the shaking caused by the Chicxulub impact. In that case, a wave surged up a narrow inlet that was the remnant of the Western Interior Seaway and a short way up a river valley. What it left was a thick deposit that is filled with glass beads and fossils and that was created just hours after the impact. Today that deposit is exposed. It is, as some have described, the surest record of conditions on the planet on the day the dinosaurs died.

◆

The location is now known as Tanis, a reference to the lost Egyptian city that is at the center of the plot in the movie *Raiders of the Lost Ark*. The name was given by Robert DePalma, curator of the Palm Beach Museum of Natural History in Wellington, Florida, who has been the main investigator at the site and an obvious admirer of the movie. Included in his list of collaborators is Walter Alvarez.

Tanis is a fossil site in North Dakota that seems to contain a stunningly detailed picture of the devastation that wrecked the planet in an instant 66.043 million years ago. It is about two thousand miles north of the Chicxulub Crater. The deposit is a jumble of debris about four feet thick. There are fossils of fish, the bodies in haphazard positions, some nearly vertical, suggesting they were caught up and died within a churning mass of mud that settled suddenly. The mud is dotted with tiny glass beads. There is even a layer rich in telltale iridium that caps the deposit.[12] Both seem to tie the deposit to the impact. Moreover, a bone fragment has been found that might be from a dinosaur.

This is a stunning discovery, already sensationalized, even though few details have yet been released about the scientific studies. And so there is caution. And anticipation.

Tanis might become the window through which many details about the Chicxulub impact are revealed, as well as those about the mass extinction that marked the end of the Cretaceous Period. But it must be remembered that the two are not necessarily the same.

After the Alvarezes made their announcement in 1980 of their discovery of an iridium layer at the K-T boundary and the discovery by Hildebrand of Chicxulub Crater in 1991—as well as all of the additional field studies and the computer simulations that revealed the consequences of a huge impact—much of the scientific community was convinced that an asteroid impact had caused a mass extinction and had ended the reign of the dinosaurs. But there was still a group within the community that was skeptical—the paleontologists. They understood that a mass extinction meant more than just the extinction of large animals.

12

Extinction

Deccan Traps: 66.4 to 65.6 Million Years Ago

As late as the eighteenth century, many Americans and Europeans held the worldview that the natural world, because it had been created by a benevolent God, was perfect and complete. It had exactly as much variety, especially in its many life-forms, as was needed to maintain a world of perfection created by a divine hand. Any addition or deletion would disrupt the harmony, would break the chain that linked all living things. And so, it was reasoned, life on the planet must never have changed: Animal species were immutable and no animal species had ever gone extinct.

Also by the eighteenth century, many Americans and Europeans had accepted the idea that the abundance of relics that were being dug out of the ground that had the familiar shape of leaves and branches or that looked like the skeletal remains of animals were exactly that: The mineralized remains of plants and animals that had lived long ago. The shells of turtles and the delicate skeletons of fish had been recovered. So had the fragmentary remains of animals that had once stood and walked and scurried around on four legs.

That these fossilized relics were found in abundance could also be explained: They were the unfortunate victims of the terrifying storms and worldwide flood that God had sent to punish those who were wicked and who were evil and that, in the process, had drowned most of the animal life. But there was a problem with this explanation: If the drowned animals had been found in such abundance, where were the bones of the people, the sinners, who had drowned in the biblical flood?

A possible resolution to the problem came in 1706 when it was reported that a fossilized tooth, seemingly similar in shape to that of a human tooth, was found along a riverbank of the Hudson River a few miles south of the frontier settlement of Albany, New York. The discoverers presented the tooth to the governor of Massachusetts, John Dudley, who recorded that it had been given to him "by two honest dutchmen." They also gave him "some other pieces of bone," some rather large. Dudley could make nothing of the bones. But of the tooth, after consulting with several local physicians, Dudley came to the "(perfect) opinion, that the tooth will agreee only to a human body." Furthermore, he suggested that it had come from someone who had succumbed to a flood, someone who had "waded as long as he could keep his head above water." But the ordeal had been too much. The sinner had become entombed in "the new sediment [deposited] after the flood," and that is where the tooth had been found.

Dudley wrote to the Boston preacher Cotton Mather about the discovery and he came to a similar conclusion, certain that this relic had come from someone "interr'd by the Flood." Mather wrote a description of the tooth and sent it to the Royal Society of London where it was published in the transactions of the society. In his description, Mather noted that the crown of the tooth was worn and on the underside were four prominent "Prongs or Roots." The tooth weighed more than two pounds and was "six Inches high" and "almost thirteen inches in circumference." That such a large tooth could be identified as a human tooth was something that Mather could readily explain. According to the Bible, a race of giants, Nephilim, lived before the flood. These people had been much larger in bodily size than modern people because they had much longer lifespans, comparable to those of the biblical patriarchs, and because modern people were a smaller, degenerate state that was the result of man losing vital forces after the fall from sin. And so, as Mather reasoned, it was to be expected that the bodily remains of those who had died in the flood would be much larger than the remains of people who were alive today.

But as the eighteenth century progressed and more and more fossils were unearthed, and as the variety of fossils increased, it was clear that many different types of animals were represented. In particular, a site with an incredible wealth of bone and tooth fossils was soon found in northeastern Kentucky.

In 1739, French Canadian Charles de Longueuil was leading an army along the Ohio River near what is now the city of Cincinnati when several members of the army, crossing through a marsh, came upon an area where huge bones were poking out of the muck. They collected a single large bone, three large teeth, and what appeared to be an elephant tusk and presented them to Longueuil. Seeing a possible value to the discovery, he had the army carry these findings with it.

They continued down the Ohio, then the Mississippi River to New Orleans, where Longueuil, now in possession of literally a boatload of treasures he had acquired in North America, including the ancient bone, teeth, and tusk, boarded a ship and sailed for France. In Paris he presented his treasures, including the fossils, to Louis XV, who had them placed in the Cabinet du Roi (the Royal Cabinet of Curiosities) in Paris.[1] The fossilized bone, three teeth, and tusk were of special note. They were the first American fossils studied by scientists. The place where they had been discovered was soon identified and inscribed on maps as *Endroit où on a trouvé des os d'Éléphant,* the "place where the elephant bones were found." Today it is known as Big Bone Lick.[2]

For decades this small trove of American fossils, as well as similar fossils collected from elsewhere in the world, was studied by French naturalists. Some suggested that these fossilized remains represented variations of a single animal species. Others suggested that these must be bones of several species. The problem was that no one had studied the anatomy of animals, even domesticated animals, such as cattle and horses, in sufficient detail to decide which anatomical features could be used to define an animal species. Into this confusion entered one of the foremost scientists of history, certainly one of the greatest anatomists of all time, the originator of the modern field of comparative anatomy, Georges Cuvier.

Swiss born, Cuvier arrived in Paris in 1795 and began work at the Muséum National d'Histoire Naturelle, where his official job was to teach, but where he spent most of his time delving into the museum's collection. Among the first specimens he examined were two elephant skulls, one from Ceylon (modern-day Sri Lanka) and the other from southern Africa. From the arrangement and the shape of the bones and, in particular, from the shape and structure of the teeth, he concluded that, contrary to popular opinion, Asian and African elephants should be considered separate species. His contemporaries soon agreed with him. He then turned his attention to the collection of American fossils that Longueuil had brought to France more than fifty years earlier.

Cuvier decided, after a careful examination, that the fossilized teeth were so different from that of living elephants, whether from Asia or from Africa—he described them as different "from the elephant as much as, or more than, the dog differs from the jackals and the hyena"—that they must represent another species. That is, they must represent an animal that had become extinct.[3]

◆

Cuvier made his proposal in 1796. Nearly a generation passed before it was widely accepted.

Thomas Jefferson was one of the many who never accepted the idea of extinction. As a patron and a practitioner of the sciences, Jefferson did make significant contributions to many fields, including astronomy, agronomy, entomology, and meteorology. He even contributed to the study of fossils, including those recovered at Big Bone Lick. As president, he set aside a room in the Presidential Mansion where he displayed and studied fossils that others had collected and had sent to him. One set of fossils, in particular, was of a mysterious animal whose remains had been found in a cave in Virginia that, after a careful examination, Jefferson had named *Megalonyx* for its "giant claw," the most distinguishing characteristic of this animal. Today this animal is known to have ranged over much of North America. Individuals stood as high as ten feet. Experts today categorize it as a type of giant ground sloth.

And while Jefferson was well acquainted with Cuvier's work, he held firmly to the idea of the completeness of nature. Every component of nature was essential to maintain harmony. Remove a single component, say, by the complete disappearance of a single animal species, and the harmony would be ruined and the planet and all of its inhabitants would descend into doom. And so, to counter Cuvier's claim of the extinction of the mastodon, he suggested that it was still alive and roaming somewhere in North America.

When he sent Meriwether Lewis on an exploration of the western half of the continent, he gave instructions that Lewis was to record anything that he might see or anything that he might be told by the native people that might be evidence that mastodons were still roaming the plains of North America. Lewis made the journey, leading the famous Corps of Discovery Expedition with William Clark. They reached the Pacific Coast and returned. When Lewis arrived back in Washington, DC, he reported in person to Jefferson, briefing him about the many accomplishments of the expedition, but having to report that he had not seen nor had anyone told him anything that could support a claim that any large animal, such as a mastodon, was still alive in North America.[4]

By the 1830s—Jefferson died in 1826—scientific opinion had shifted in favor of Cuvier's idea about the extinction of animals. The conversion was especially quick in Britain, where, as in the United States, many naturalists had originally opposed the idea. Now most embraced it. And one who did so, and whose other ideas and whose writings would have a commanding influence on the next few generations of geologists, was Charles Lyell.

Whenever the name of Charles Lyell is mentioned, one is immediately drawn into a world where the geologic forces that can be seen today operate slowly and steadily over immense period of time to cause mountains to rise and to erode, to produce vast deposits of sediments on the ocean floor, and to occasionally throw up a small volcano. And that was the subject of the first volume of Lyell's book, *Principles of Geology*, published in 1830. In the second volume, published two years later, he followed the same idea of slow and steady change and applied it to the biological world.

Lyell went to Paris and met with Cuvier a number of times. One of the subjects they talked about was the dodo, a flightless bird once endemic to the island of Mauritius in the Indian Ocean, that some of their colleagues, in order to avoid the idea of extinction, insisted never existed. But Cuvier knew of fossil bones of the bird that were in a collection at his museum in Paris. And Lyell had seen a foot and a head displayed at a museum in Oxford that could not have belonged to any other bird. And so both men were convinced that the dodo had existed and that extinction was a reality, even if it had been caused by overhunting by men. Furthermore, the disappearance of this single animal species had not caused a downward spiral of the planet's life-forms into an irreversible doom.

They also agreed that other animal species were now extinct. A woolly rhinoceros found in Siberia, a large crocodile found in the region of Le Mans in northwestern France, a large bear—what would now be known as a cave bear—from Ansbach Germany, and a deer, often misidentified as an elk, with enormous antlers found in Ireland—all of these animals had once been plentiful and all had disappeared, only their fossil bones remained. Furthermore, the extinctions had been by natural forces. From that, the two men were in complete accord. But there was an important point of disagreement between the two men.

In the second volume of *Principles of Geology*, Lyell acknowledged that extinction was part of the "regular and constant order of Nature," that it had occurred because the planet was "in a state of continual fluctuation, the igneous and aqueous agents remodelling [*sic*], from time to time, the physical geography of the globe." But according to Lyell, the "remodelling" had proceeded at a slow and steady pace. Hence, extinction must also have occurred slowly and steadily. Any abrupt disappearance of an animal species in the fossil record was an illusion, owed to one of the many inevitable gaps in the fossil record.

Cuvier saw it differently. In a study of the rock strata in the Paris Basin, he had identified several abrupt changes in the strata and saw that each one was associated with an abrupt change in the type of fossils. To him, the meaning was clear: Here was evidence of geologic catastrophe on a scale and intensity unprecedented in human history. To Lyell, who examined the same rock strata, here were more of the inevitable gaps in an imperfect fossil record.

And so began one of the great debates in geology: Cuvier's catastrophism versus Lyell's uniformitarianism. It was a debate that would not be fully resolved until it was possible to determine with great precision the absolute ages of individual geologic events. And so for more than a century opinions dominated the debate, not definitive facts.

◆

At first, Cuvier's opinion prevailed, even in Britain, where there were many diluvialists—that is, those who held to the belief that great floods had been responsible for the major geologic features. Among those who advocated such a belief was William Buckland, the first professor of geology at the University of Oxford, William Conybeare, who had authored *Outlines of the Geology of England and Wales*, and Adam Sedgwick, a fellow of Trinity College, Cambridge. It was an impressive trio. Each man was a fellow of the Royal Society, the foremost scientific organization in Britain. And each was an influential clergyman of the Anglican Church.

Lyell never rose to a high academic position nor did he ever serve as a church official. He began his career as a barrister, but, because he came from a family of means, he was able to indulge in travel. And it was his traveling that shifted his attention to geology.

He examined the rocky outcrops of England, Scotland, and Wales. In France he met Cuvier and the two men dug fossils from the stratified beds of the Paris Basin. In Italy he took note of where fossil seashells could be found along the coast and how those locations were related to nearby mountains. Near Naples he climbed Vesuvius and watched the roiling and boiling of lava within the summit crater. On Sicily he followed a sheet of lava as it descended down the side of Etna.

He saw all of this through a new lens, one created a few decades earlier by James Hutton, who, from his own travels, ascribed the arrangement of rocks of Britain and of Europe to the passage of an immense period of time. From this, Lyell knew he was assured of sufficient time for slow processes, such as those happening on the planet today, to produce huge changes. That was the basis for his three-volume work, *Principles of Geology*. And that included the slow processes that led to extinctions.

Buckland, Conybeare, and Sedgwick accepted none of this. Each one criticized Lyell's *Principles of Geology* because it had omitted the important element of great floods. And they stood by Cuvier and maintained that the past had been different from the present.

The shift in opinion about the rate of geologic change—and the means of extinction—did eventually come from Lyell's *Principles of Geology*, but in an indirect way. It was one of the books that Darwin took with him when he left England in 1831 on a five-year voyage that would take him around the world. He came back convinced of Lyell's uniformitarianism. From that, he would be led to the idea of natural selection and that, in turn, would lead him to write the highly popular and influential *On the Origin of Species*.

The publication of *On the Origin of Species* in 1859 set the tone for biology, including ideas about extinction. Lyell and Darwin were now scientific orthodoxy. And extinction had become a fact. It had occurred slowly. It was the inevitable fate of all species, a passive end to a struggle for survival.

For the next hundred years, extinction was a moderately interesting topic, one that attracted only the occasional scientific commentary. But, as more and more of the fossil record was filled in, it was gradually realized that there had been episodes when entire groups of animals had been wiped out. But what was the mechanism for such large-scale destruction? That was the challenge.

◆

Cuvier, Buckland, and other catastrophists had initially invoked floods. Later in life, Cuvier was more specific and suggested that a rapid rise or a sudden drop in sea level had either drowned land animals or deprived marine animals of nutrients, thus leading to extinctions. Others invoked the sudden growth of huge mountain ranges as a way to disrupt ecological niches and, thus, cause extinctions. But, as more and more of the geology of the world was understood, neither sea-level change nor the building of mountains seemed adequate to explain the elimination of a wide range of animals.

A cosmic origin was proposed in the 1950s, inspired by discoveries being made in physics and astronomy that showed that, occasionally, the Earth must have been bombarded by potentially lethal radiation that originated from nearby cosmic explosions, such as supernovas. It was also suggested that the two largest extinction events—the two that John Phillips had recognized in the 1840s and used to define the division between the geologic eras—that divide the Phanerozoic Eon into the Paleozoic, the Mesozoic, and the Cenozoic eras had been so severe that there may have been multiple causes. Hence, these largest-of-all extinction events might be too complicated to ever be understood.

Then came Shoemaker and his proof that the Earth had been bombarded by large meteors. And then came the Alvarezes and their discovery of a worldwide iridium-rich layer that corresponded to Phillips's second extinction event, the one that coincided with the largest-known meteor to hit the planet in the last two billion years and seemed to coincide with the end of the dinosaurs. And that reawakened an interest in large extinctions.

In 1982, a reexamination of the fossil record was done, much along the same lines done initially by Phillips. It was found that there had been five major mass extinctions. There were the original two, which were the largest of the five. And there were three more: one at the end of the Ordovician Period, another at the end of the Devonian Period, and the third at the end of the Triassic Period. These have now been canonized in the scientific literature and in the popular press as the *five major mass extinctions.* Since 1982, the fossil record has been scrutinized with more care and many more fossils have been discovered and recovered so that there are now additional major mass extinctions.

The Permian extinction has been downgraded from a loss of 95 percent of animal life down to 60 to 80 percent. And what was once thought to be a problem in the preservation of fossils is now the Guadalupian extinction when 60 percent of animal life became extinct, which qualifies it to be one of the major mass extinctions. There were also at least two major mass extinctions during the Cambrian Era, known as the Botomian and the Dresbachian extinctions, though little is known about either of them due to a lack of fossil evidence so early in the evolution of animal life, but initial estimates are that the extinction rates may have been as high as 80 percent. And then there were events that included rapid shifts in the type of animal life, as during the Carnian Pluvial Event and the Carboniferous Rainforest Collapse, and these, too, should be considered as types of mass extinction.

In short, these brief, global die-outs of most of the animal life have occurred many times since animals first appeared on the planet several hundred million years ago. Each time another mass extinction is identified, there is a rush to see if an iridium anomaly or some other telltale sign of a major meteor impact might be associated with it. There is often an initial report of such an association, but after additional work, the early report has always fallen apart, except once, the mass extinction that occurred at the end of the Cretaceous Period, which is still tied to a large meteor impact and the creation of the Chicxulub Crater.

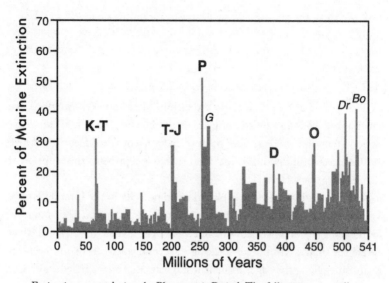

Extinction events during the Phanerozoic Period. The following are usually
regarded as the *five* major mass extinctions: K-T, Cretaceous-Tertiary;
T-J, Triassic-Jurassic; P, end-Permian; D, end-Devonian; and O, end-
Ordovician. Recent research has shown there are at least three other major
mass extinctions: G, Guadalupian; Dr, Dresbachian; and Bo, Botomian.

That has prompted a reconsideration of the causes of all mass extinctions.
Might all mass extinctions, including the one that marks the end of the Cretaceous Period, be explained by the same type of catastrophic event, one that
was global in its effect and that eliminated most of the animal life on the planet?

Many times in the history of the planet, the ground has opened up and
huge volumes of molten lava have poured out and onto the surface. These
were once thought to have taken many millions of years to occur, but
improvements in the age determination of rocks show that some of these
sudden outpourings occurred in less than a million years. Some have happened in less than a hundred thousand years.

Where these voluminous outpourings have occurred on land, the broad
volcanic feature that is formed is known simply as a flood basalt. Where
they have occurred on the ocean floor, the resultant feature is called an
oceanic plateau. There is also a more general term that applies to both:
Where a voluminous outpouring of lava has occurred over a brief period of

geologic time, the feature is known, somewhat unimaginatively, as a Large
Igneous Province, or LIP.

◆

Wherever one attempts to cross the terrain of eastern Washington, one
invariably encounters a series of giant steps. The heights of most are measured in tens of feet. Some have risers more than a hundred feet high. There
are hundreds of individual steps. And each one represents the eruption of
a single large lava flow.

A series of giant steps is displayed in grand style in the Grande Ronde
Valley in southeastern Washington. Another spectacular series can be
found along the downstream side of Palouse Falls where the Palouse River
enters the Snake River. The stacking of several steps can be followed for
many miles along the edges of ancient spillways that once carried enormous
floods and that formed what are known today as Grand Coulee and Moses
Coulee in central Washington.

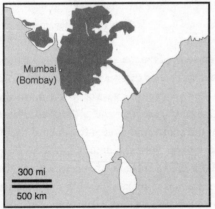

Deccan Traps, 66 million years
India

Columbia River Basalts,
15 million years
Washington and Oregon

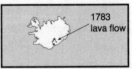

Laki eruption, 1783
Iceland

Comparison of the lava flows erupted by the Deccan Traps, the
Columbia River Basalts, and the Laki eruption in Iceland in
1783. The 1783 eruption was the most voluminous eruption of
lava in history. All three maps are drawn at the same scale.

Start in the north-central part of the state at Grand Coulee Dam and follow the Columbia River to the Pacific Ocean. In dozens of places there are a distinct series of steps that define the riverbank. One of the more intriguing is near Vantage, Washington, where a high wall of vertical columns defines the riser of each step. The same is true at the several steps seen at Wallula Gap where the river makes a sharp turn to the right just south of the twin cities of Kennewick and Pasco.

After another hundred miles one is within the narrow confines of the Columbia River Gorge. Here the massive stairstep lava flows stand in high relief as Table Mountain and Hamilton Mountain and as a spectacular colonnade at Cape Horn, all on the northern side of the river. Along the southern side the giant steps run continuously from the city known as The Dalles to a famous viewpoint called Crown Point. In between, four separate steps, again each one representing an individual lava flow, are exposed in the six-hundred-foot drop at Multnomah Falls. Farther down the river, these same stairstep flows, much muted by the heavy vegetation, form the Tualatin Hills west of Portland. At the Pacific Ocean, these same flows form the rocky exposures found from Seal Rock in Oregon to Grays Harbor in Washington, a distance of almost two hundred miles. This includes Tillamook Head and the pinnacle of Haystack Rock that stands partway out in the Pacific Ocean near Cannon Beach.

Each of these massive lava flows erupted near the Washington-Idaho border and then poured west. The longer ones have lengths of several hundred miles. The cooling history of these flows—indicated by the mineral content—shows that these flows covered that distance in about one week. They are *flood basalts*.

These flows lapped up against the Okanogan Highlands to the north and pushed the course of the ancestral Columbia River out of the basin and up against the highlands. The flood basalt poured across what is now northern Oregon and over the place where Mount Hood now rises. (That volcano would not form for another fifteen million years or so.) Some of the molten rock also flooded southwestern Washington, leaving thick lava flows as far north as Grays Harbor. In the Willamette Valley lava flowed as far south as Salem. It found passages and reached the sea at Cape Disappointment, where a lighthouse now stands atop one of the massive flows. The remnants of other massive flows can be seen at Seal Rock and at Yaquina Head near Newport, Oregon.

In places, these flood basalts are more than two miles thick, more than twice the depth of the Grand Canyon. They cover nearly a hundred thousand square miles. Most of the lava was erupted in less than a million years, perhaps in as little as a few hundreds of thousands of years. This is a geologic entity known as the Columbia River Basalts. It is the youngest and most intensely studied of the world's many LIPs.

Spend time traveling across the Columbia River Basalts, climbing the steep cliff faces and marveling at countless colonnades and the many waterfalls produced by this volcanic landscape. And then realize that this is the *smallest* LIP on the planet.[5]

Remnants of one of the *largest* on the planet is found on the eastern side of North America. The Central Atlantic Magmatic Province, or CAMP, has already been described in connection with the Late Triassic extinction. That brief series of eruptions produced more than ten times the volume of lava as did the eruption of the Columbia River Basalts. And it, too, was erupted in less than a million years.

And so there are at least two types of punctuation marks in the planet's geologic history. There are the great meteor impacts. And there are the sudden outpourings of huge volumes of lava.

Of the latter, every mass extinction since the Carboniferous, that is—since the last 250 million years—can be associated with an LIP.[6] Only one can be associated with a meteor impact, the one that occurred at the end of the Cretaceous Period. And so eyebrows should be raised. There is also an LIP at the end of the Cretaceous Period, the eruption of the Deccan Traps in India. Might this sudden outpouring of lava by responsible for the extinction of the dinosaurs?

◆

As late as the 1960s general opinion still held that dinosaurs had disappeared gradually, an opinion that reflected the continued influence of Lyell and Darwin. And so when the Alvarezes proposed in 1980 that a giant meteor impact had done the killing, their proposal was taken as a challenge. A concerted effort was then made to find evidence of the last surviving dinosaur and determine when it had lived.

That has proven to be a more difficult task than one might imagine. In the Hell Creek Formation, where the bones of *Triceratops* and *Tyrannosaurus* and many other colossal dinosaurs have been found in abundance, there is a noticeable absence of fossil bones in the upper ten feet or so of the formation. Some experts have proposed that this absence is due to a local change in environment, that the manner in which sediments were laid down made it unlikely for dinosaur bones to be preserved. But the same is found in the Raton Basin of Colorado, where the occasional fragmentary remains of fish, crocodiles, and turtles have been found, but only one bone that might be attributed to a dinosaur has ever been recovered. The only thing that can be said for certain about dinosaur bones and the K-T boundary at these two locations is no one has ever found a dinosaur bone above the K-T boundary.

Dinosaur footprints are more plentiful in the geologic record than dinosaur bones, and so they might be a better indicator of when the last dinosaur lived. But this, too, has fallen short of reaching the K-T boundary. Thousands of dinosaur footprints have been found in Utah. And there are several places scattered across the Wasatch Plateau in the central part of the state where sets of footprints, as well as fragments of dinosaur eggshells, have been found that are within several feet of the K-T boundary, but still significantly below the boundary. In the Raton Basin, impressions and natural casts of dinosaur footprints have been found slightly more than a foot beneath the boundary. That is, these impressions and natural casts were made hundreds of thousands, perhaps more than a million, years before the extinction event. And so all that can be concluded is that large dinosaurs were walking across Utah and Colorado until shortly before the end of the Cretaceous Period, but no evidence has yet been found of a dinosaur that died during the extinction event—that is, no one has uncovered a dinosaur killing field.[7]

Whether that will ever happen is unknown. One can only look across the arid hills and down into the steep ravines of Montana and North Dakota and wonder what key discoveries are yet to be made. But it is important to remember that a mass extinction is about more than the demise of large animals. Large animals are more susceptible to dying away during any extinction event, even minor ones. The key to recognizing and understanding what caused a mass extinction is the disappearance of the less vulnerable animals, especially the microscopic ones that are plentiful and that can be found over wide areas. In fact, Walter Alvarez's interest was drawn to the K-T boundary in Italy because he saw that dozens of different types of microscopic marine animals, known as foraminifera, lived before the extinction event and only one type survived and continued to thrive after.

The cause of an extinction is determined from the timing of events. In the case of the one that occurred at the end of the Cretaceous Period, three key questions must be answered: When did the massive lava flows of the Deccan Traps erupt? When did the Chicxulub impact occur? And when did animals species, such as foraminifera, disappear? When the Alvarezes made their announcement of an impact-driven extinction, the timing of any of these three events was known to only a few million years. Today it is possible to resolve the timing down to a few tens of thousands of years.

The eruption of lava of the Deccan Traps occurred in three phases. The first phase preceded both the impact and the extinction by three hundred thousand years. Then there was a hiatus between the first and second phases of about one hundred thousand years. It was during the hiatus that the asteroid hit.

The Chicxulub Crater formed 66.043 million years ago. Within a few tens of thousands of years afterward, the second and main phase of the Deccan Traps began. Lava erupted for more than a million years and covered an area of more than half a million square miles, equal to the combined areas of California, Oregon, Washington, Idaho, and Nevada. The longest lava flows on the planet formed during these eruptions, more than a thousand miles long. The total accumulation of lava was more than a mile thick.

But when exactly did the extinction of animal species occur? Was it immediately after the impact or during the eruption of voluminous lava flows? That is at the center of the debate today.

First, it should be realized that there is a North American bias—some have called it a "chauvinism"—to the study and understanding of the end-of-the-Cretaceous extinction. The structure of the Chicxulub Impact Crater indicates the asteroid entered the atmosphere and struck the planet as it was traveling north.[8] And so, because it hit near the southern edge of the continent, the most destructive effects of the impact—that is, much of the ejecta and the more intense local firestorms—would have been felt in the down-range direction, or across North America.

And that is indicated when different sections of the K-T boundary that are found around the world are compared. More than a hundred such sections have been found. And the environmental effects across the boundary are much less severe for sites, say, in Antarctica or in South America than in North America or around the Caribbean.

Moreover, the global fire that was expected to have followed the Chicxulub impact has not been clearly identified in the rock record. That is, evidence for a global fire at the K-T boundary has not been found anywhere.

And so what was the sequence of events near the K-T boundary? Geologists have examined those hundred or so sections with the geologic equivalent of a fine-tooth comb. These include the section in Texas where the Brazos River turns into rapids where it passes over a hard sand bed, which is, in part, a remnant of a huge tsunami. And dozens of sections in India where the timing can be seen between the sudden disappearance of foraminifera and the volcanic pulses. The conclusion is that the extinction did not occur immediately after the impact. There was the passage of hundreds, thousands, or, perhaps, tens of thousands of years.

The asteroid struck the Earth in a shallow tropical sea. The floor of the sea was covered by a thick layer of limestone. The limestone vaporized, filling the atmosphere with carbon dioxide. The asteroid strike also caused great clouds of small beads of rock and of dust to be thrown high into the air. The reentry of the beads heated the atmosphere, and so for days, weeks, maybe months, most of the planet was an inferno. But as the planet cooled, the amount of cooling intensified as the blanket of dust caused most of the sunlight to be reflected.

There was a yearlong period of darkness. Photosynthesis stopped. A major die-off of animals occurred. Then a greenhouse warming induced by the additional carbon dioxide reversed the trend and the planet began to heat.

Then, after tens of thousands of years, the main phase of the Deccan Traps erupted. Massive amounts of both sulfur and carbon dioxide were injected into the air. The sulfur combined with water vapor and formed aerosols that again blocked sunlight from the planet. Again photosynthesis stopped. And again there was a die-off of animals.

Thousands of years were required for the aerosols to clear. Then the great amount of carbon dioxide in the atmosphere—the Deccan Traps had put thirty to a hundred times as much carbon dioxide into the atmosphere as the Chicxulub impact—caused the planet to reheat, this time to a more extreme level. High planetary temperatures lasted for thousands of years. And more animals died.

And so did an impact or an eruption kill the dinosaurs? It seems that it was both, the two events acting in concert. An impact on a scale that could produce the Chicxulub Crater happens once every few hundred million years. An eruption on the scale of the Deccan Traps happens two or three times every hundred million years and lasts only a million years or so. And so the chance that two rare events would occur during the same short interval of time is low—but it is not zero.

13

How the
Mountains Grew

An Orogenic Interlude[1]

Here is the challenge. Travel across the major mountain ranges of North America, examine the rocks, taking special care to notice how they are arranged, then decide how each range formed.

Start in the Appalachian Mountains. To move quickly across the range, follow the interstate that connects Frederick, Maryland, and Morgantown, West Virginia. The four-lane highway runs across a series of parallel ridges and valleys. The ridges are high because they are capped by erosion-resistant sandstone. The intervening valleys are underlain by soluble limestones and easily eroded shales. Each layer of rock accumulated on the floor of an ancient sea.

Stop at any of the gaps and look at the cutaway view of the interior of a ridge. The layers of rock either arch upward or bow downward. In a few places, one can see both. The cutaway view at Sideling Hill, west of Hancock, Maryland, is the most famous. It is one of the most visited and photographed rocky outcrops in North America. When the interstate was first built, so many people were stopping to look at the giant downward-folded rock that the state of Maryland built a visitor center to accommodate the interest—and to maintain safety.

The exposed rock at Sideling Hill is more than eight hundred feet high and forms a huge *U*. It is typical of what can be seen throughout the

Appalachians: layers arched and folded in the manner that a carpet buckles when pushed up against a wall. It should be obvious that the Appalachians formed by compression.[2]

Rock deformed into a broad inverted arch by the Alleghenian Orogeny. The road cut is Sideling Hill along Interstate 68 west of Hancock, Maryland (photo courtesy of Acroterion).

Now jump to the Rocky Mountains. Take another drive along an interstate, this time from Denver to Grand Junction. Just west of Denver is an upended slab of the Dakota Sandstone. The place is known as Dinosaur Ridge for the many imprints of dinosaur tracks in the sandstone. Those who are more interested in how the Rocky Mountains formed know it as the Dakota Hogback.

Drive a bit farther west and one is soon out of the sandstone and into hard crystalline rock that was once deep inside the Earth and has been pushed upward. It should be apparent now that the Rockies are not just different in scale from the Appalachians, they are also fundamentally different in appearance, lacking the corrugations that come with compression.

Continue west and drive through Idaho Springs and Georgetown where the highway crosses the Colorado Mineral Belt, another feature that distinguishes the Rockies and the Appalachians. Not until one reaches the popular ski resort of Vail does one begin to see great folds in the rock. A few more miles west is Gypsum, the name derived from the rich layers of the mineral gypsum exposed in the cliffs, an indicator that an ancient sea has dried up. Farther west is a sequence of older sedimentary rocks—a Cretaceous limestone, a Jurassic sandstone, a Triassic siltstone, and Pennsylvanian shales.

At New Castle, the highway runs through another high ridge, the Grand Hogback, of upturned slabs, a mirror image of the Dakota Hogback near Denver. This is the western limit of the Rocky Mountains.

Now reflect on what has been seen on this two-hundred-mile transect. In the center is a mass of crystalline rock—ancient igneous and metamorphic rock—that forms the core of this section of the Rocky Mountains. This core was once covered by thick sedimentary deposits that accumulated on a seafloor or along coastal plains. Look closely and one can see that, in places, the crystalline rock has been thrust over and now lies over the former seafloor and coastal rocks. And, at the edges of the Rocky Mountains, the sedimentary rocks are turned almost vertically forming the two hogback ridges. In short, every rock structure one can see within these mountains is about thrusting and some compression.

Continue west through Grand Junction and Green River and find a way to Deseret in west-central Utah. Deseret is at the eastern edge of yet another major mountain range in North America. It is one of the most desolate regions of the continent. The highway that runs through it is regarded by some travelers as "the loneliest road in the world."

Nevada State Highway 50 is the proverbial open road. It winds along in great curves across a vast and barren landscape beneath an ever-present, clear blue sky. The few towns that do exist have populations that usually number in the single digits. But there is geologic and geographic distinction here. As one winds along the great curves, one soon realizes that the road is forced to curve because of a series of evenly spaced parallel mountain ranges separated by equally parallel and broad basins.

Known as the Basin and Range Province, this region of occasional high relief—a few peaks rise above thirteen thousand feet—covers all of Nevada and most of western Utah. It begins in southern Oregon and Idaho, includes the easternmost part of California, including Death Valley, is found in southern Arizona and New Mexico and in West Texas, and continues south into Mexico.

The northern section, the one that State Highway 50 crosses, is bound on the east by the Wasatch Mountains in Utah and on the west by the Sierra Nevada in California. Between are other major mountain ranges, including the Snake Range, the Toiyabe Range, and the Panamint, all running north-south.

The north-south alignment of basins and ranges indicates great fractures are in the crust, fractures that appear as long parallel geologic faults where large crustal blocks have shifted. Where a block has dropped down, a basin has formed. Where a block is still standing high, a high ridge runs. To produce such an alignment is quite simple: Unlike the Rocky Mountains, where there is thrusting and compression, the Basin and Range is all about a great stretching of the crust.

And so, of the three major mountain masses of North America described here, each one has had a different origin. Mountain masses are not simply land that has been raised. They have a collective story that reveals something fundamental about the Earth. It has taken centuries to understand what that story is.

◆

In the winter of 1320, the same year that he completed *The Divine Comedy*, Italian poet and moralist Dante Alighieri addressed members of the court of Verona and his patrons about a solution to a general question: Why was the land higher than the sea? He offered what was then a standard explanation: The land had been drawn upward by an attractive force that radiated from the stars, a force that acted in a manner that was similar to and as mysterious as the one that caused iron to be attracted to magnets.

This standard explanation dated back at least to the twelfth century, when Albertus Magnus, who is also known in history as Saint Albert the Great and as Albert of Cologne, simply supposed that more stars hung over mountains than over other regions of the planets. Hence, the positions of the stars must be responsible for the presence of mountains. A generation later, the Italian monk Ristoro d'Arezzo, who was intensely interested in minerals and in general questions about the Earth, elaborated on Magnus's suggestion by supposing that forces internal to the Earth operated in such a way as to cause the arrangement of mountains to mimic the pattern of constellations. Dante merged these ideas with those offered by the ancient philosopher Aristotle and proposed that Ristoro's internal forces were actually an "elevating virtue" (*virtus elevans*) that was produced by the stars, releasing, as Aristotle and many other ancients had proposed, strong internal winds that were responsible for all topography—not only high mountain peaks but also small hills.

The first person of record who proposed that mountains had a history that could be studied and revealed, and, hence, they must have had a physical origin and an evolution, was the English theologian Thomas Burnet. In 1672, he crossed the Alps at Simplon Pass from Switzerland to Italy on his way to see the celebrated ruins of Rome. But it was not the ruins in Rome that caught his attention. Instead, as he later wrote in a book entitled *The Sacred Theory of the Earth*, it was "the ruins of a broken world" that he saw in the Alps.

As late as the seventeenth century, people considered mountains to be places that should be avoided, places of foreboding. It was in mountainous realms that demons lurked, where the storms were common, where food was scarce, where sudden storms and the combination of perpetual cold and wind could quickly end one's life. But when Burnet passed through the Alps he was fascinated by what he saw. "There is something august and stately in the Air of these things," he wrote after the Simplon crossing, "that inspires the mind with great thoughts and passions." Moreover, as he continued, "We must therefore be impartial where the truth requires it, and describe the earth as it is really in itself."

Burnet was aware that mountains were not mentioned in the first few chapters of the Book of Genesis. He accepted that as evidence that God had created the Earth as a smooth and featureless ball. But, as he also noted, mountains are mentioned later when, in Genesis 7:19, "the waters prevailed so mightily on the earth that all the high mountains under the whole heaven were covered" and, in Genesis 8:4, when "the ark rested . . . upon the mountains of Ararat" after the Flood. And so, he reasoned, it was the flood that had brought imperfections to the planet, had scarred the face of the Earth by the growth of mountains, a reminder of the punishment for man's initial transgression.

But Burnet's contemporaries were not pleased with his reasoning, his suggestion that the Earth's surface had drastically changed since its origin.

That led to him being forced to end his ecclesiastic duties. Additional criticism came a century later when early geologists, such as Hutton and Lyell, who were arguing that mountains were recent constructs on the planet, described Burnet's reasoning as a "fantasy" and a "dream."

Nevertheless, Burnet must be credited with playing an essential role in the early development of geologic thought because it was he who began the erosion of biblical orthodoxy and stated, quite publicly, that since the moment of its creation the Earth had changed in dramatic ways. It was a desire to explain the origin of mountains that inspired him.

The feelings of avoidance that Burnet's contemporaries had for mountains were reversed in those who lived in the next centuries. By the late eighteenth century and throughout the nineteenth century, writers of the Romantic Age sought out mountains for the remoteness and for the solitude and as a way to embrace nature in a wild and undisturbed state. Samuel Taylor Coleridge described himself as becoming "much addicted" to the thrills of mountain climbing and was the first to use the word "mountaineering." John Keats dreamed of climbing in the Alps and did reach the top of Ben Nevis, the highest point in the British Isles. William Wordsworth was a self-proclaimed "rover . . . / In the high places, on the lonesome peaks, / Among the mountains and the winds." And among these climbers (and poets) were those who wanted to understand how mountains had formed.

One of the early climbers and seekers of the origin of mountains was Swiss naturalist and mountaineer Horace Bénédict De Saussure. During the 1780s and 1790s, he crossed the Alps at least fourteen times and ascended many of the major peaks, including the highest, Mont Blanc. When he made these crossings and ascents, he also took considerable time to examine the rocks, convincing himself that the giant wavelike ribbons of limestone that ran through the Alps could not, as was then generally believed, have precipitated from and grown upward like giant crystals during the drying up of a global ocean. Instead, based on his field observations, Saussure argued that these limestone beds had formed horizontally on the floors of ancient oceans, then had been contorted, by some unknown force, into their current wavelike forms. But what had raised the limestone? Why do the Alps stand so high?

Many answers were proposed. One of the more popular ones was offered by German geologist Leopold von Buch, who, after a trip to the Canary

Islands in 1815, was convinced that the buildup and release of pressure in underground chambers of molten lava was the key to understanding how volcanoes grew. He then traveled to the Dolomites in the southern Alps, noticed that great bodies of granite—rock that had once been molten—formed the core of those mountains, and concluded, applying his experience from the Canary Islands, that it was when such great volumes of rock were molten and expanding that mountains grew—and the Alps had formed.

Counter to his ideas were those of French geologist Léonce Elie de Beaumont. Beaumont thought the key was realizing that not only the Alps but also the Rockies, the Appalachians, and the Andes consisted of a series of long parallel ridges. From that, he suggested that if the Earth had once been molten and had been slowly cooling, then it should also be shrinking. And as the planet shrank, the surface would become wrinkled and folds would form in the rock strata and produce the parallel ridges. (The analogy that was often used was that of a withering apple.)

As the 1820s closed, Beaumont's idea of a shrinking and wrinkled Earth and Buch's "craters of elevation" theory, as the later came to be known, were at the center of a debate about the origin of mountains. During the subsequent decades, across the Atlantic, geologists in North America began to contribute to the debate after the completion of the first geologic surveys of the Appalachian Mountains. From those surveys a multitude of new theories would be offered.

◆

The first such surveys were done by two brothers, Henry Darwin Rogers and William Barton Rogers, who produced geologic maps of large sections of Pennsylvania, New Jersey, Virginia, and West Virginia in the 1830s and 1840s. Their work was financed by legislatures of the individual states who were interested in knowing what mineral resources might exist in their states, to evaluate whether soils were suitable for farming, and to determine what resources may be available in surface water or groundwater.

The Rogers brothers became especially interested in the Appalachian Mountains, in the arrangement of the ridges and in the origin of the many folds. They produced the first detailed mapping of a mountain range anywhere in the world. The Alps did not receive comparable attention until late in the nineteenth century. And the brothers had a novel theory to explain how the mountains had originated.

Their theory argued that mountains were built during violent upheavals of the land. The upheavals were caused by great wavelike undulations of the crust that occurred because a vast sea of underlying molten rock was sloshing and churning with a wavelike motion. The Rogers' wave theory of mountain building had a heyday of a few years. It was soon replaced by another, also flawed theory.

Geologist James Hall of New York was vigorously opposed to the "undulating" theory proposed by the Rogers brothers, objecting mainly to its catastrophic tone. He traveled through the Appalachians and made his own field observations, also studying the alignment of ridges and the many folds.

From that, Hall was inspired to construct a device that he called a "squeeze-box." It was a rectangular trough that he filled with layers of sand and mud. He then applied pressure to the box by means of a vise, compressing the contents. After several such experiments, he was satisfied that he could reproduce on a small scale the giant folds that he had seen in the Appalachians. No vast sea of pulsating molten rock was needed.

Moreover, from his own study of the Appalachians, he pointed out a key element that the Rogers brothers had ignored: The total thickness of sedimentary layers in the Appalachians was more than eight miles. How could so much sediment have accumulated when no ocean in the world was this deep?

The region where sediment had accumulated continued to sag under the additional weight as more and more sediment was added. The folds had formed because the shape of the basin where the sediment had accumulated had caused compression of the lower layers, contorting them into giant folds. But what had raised these sediments to form the mountains? Hall had an answer—or so he thought.

At the 1857 meeting of the American Association for the Advancement of Science, Hall, as president of the association, was allowed the honor of giving the key address. He chose the formation of mountains by sedimentation as his subject. He told his audience "elevation is due to deposition" and "uplift has not produced elevation, but had rendered the strata liable to degradation." He continued with such absurdities for more than an hour and his audience became more perplexed.

Years later, Joseph Le Conte of the University of California recalled being at the lecture. Arnold Guyot of Princeton University was sitting behind him. At one point, Guyot leaned forward and whispered in Le Conte's ear, "Do you understand anything he is saying?" Le Conte whispered back, "Not a word." Nor did anyone else. Hall's idea about mountain formation simply faded away.

But the question remained: Why did thick accumulations of sediments rise upward and, from them, how did mountains form? As the second half of the nineteenth century was beginning to close, geologic studies had also been done in the Alps and in part of the Rocky Mountains, the Andes, the Caucasus, and the Himalayas. And another answer was offered, this time by James Dwight Dana of Yale College.

By the late nineteenth century, Elie de Beaumont's claim of a cooling and shrinking Earth was widely accepted. Dana built on this to propose that as the Earth cooled and shrank, broad depressions formed in the Earth's surface. Those depressions were the ocean basins. Because the basins were low, they filled with sediments. The basins continued to deepen, and the sediments continued to accumulate. The result was a feature that Dana called a *geosyncline*. Then, according to Dana, something remarkable happened. The basins would become so deep and would so strain the Earth's crust that deep fissures would form. And up through those fissures molten rock would flow, causing the entire region to rise—reminiscent of von Buch's idea—and reversing the deepening of the basins.[3]

Dana's geosynclinal theory of mountain building remained at the heart of geology for more than a century. From this, a complex theory of mountain formation was built.

By the mid-twentieth century, geologists were discussing miogeosynclines, eugeosynclines, monogeosynclines, and polygeosynclines, all variations on Dana's original idea. But the key remained the contraction by cooling that gave rise to subsidence that, in turn, produced lateral strain and fissuring and caused molten rock to rise and push the crust upward to form mountains.

And, yet, there was a fundamental problem with the idea: It could not account for even a modest amount of thrusting of an older layer of rock over a younger one, as is dramatically displayed at Lone Rock Point along the eastern shore of Lake Champlain in Vermont, or along the North Mountain Fault in Virginia that runs between the Blue Ridge Mountains and the Valley and Ridge Province. And there was no accounting for the hogback ridges seen at opposite ends of the Rocky Mountains or the great amount of thrusting seen within those mountains.

All of this—the centuries-long pursuit of the origin of mountains by many distinguished and innovative thinkers, like Burnet and Buch, Elie de Beaumont, the Rogers brothers, Hall, and Dana—illustrates how difficult the problem was. And, yet, many geologists continued to march ahead. By the 1960s, Dana's idea about geosynclines had developed into a highly complex theory, one now populated with terms like zeugeosynclines, taphrogeosynclines, and paraliageosynclines. But it abruptly stopped with the development of the theory of plate tectonics.

The overriding problem in understanding the origin of mountains—and of the Earth and its history, in general—was the long-held assumption that the crust had moved vertically. Horizontal movements of any significant amount were impossible. The theory of plate tectonics removed that constraint. Continental masses had shifted thousands of miles. During the shifting, they had collided—and mountains had formed.

That explained the uplift, the folds, and the thrusts in the Appalachian Mountains. And in the Alps and the Andes and the Apennines. It explained how the great mountain masses of the Caucasus and the Himalayas had formed.

But if one looked at North America, there was still a problem. The main mountainous mass of that continent was the Rocky Mountains. The highest peaks are not along an obvious plate boundary, but are scattered across the interior of the continent in Wyoming and Colorado. And so, even as the theory of plate tectonics was revolutionizing an understanding of the Earth, a problem remained: How did the Rocky Mountains form?

◆

Oyster Ridge is a low, narrow ridge that runs for more than forty miles north of I-80 in southwestern Wyoming. The name was given by Ferdinand Hayden, who found fossil oyster shells along the ridge during his geologic survey of the area in 1872. The ridge and the oysters that are contained within it are commemorated in the Oyster Ridge Musical Festival that is held annually in the small town of Kemmerer, Wyoming, that sits close to the ridge. The particular style of guitar-picking music played at the festival is known, quite appropriately, as oyster grass.

The oysters whose remains lie along Oyster Ridge were alive and thriving ninety million years ago along what were then tidal flats and river deltas that lined the western shoreline of the Western Interior Seaway. During the next few tens of millions of years, the region was raised during a mountain-building event known as the Sevier Orogeny. In addition to Oyster Ridge, several other parallel ridges were formed. These can be followed from near Salt Lake City, Utah, to Jackson, Wyoming. In form and in spacing they resemble the sets of parallel ridges seen along the Allegheny Mountains of Virginia in the Appalachians, ridges that resemble the corrugated folds and buckles that are produced when one slides a carpet up against a wall.

More travel shows that the Sevier Orogeny produced mountain ranges that ran for more than a thousand miles. The southern end is at least as far south as Las Vegas, Nevada. From there, the folds produced by the Sevier Orogeny can be followed northward through Utah and western Wyoming, across eastern Idaho and western Montana, and into British Columbia.[4] The spectacular contortions seen in rocks exposed in the Lemhi Mountains of Idaho were produced by the Sevier Orogeny. So were the gigantic rhythmic folds in the mountains of the Sawtooth Range in Montana and the huge faulted blocks seen along the famous Going-to-the-Sun Road in Glacier National Park. All of this was a result of a head-on, plate-to-plate collision.

Those who have studied the many folds, buckles, and faults produced by the Sevier Orogeny have determined that this was a "thin-skinned" event, meaning it was rocks near the surface, mostly sedimentary, that were deformed and contorted into ridges and into ranges.[5] It is a distinction that means the Lemhi Mountains of Idaho and the mountains of the Sawtooth Range in Montana are *not* part of the Rocky Mountains.

To be in the Rocky Mountains, one must travel through the Uinta in Utah; the Wind River, the Big Horn, the Absaroka, and the Beartooth in Montana and Wyoming; the Front Range of Colorado; or the Black Hills in South Dakota. These ranges were produced by the Laramide Orogeny. And their origin lies deep inside the Earth.

To see the distinction between the Sevier and Laramide orogenies, after investigating the fold at Oyster Ridge, run up to southern Montana and travel the steep highway between Red Lodge and Cooke City. Near mile marker 62, along a series of long switchbacks is a light cream-colored rock with a sandy texture. This is the Quad Creek Quartzite. Zircons recovered from this rock show that it has an age of 3.25 billion years, a few hundred million years younger than the Morton Gneiss. The zircons and other minerals also indicate that this quartzite was once tens of miles beneath the surface. Rocks found in mountain ranges elsewhere in Wyoming—the Bighorn Mountains, the Wind River Mountains, the Laramie Mountains, the Sierra Madre, and the Medicine Bow Mountains—tell the same story of once residing at great depths, then being pushed up to the surface.

That is what distinguishes the mountains produced by the Sevier Orogeny from those produced by the later Laramide Orogeny. During the Laramide Orogeny rocks were raised from deeper levels. But what changed so that deeper forces would be at work?

For many years it was suggested that the shift in how mountains were raised during the Sevier Orogeny then the Laramide Orogeny was due to a change in the condition of the Farallon Plate. As an oceanic plate slides beneath a continent, it does so at an angle of thirty degrees or more. That was the situation during the Sevier Orogeny. But during the Laramide Orogeny, so it was suggested, the angle of subduction was shallow and the Farallon Plate had slid along the base of the North American Plate. That had the dual effect of causing the Rocky Mountains to form far inland and to push up rocks from deep inside the Earth.

But where was the proof? What was the evidence that the Farallon Plate had shallowed its angle of subduction?

It was a long time in finding the proof. And when a crucial piece was found, it was lying on the seafloor of the northeastern Pacific Ocean as a feature known as the Shatsky Rise.

◆

Ocean crust is created along the Mid-Ocean Ridge, a sinuous feature that runs along the floor of the world's oceans and where the seafloor is literally spreading apart. As it spreads, molten material wells up, solidifies, and is rafted away slowly from the ridge, much in the manner of a conveyor belt.

Most of the seafloor has been created in this way. But a small percentage has had a dramatic origin, an occasional surging of lava upward along short sections of the ridge, a pouring out of lava onto the seafloor, and the building up of a large undersea plateau. That is how the feature known as the Shatsky Rise was formed.

Located today about a thousand miles southeast of Japan on the floor of the Pacific Ocean, the Shatsky Rise covers an area about equal to the area of Montana. Its highest point stands about two miles above the normal seafloor, which means its upper reaches are still about two miles below the surface of the ocean. It is a part of the Pacific Plate that formed when a surge of lava erupted from the Mid-Ocean Ridge.

One can take the current location of the Shatsky Rise and, using a variety of indicators, such as the pattern of magnetic anomalies on the ocean floor, run the movement of the Pacific Plate backward and determine that the Shatsky Rise formed about 145 million years ago along a section of the Mid-Ocean Ridge that was part of the boundary between the Pacific and the Farallon Plates.

But the Shatsky Rise is only half of the story. It is only half of the lava plateau that formed 145 million years ago. Eventually, the surge of lava ended, the Mid-Ocean Ridge resumed the usually slow and steady production of lava, and the Pacific and the Farallon Plates continued to move away from each other. The Shatsky Rise was carried away by the Pacific Plate. But there should be a mirror image of the Shatsky Rise that was carried away by the Farallon Plate, a so-called *Conjugate* Shatsky Rise. Where is it? The quick answer is: It lies under North America and gave rise to the Rocky Mountains.

For tens of millions of years, the Mid-Ocean Ridge produced what could be called "normal" ocean crust that was carried by the Farallon Plate up to the edge of the North American Plate, then down and under the continent at a steep angle. One result of the steep subduction was the Sevier Orogeny.

Then about eighty million years ago, the part of the Farallon Plate that carried the Conjugate Shatsky Rise arrived at the edge of the North American Plate somewhere near where southern California is today. Then the Conjugate Shatsky Rise slid down under the continent. Because it was an unusually thick pile of lava, it was more buoyant than normal ocean crust and so it slid down under North America at a shallow angle, in essence, scraping its way along the base of the North American Plate, forcing huge blocks of crust from deep in the Earth upward and forming the Rocky Mountains.

There is also other evidence of this movement of the Conjugate Shatsky Rise beneath North America. It can be seen in the migration of volcanism across the continent.

When an oceanic plate slides into the Earth, it includes not only lava produced along the Mid-Ocean Ridge but also water-rich sediments that have accumulated on the seafloor. Some of these sediments are also taken down into the Earth. And as they descend, they are subjected to increasing temperature and pressure, so that, at a depth of about sixty miles, the water seeps into and mixes with the hot mantle. That lowers the melting temperature, and molten rock is produced.

The molten rock then rises, much of it solidifying near the surface as huge bodies known as *batholiths* and some of it reaching the surface and erupting. That is why a long arc of volcanoes is often associated with a subduction zone. The volcanic islands of the Aleutians owe their existence to the sliding, or subduction, of the Pacific Plate beneath Alaska and the Bering Sea. The volcanoes of South America that lie along the Andes formed in a similar way.

When the Farallon Plate was sliding at a steep angle beneath North America a string of volcanoes did form along a long arc close to the western edge of North America. And a long series of batholiths formed. The ash and lava erupted from the volcanoes can be hard to find—much of it has been eroded away—but the large batholiths are easy to find.

Pick up a rock in Boise National Forest in central Idaho and it is almost certainly a piece of the Idaho Batholith. Half Dome in Yosemite National Park in California is one of the many bulbous masses of once-molten rock that rose up and solidified and that now comprises the Sierra Nevada Batholith. Draw a line between Los Angeles and San Diego. East of that line are rocks of the Peninsular Ranges Batholith, including those that comprise Mount Palomar, where one of the largest telescopes in the world is located. All three batholiths solidified from huge volumes of molten rock produced when the Farallon Plate was sliding at a steep angle beneath North America between 140 and 80 million years ago.

Then, as the Conjugate Shatsky Rise slid under North America and the subduction angle of the Farallon Plate shallowed, the place where water-rich minerals mixed and caused the hot mantle to melt and shifted eastward. And so did the volcanism.

The farthest east that mountains were formed during the Laramide Orogeny was at the Black Hills in South Dakota. The farthest east the volcanism migrated is also near the Black Hills. This includes Devils Tower in Wyoming, a thousand-foot-high rocky sentinel, and the nearby Missouri Buttes.

After tens of millions of years of sliding beneath North America, the Conjugate Shatsky Rise finally fell victim to the constant high pressure and high temperature of the mantle and the mineral content of its rocks changed. The change was in such a way that denser minerals formed. And so the Conjugate Shatsky Rise started, at last, to sink deep into the hot mantle. When it did, the Laramide Orogeny ended. And so did volcanism around the Black Hills. Moreover, the sinking became so fast that it tore apart of the rest of the Farallon Plate. That left a void—and hot mantle material moved in to fill it. And that set off a new round of volcanism to the west.

There was an *ignimbrite* flare-up across Idaho, in Utah and Arizona, and in Colorado and New Mexico. An ignimbrite is a huge volume of volcanic ash that is poured out during highly explosive eruptions. The Mogollon-Datil volcanic field in southwestern New Mexico formed at this time. So did the San Juan volcanic field in southwestern Colorado. That was thirty million years ago.

Volcanic ash is friable and easily weathered. And so when the eruption of such huge volumes of volcanic ash is laid down—the thickness of such deposits are commonly measured in tens of feet—and subject to erosion, strange landforms often result. The Wheeler Geologic Area in the San Juan volcanic field near Creede, Colorado, is a wonderland pink-and-white spires and cones that are so closely bunched together that visitors to the area often describe the scene as "an army of gnomes." In New Mexico the erosion of volcanic ash has produced many shallow caves. Because these thick deposits of volcanic ash are easy to dig into with stone tools, for centuries people have been enlarging and reshaping the caves to make them into dwellings. Such is the origin of the Gila Cliff Dwellings that were occupied at least as early as the sixth century C.E.[6] Both the fields of gnome-like figures in Colorado and the cave dwellings of New Mexico are, in a distant sense, consequences of the accelerated sinking of the Conjugate Shatsky Rise.

And what of the majesty of the Rocky Mountains? That is owed to their youthfulness. It is the excitement of youth that brings people to these mountains to scale them and to take in the spectacular scenery. Every distinct range has a story to tell, but the story is often complicated by the raising of additional nearby ranges. To best understand these mountains, it is best to seek out the youngest and most isolated. And that is the Black Hills of South Dakota.

The story of the rocks of the Black Hills begins a few billion years ago in the Archean Period. At the start, for more than a billion years, a wide variety of rocks were laid down on the floors of different ancient oceans. Then, about 1.7 billion years ago, as early continental masses in the form of cratons began to come together and form the core of North America, molten rock was produced along the seams where those cratons were colliding. One of those seams runs along the western extreme of South Dakota. The molten rock rose and heated the Archean sediments and, from those sediments, a new suite of rocks was metamorphosed. The molten rock solidified. It is the Harney Peak Granite that the huge portraits of four presidents are carved into at Mount Rushmore. The multiple layers of metamorphosed Archean rocks can be difficult to follow but one, in particular, does stand out: the band of dark contorted rock that lies below the portrait of George Washington. Then there was another billion years of deposition and erosion, of which little of it has been yet deciphered and understood, in the rocks at the Black Hills.

The next notable event recorded in the Black Hills is the deposition of a yellow limestone, usually known as the Madison Limestone, that formed in a warm, shallow sea. It is composed of sediment and shells that formed during the Mississippian subperiod of the Paleozoic Era. Lying atop this is the Spearfish Formation of the Permian and Triassic periods. It is a soft red shale that can be seen east of Rapid City, South Dakota, and followed by driving along I-90 northward to Sturgis then to Spearfish, both in South Dakota. Above the soft red and easily eroded shale of the Spearfish Formation is a white sandstone, the Lakota Formation. It formed in the Cretaceous Period along the eastern edge of the Western Interior Seaway. The sandstone is well-packed and erosion-resistant and appears as a long hogback that runs between Rapid City and Sturgis. These three units—the limestone, the shale, and the sandstone—were raised and deformed and slightly cracked. And that was by the Laramide Orogeny.

Imagine that it is fifty million years ago and you are standing where Rapid City or Sturgis or the town of Spearfish is located today. The entire center of the continent has been raised recently and there is no seawater that remains from the Western Interior Seaway. The raising and the disappearance of the Western Interior Seaway is courtesy of the Conjugate Shatsky Rise.

Now look to the west and wait a few million years. The land around Rapid City, Sturgis, and Spearfish starts to rise. But there is no buckling or folding of the rocks. Instead, the land rises upward in a great arch. As it rises, the limestone, the shale, and the sandstone are all pushed upward and outward. Cracks develop in all three. The cracks are hard to see today in the sandstone. They never left much of an impression in the soft shale. But there is a record of the cracking seen in the limestone. The limestone dissolves in rainwater, which is slightly acidic. And the dissolution is most efficient along the cracks, leaving systems of caves. Today there are hundreds of cave systems in the Madison Limestone around the Black Hills. Two in particular, Wind Cave and Jewel Cave, have more than a hundred miles of passages. These cave systems with their many intricate connections are a testament to the arching of the land during the Laramide Orogeny.

Now watch the Black Hills grow. As the land rises, rain and wind erode the surface, laying bare a succession of deeper and deeper geologic layers. The material that is eroded is transported eastward by streams and rivers and forms new rock layers that are now being eroded again at Badlands National Park.

The upward arching was so great and the erosion dug down so deep that the Harney Peak Granite is exposed. Around it is a ring of metamorphosed ocean sediments of the Archean. Moving outward away from that ring, exposed today as a skewed bullseye are the thick sedimentary layers—the Madison Limestone, the soft shale of the Spearfish Formation, and the hard sandstone of the Lakota Formation—that formed beneath seas during the Paleozoic and the Mesozoic.

The arching occurred when the next geologic eon came to South Dakota, the Cenozoic. By then, the Conjugate Shatsky Rise had broken away from the rest of the Farallon Plate and continued to slide deep into the Earth. As it tore away, the Farallon Plate recoiled—and another series of mountain ranges formed, those of the Basin and Range.

◆

The entire state of Nevada has been stretched from east to west. It has been stretched so much that, essentially, everything seen between Reno and Salt Lake City was once piled up in half the distance.

It is from this immense amount of stretching and, by consequence, a thinning of the crust that, paradoxically, hundreds of mountain ranges and intervening basins formed. But how did the stretching and thinning of the crust, which should have greatly lowered the land, produce so many high ranges?

For the answer, one has to examine the aftermath of the Laramide Orogeny. As the thick and buoyant section of the Farallon Plate known as the Conjugate Shatsky Rise slid beneath North America, a broad plateau of mountains formed, one that extended from the western edge of North America to the middle of the continent. A study of the geologic remains of this plateau indicates that it was similar in size and in general appearance to the Altiplano of South America, the high plateau that currently forms much of Bolivia and northern Chile. For that reason, the North America plateau, which included much of what is today the western United States, is called the *Nevadaplano*.

It was a formidable mountain mass, the highest peaks rising more than three miles. And much of the eastern half still exists and still stands at high elevation. That eastern half is the Rocky Mountains. The western half has been lowered considerably, not by erosion of the ground surface but by what was happening inside the Earth.

As the Conjugate Shatsky Rise broke away and slid farther east and deeper into the Earth, the section of the Farallon Plate that was still under North America and that lay to the west recoiled. It recoiled in such a way that the Farallon Plate, as it continued to subduct, swung downward, much in the manner of the opening of a trap door, and resumed a normal steep angle of descent. Then, as a consequence, the place where the plate was entering the Earth rolled back to the west. It was this combination—the swinging downward and the rolling back—that stretched the crust.

The stretching caused sets of parallel cracks to form in the Earth's crust. And these cracks developed into faults that define the edges of individual crustal blocks. As the stretching continued, the blocks slid along these faults—and the blocks shifted and tilted. It was this stretching and sliding and shifting and tilting of crustal blocks that has given rise to the distinctive pattern of parallel basins and mountain ranges in the Basin and Range. But the story does not end here.

There has been a remarkable amount of stretching across the Basin and Range. For example, the rocks of Red Rock Canyon in the Spring Mountains west of Las Vegas formed originally a hundred miles to the east of where they are today. The great masses of granite exposed in the Sierra Nevada were originally near the Nevada-Utah border and have been moved westward hundreds of miles. Given the great amount of stretching, this region of what was the Nevadaplano should now be near sea level. But much of Nevada, even the floors of the basins, lie more than a mile above sea level. What kept this region from dropping farther? The answer lies with the Farallon Plate.

The outer rind of the Earth can be divided two different ways. It can be divided into a crust and a mantle, the two distinguished by mineral content; the crust is richer in quartz and feldspar than the mantle. It can also be divided based on mechanical behavior. There is a cold, rigid outer part, the *lithosphere*, composed of the crust and the uppermost part of the mantle. Beneath is a hot, pliable layer, the *asthenosphere*, or "weak" layer, that, though solid, can flow slowly, similar to how molasses flows. The cold lithospheric mantle is heavier than the asthenosphere, and so when disrupted, the entire lithosphere will drop down into the hot asthenosphere.

The tectonic plates that move around and shift the positions of the continents are part of the lithosphere. Where two plates collide and one plate is sliding beneath the other, a cold slab of the lithosphere slides into the Earth. That is what is happening today as the Pacific Plate is moving northward and sliding beneath Alaska and eastern Russia. That is also what happened as the Farallon Plate slid beneath the western coast of North America. But there is another way for a slab of cold lithosphere to descend into the mantle. It can peel off the bottom of a tectonic plate. It was the peeling off of cold lithosphere into hot asthenosphere that kept the region of the Basin and Range from dropping down to sea level.

Again that was a conjecture. Where was the proof? The surface geology could suggest that this was happening, but one needed to peer into the Earth to see if it was right. Only recently has that been possible—after the completion of a program called EarthScope.

For fifteen years, from 2003 to 2018, EarthScope operated hundreds of highly sensitive seismometers, closely spaced, across the United States. Those instruments could record every slight vibration of the ground surface. From those data, detailed images of the Earth's interior have been made, based on seismic tomography, revealing what lies deep beneath the surface.[7]

Beneath Nevada is a cold cylindrical slab—some have called it a "lithospheric drip"—in the asthenosphere. Around it is a larger feature, the Nevada swirl, that is probably the hot asthenosphere flowing around the cylinder. Together they indicate that a slab has, in fact, peeled off the bottom of the North American Plate and is now slowly descending into the Earth.

When this slab peeled away, two things happened. The slab was like the heavy ballast of a ship: As it peeled away, the rest of the North American Plate rose. Also, the movement of the slab downward caused hot mantle material from the asthenosphere to flow to fill the void. The net result was that the ground surface remained high, and the basins and the mountain ranges of the Basin and Range remained high above sea level.

The peeling away of a lithospheric slab and its descent into the Earth is known as *delamination*. It was first recognized to be occurring beneath the Basin and Range. Because it was produced by what seemed to be an unusual set of circumstances—the breaking away of the Conjugate Shatsky Rise and the swinging downward and rolling back of the Farallon Plate—one might think that delamination is not an important process in understanding the Earth, especially in understanding how mountains form. But, as more and more tomographic images have been made of more and more regions of the Earth, it has become apparent that delamination has played an important role. In fact, it seems to be the crucial element in how mountain ranges continue to grow.

Heavy slabs that have separated from the lithosphere and are now descending into the asthenosphere have been discovered lying deep beneath the Andes, the Apennines, and the Alps. They have been found beneath the Atlas Mountains in northwestern Africa and the Carpathian Mountains of Eastern Europe. A large descending slab may be responsible for producing a recent "pulse" of mountain building that gave rise to the Iberian Massif, the set of rugged mountains of northern Portugal and western Spain that are far from an active plate boundary. A lithospheric slab has also been discovered recently to lie beneath one of the oldest ranges of mountains in North America: the Appalachians.

The ranges and valleys of the Blue Ridge Mountains of the Appalachians are exceptionally rugged and steep. Consider the profile of the Cullasaja River in North Carolina. At Highland Falls the river drops more than a hundred feet. A few miles downstream at Dry Falls, where visitors can walk under an overhanging bluff and stand behind the falls, the drop is sixty-five feet. And a few miles farther, after a series of rapids, the river drops an additional 250 feet. Such a steep gradient should have been eroded into a steady decline in the hundreds of millions of years that have passed since this section of the Appalachians first formed. The steepness of the river profile and the existence of multiple falls are strong indicators that the Blue Ridge Mountains have continued to rise.

Now consider the location and the age of the most recent volcanic eruptions in the Appalachians. They occurred in what is now the Shenandoah Valley near Harrisonburg, Virginia. In particular, Mole Hill just west of the city is a volcanic neck, the same type of erosion-resistant feature as Devils Tower in Wyoming—that is, the remnant of a magma chamber that once fed a volcano. Mole Hill and Devils Tower are also, coincidentally, similar in age, about forty-five million years.

Earthquakes occur at a much lesser rate in the Appalachians than in the Rocky Mountains or almost anywhere in the western third of North America, but significant events have occurred. The most recent one was on August 23, 2011, a magnitude 5.8 earthquake that was centered beneath Mineral, Virginia, about sixty miles southeast of Harrisonburg. When an earthquake of that magnitude occurs beneath Los Angeles, it is seldom reported to have been felt more than a few hundred miles away. But the crustal rocks of the eastern half of North America are different from those in the western half in that the eastern half of the continent is much colder—and so the energy radiated by earthquakes as seismic waves is not attenuated as quickly.

Shaking from the 2011 earthquake was felt across much of the eastern United States and southeastern Canada. In Washington, DC, it caused cracking and spalling of the Washington Monument, it shifted the red sandstone blocks on the turrets of the Smithsonian "Castle," and it broke spires and stone ornaments on the Washington National Cathedral. In New York City, three hundred miles from the epicenter, about the same distance from Los Angeles to San Francisco, the swaying was so great that some buildings were evacuated.

The 2011 earthquake occurred in a region of persistent seismicity known as the Central Virginia Seismic Zone, a place where other notable earthquakes have occurred. In 1875, a magnitude 4.8 event shook bricks from chimneys, broke plaster and windows, and overturned furniture at several places. In 2003, an earthquake doublet—that is, two identical earthquakes, in this case, both of magnitude 4.4—were separated in time by a mere twelve seconds and caused minor damage in the region.

These three indicators—rugged topography, volcanic activity within the last few tens of millions of years, persistent earthquakes—dispel the notion that the Appalachians are a geologic dead zone, a place where no significant tectonic activity is occurring today. And if one takes a much broader view of the geological region, this becomes even more obvious.

Early European explorers of the East Coast were stopped from penetrating far into the continent by a geologic feature. They would sail up tidal estuaries and quiet rivers and be halted in their progress by waterfalls or steep rapids. It was the same all along the coast. It was at these places, the heads of navigable rivers, that settlements, then cities grew. In Washington, DC, it is the Potomac River; in Richmond, Virginia, the James River; and in Augusta, Georgia, the Savannah River. This was the geologic boundary between the hard consolidated rocks of the interior that make up the Appalachians and the sandy, flat outwash plain known as the Piedmont.[8] Why such a boundary existed puzzled geologists for a long time.

And then there are the mountains themselves. Three hundred million years have passed since a collision between Africa and North America caused the Alleghenian Orogeny, the final episode in the building of the

Appalachians. Chemical and mechanical weathering should have eroded those mountains down to a flat plane long ago. And yet, we see mountains today.

What is needed is a process that could produce uplift, volcanism, and seismicity long after an orogeny has ended. And that can be accomplished by delamination.

In 2019, seismic tomography showed that a cold slab does lie in the asthenosphere beneath the Appalachians. How far the slab extends is still being worked out, but it seems that delamination is a process that can operate long after plates have collided and then pulled apart.[9]

Delamination has also provided a solution to yet another long-term problem in mountain building: the growth spurt of mountains. A prime example is the Sierra Nevada.

At the end of the Laramide Orogeny, what is now the Sierra Nevada formed the high western flank of the Nevadaplano. As the crust stretched and the Basin and Range formed, the rocks of the Sierra Nevada moved westward to their present position. Because of the far western movement, they also dropped considerably, so that what is seen today as high jagged peaks was once no more than a series of long hills a few thousand feet high. Then, in the last few million years, high mountains have sprung up as another slab of the lithosphere started to peel away, this one directly beneath the extreme western part of the Basin and Range.

The peeling away and dropping of the slab caused the crustal block that includes the Sierra Nevada to rise and to tilt downward to the west. These are the mountains seen today, their summits still rising about a tenth of an inch each year.

The tilting and continued rise has also had another effect: It has allowed hot asthenospheric material to rise close to the surface and melt some of the continental crust rocks. From that, a line of volcanic centers have formed. At the southern end are lava domes and cinder cones and numerous hot springs, steam vents, and boiling pots in the Coso Volcanic Field. Farther north is the explosive volcano of Long Valley, where a major eruption occurred about two hundred thousand years ago.[10] Just to the south of Long Valley are the thick basaltic flows of Devils Postpile National Monument that erupted in the last one hundred thousand years. And just north is a line of recent volcanic features that lead to Mono Lake. Here one finds several funnel-shaped craters partially filled by small ponds and several large domes made of frothy volcanic pumice and black volcanic obsidian glass. All of this is courtesy of the continued peeling away of the lithospheric slab that lies deep beneath the base of the Sierra Nevada.[11]

But it does illustrate an important point: While the theory of plate tectonics emphasized the importance of plate collisions in the formation of mountains, that was not the complete story. Events happening deep in the Earth have also played a role. To fully understand how mountains are built—and the geological history of North America—one must look deep in the Earth. And that begins by looking to see where the Farallon Plate is today.

◆

Drive almost anywhere across the contiguous United States and, with few exceptions, hundreds of miles beneath you is the Farallon Plate. The few exceptions are at three of the four corners. A slab of the Farallon Plate probably, but not definitely, lies beneath Maine and almost certainly does not lie beneath the large peninsula of Florida; that is, the part of Florida known as the Suwannee Terrane that was once part of Africa. And it certainly does not lie beneath that part of California south of Los Angeles or that runs as a strip along the coast from Los Angeles to San Francisco. But everywhere else, if one could drill deep into the Earth, one would find a piece of oceanic lithosphere that once lay at the bottom of the Pacific Ocean—though then it was the Panthalassa Ocean. That is the Farallon Plate.

Today the Farallon Plate has been torn into a few large slabs and many small ones. The Conjugate Shatsky Rise is currently under the eastern region of the Great Lakes and may extend as far east as New England, which suggests that the slow movement of this ancient oceanic plateau through the asthenosphere might be the ultimate cause of the earthquakes that are scattered across this broad region, including the persistence of earthquakes beneath Charlevoix, Quebec, and the occurrence of two magnitude 6 events beneath New England, one in 1638 and another in 1755. The large earthquakes that occurred beneath New Madrid in southeastern Missouri in 1811 and 1812 might have a similar origin because a small slab of the Farallon Plate is descending beneath this region of the mid-continent.

For more than a hundred million years the Farallon Plate collided with and slid beneath the North American Plate. As it did, it carried several large terranes that are now securely accreted to North America. Both the Wrangellia and the Alexander Terranes, which include most of Vancouver Island and the western half of British Columbia and continue into southeastern Alaska, were once part of the Farallon Plate. So was the horseshoe-shaped Crescent Terrane that runs along the northern, eastern, and southern edges of the Olympic Peninsula. Individual lava that erupted from the Mid-Ocean Ridge flows can still be recognized in many parts of the Crescent Terrane. Such flows are visible at Portage Head just south of the northwestern tip of Washington State at Cape Flattery, at Tongue Point near Port Angeles, and at the Black Hills east of Hood Canal, as well as from the state capitol building in Olympia.

South of the Crescent Terrane is another large, accreted terrane, Siletzia, that includes essentially all of southwestern Washington and western Oregon. Bald Mountain north of Newport, Oregon, consists of submarine lava flows that were part of the Farallon Plate. So are the dark rocks of the Tillamook Highlands and the platform of volcanic rocks that surround Roseburg, Oregon. These were all once part of an ancient seafloor that was part of the Farallon Plate.

But most of the Farallon Plate has disappeared from sight and now lies hundreds of miles beneath North America. Then, at the eastern end of the continent, it takes a plunge of more than a thousand miles into the Earth, possibly reaching as far as the top of the Earth's core.

◆

But why does the Farallon Plate follow the path that it does into the Earth: an initial steep plunge from the surface to a depth of about four hundred miles, then a leveling out as it proceeds for thousands of miles beneath North America, and finally a second steep plunge that sends it down to the deepest part of the Earth? And that leads to a more fundamental question, one that geologists have wondered about since the theory of plate tectonics was first proposed: Why do the tectonic plates move? The answer to both is simple: It is gravity that is pulling on the heavy oceanic plates and dragging them down into a hot and pliable Earth.

An oceanic plate—that is, the oceanic lithosphere—is colder and denser than the hotter asthenosphere beneath it. And so when an oceanic plate begins to descend, gravity pulls it downward into asthenosphere that, over geologic time, flows like a liquid. In turn, as the slab descends, it pulls on that part that is still on the surface. And so the tectonic plates are moving and shifting their positions because their edges are being *pulled* downward by descending slabs of oceanic lithosphere.

In particular, the Pacific Plate is moving slowly to the northwest because it is being pulled along its western and northern edges by huge cold slabs of oceanic lithosphere that are descending beneath Mariana Islands, Japan, the Kurile Islands, and the Aleutian Islands. The Juan de Fuca Plate—one of the remnants of the Farallon Plate that still forms part of the seafloor—is moving eastward toward North America and descending beneath the continent because it is being pulled by a slab of the Farallon Plate that is currently sliding downward beneath British Columbia, Washington, Oregon, and northern California.[12]

There is an important corollary here. As the theory of plate tectonics was originally envisioned, it was proposed that the movement of the plates was being driven by huge convection cells within the Earth's mantle. An analogy that was often given to illustrate this movement was that of a boiling pot of oatmeal. Heat applied to the bottom of the pot created a series of convection cells that caused the oatmeal to rise from the bottom of the pot, turn over at the top, then descend. That view has greatly changed. The Earth's deep interior can still be viewed as a pot of hot oatmeal, but one that is being stirred by large cold spoons—the spoons being the cold slabs of descending oceanic lithosphere, such as the Farallon Plate.[13]

But why doesn't the Farallon Plate simply continue its initial steep descent and plunge deep into the Earth? Why does a long section of it move horizontally beneath North America? Again, it is about the pull of gravity.

By 1900, it was understood that the Earth's interior could not be homogenous, that it must have some type of structure. The most obvious reason was a comparison of the density of rocks at the surface and the density of the entire planet. Surface rocks have a density about three times that of water. The average density of the entire Earth—known from the radius and the mass of the Earth—is about twice that of surface rocks. For various reasons, it was argued that the greater density was due to a large iron core.

In essence, that was all that was known about the Earth's deep interior until the beginning of seismology.

By the 1940s, seismologists, studying the slight wiggles traced out on paper by earthquake waves that had passed through the Earth, determined that the Earth's interior consisted of concentric shells. The outermost shell was a thin crust that varied from several miles to a few tens of miles in thickness. Beneath it was the mantle that is nearly two thousand miles thick and contains the bulk of the planet's mass. And beneath the mantle is an iron core that consists of two parts: a liquid outer core and a solid inner core.[14]

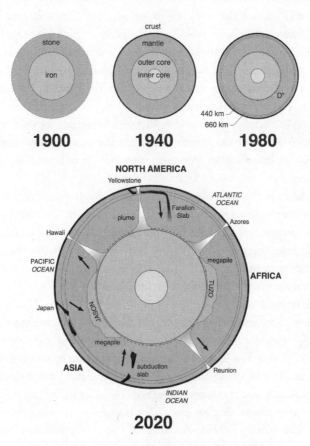

History of ideas about the Earth's interior. (Modified from a drawing by Ed Garnero, Arizona State University)

By the 1980s three important boundaries had been discovered within the mantle. Above a depth of 270 miles, or 440 kilometers, was the *upper* mantle. Beneath it was a *transition zone* that continued down to 410 miles, or 660 kilometers. And beneath the transition zone was the greater part of the mantle—the *lower* mantle—that continued down to within a few miles of the core-mantle boundary. And beneath the lower mantle and immediately above the core was a thin layer that is still being defined and characterized and that is known, for historical reasons, as D″.[15]

The upper mantle and the transition zone are of interest here because it is the nature of these two boundaries within the mantle that explains why the Farallon Plate does not make an immediate and continuous plunge down into the Earth. As one descends down into the Earth, the pressure is so great in the transition zone that the arrangement of atoms shifts into tighter configurations, which means there is an abrupt increase in density across this zone.[16] And so the Farallon Plate cannot immediately plunge through and enter the lower mantle. Instead, it continues to slide across the top of the 420-mile boundary for thousands of miles until—through the constant high heat and pressure of the mantle—the density of the Farallon Plate increases enough to break through the boundary and continue its descent into the Earth.

All of this has become clear only recently through seismic tomography, by the ability, in the last few decades, to produce detailed images of the Earth's interior. Such images show that it is not only the Farallon Plate that has been interrupted in its dive into the Earth, but as the Pacific Plate slides beneath Japan and eastern Asia, its downward movement has also been interrupted, causing it to slide horizontally for thousands of miles beneath Asia.

Looking deeper into the Earth, seismic tomography has recently revealed two large irregular blobs that lie atop opposite sides of the core-mantle boundary. These two blobs are the size of continents in area and rise for hundreds of miles above the core-mantle boundary. One is positioned beneath Africa and the other beneath the southwestern Pacific.[17] Their origin is still debated, but one possibility is that they are the graveyards of oceanic slabs that have broken through the 410-mile boundary and have built up huge piles—*megapiles*—atop the core-mantle boundary. Whether this is true is yet to be seen. But it is important to note that plumes of hot mantle material are rising along the edges of each of these two megapiles, plumes that—now clearly imaged by seismic tomography—are feeding magma to so-called "hot-spot" volcanoes such as Hawaii, Yellowstone, the Azores in the Atlantic Ocean, Reunion in the Indian Ocean, and dozens of others.

From this, a new view of how the Earth works and how it has evolved has emerged. The great slabs of cold oceanic lithosphere are the key to the planet's evolution. As they slide into the Earth and are prevented from diving immediately deep into the Earth, these slabs accumulate along the 410-mile boundary. Then, in a brief period of time, perhaps as short as a few tens of millions of years, these slabs cascade down into the lower mantle as a series of *mantle avalanches.*

The idea of mantle avalanches was introduced in the 1990s and initially showed much promise. But then, the mechanism that was proposed for why these slabs have suddenly broken through the 410-mile boundary and down into the lower mantle could not be supported by laboratory measurements.[18] And so the idea was abandoned. But new research has seen a revival.

When Wegener originally proposed his idea of continental drift, he suggested that the continents pushed their way across ocean basins in the manner of an icebreaker making its way across a field of thick ice. As many geologists realized, that was physically impossible—and so his theory was strongly criticized. Then in the 1960s, after many new observations were made that indicated the continents had moved, a possible mechanism was proposed—and the theory of plate tectonics was accepted.

If mantle avalanches have occurred, then the timing of such events was probably not random. They most likely occurred after the formation of a supercontinent when a single large continental mass existed and around its edges large slabs of oceanic lithosphere were sliding into the Earth. (In fact, the two megapiles *might* be piles of subduction slabs that accumulated after mantle avalanches, the one beneath Africa after Pangea and the one beneath the Pacific after Rodinia.)

The cascading of slabs down into the lower mantle and onto the core-mantle boundary would have been balanced by a sudden upward flow of hot material that in turn could have produced widespread volcanic eruptions and formed large igneous provinces that led to the disintegration of the supercontinent. The eruptions would have released huge amounts of carbon dioxide and altered the climate and leading to major mass extinctions, thus, possibly tying what occurred deep inside the Earth with biological evolution.

Such connections are yet to be made. But that has not stopped one from wondering. Look at the Grand Teton south of Yellowstone National Park in Wyoming. These mountains are steep and rugged, indications of their youth. Field studies show that they took their present shape less than ten million years ago. The simple explanation for these mountains is that Jackson Hole is dropping and the Teton Range is rising as this section of the continent is being transferred from the North American Plate to the Pacific Plate.

Look at the peaks. These are granites and gneisses and schists that are more than two billion years old. Look near the top of one of the highest peaks, Mount Moran. There is a vertical black streak near the summit. This feature formed nearly eight hundred million years ago, near the end of the Paleozoic Era, long before the rise of the Teton Range.

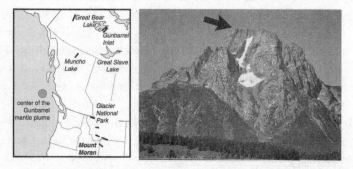

Left: Center of the Gunbarrel mantle plume and segments of radiating fractures. Right: Mount Moran, Teton Range, Wyoming. The 150-foot-wide dark dike near the top of Mount Moran, indicated by the arrow, was formed by the Gunbarrel mantle plume (photo courtesy of Paul Hermans).

This vertical black streak is a volcanic dike, about 150 feet wide, that marks a pathway that molten rock followed as it flowed horizontally through a rigid crust. What is remarkable is the source of the molten rock: It was several hundred miles away at a point off the western coast of Vancouver Island near Neah Bay in Washington. This point of origin was above a mantle plume that was rising up and pushing against the base of the lithosphere. That caused the lithosphere to fracture along radial lines. At that time, about eight hundred million years ago, North America was part of the supercontinent of Rodinia. What one sees near the top of Mount Moran is one of the effects of the rise of the plume and fracturing of the lithosphere that ultimately led to the splitting apart of Rodinia.

Similar black streaks—that is, volcanic dikes and molten rock sent off radially from the same mantle plume—have been found across the northwestern section of North America. Keeping within the Teton Range, they are visible from Teton Park Road on the eastern face of the Middle Teton and the southeastern flank of the Grand Teton. Farther afield, volcanic features of the same age and of the same origin are exposed as a dark horizontal band across the vertical rocky cliffs at Logan Pass in Glacier National Park and at Gunbarrel Inlet on the southeastern side of Great Bear Lake in the Northwest Territories of Canada.

The point is that mountains and the geologic history recorded within them are far more complex than the simple raising or lowering of the land surface. And the forces that create mountains are not just the product of plate collisions. Look at the Grand Teton or any other range. The history of supercontinents and mantle plumes, the possible occurrence of mantle avalanches, and the ongoing interaction of tectonic plates are all recorded here.

14

The Great Lakes of Wyoming

Early and Middle Cenozoic Era: 66.043 to 2.580 Million Years Ago
Paleogene and Neogene Periods

I magine the final days of the last *Tyrannosaurus rex* to walk upon the Earth. It was probably a young adult, the vigor of youth giving it the stamina and the endurance to survive longer than any other member of its species during the harsh changes that were to come. It was probably living in Montana or North Dakota or Alberta or Saskatchewan, somewhere that was near the northern limit of where its species ranged, as far as could be expected for a *Tyrannosaurus rex* to be from the point of the meteor impact that would lay waste to most of North America.[1]

At the time of its death, the Rocky Mountains had been rising for a few million years. This last survivor would have been following herds of other dinosaurs, tracking and hunting them in the forests and along the muddy banks of streams and rivers that were coursing down the steep slopes of this new mountain range. But the end was inevitable. No degree of evolutionary perfection could have guaranteed survival. The extraordinary sense of smell and the keen eyesight and the complex brain were not enough for this huge carnivore to survive one of the planet's great geologic catastrophes.

According to the definition agreed to by members of the International Commission on Stratigraphy, the Mesozoic Era ended at the moment that a large rocky asteroid touched the surface of the Earth along the northern edge of the Yucatán Peninsula and the Chicxulub Impact Crater started to form. And so technically, dinosaurs did live into the next geologic era, the Cenozoic, but most did not survive for very long. The last *Tyrannosaurus rex* probably lived for at least a few hours. It may have been a few months if it was able to search and find carcasses of dead animals and scavenge through their remains. But the environment had become too harsh. And so there was a moment in the history of the Earth when the last *Tyrannosaurus rex* took its last breath and died.

Part of the reason for its quick demise was that it was in North America where the effects of the impact were the harshest. Those animals, including dinosaurs, that lived on the opposite side of the Earth may have continued to live and reproduce and the various species survived for as long as another ten thousand years. But their end, too, was inevitable. The eruption of the massive lava flows of the Deccan Traps further fouled the air and spoiled the water, and that limited the ability of plants to photosynthesize and disrupted the food chain.

The populations of microorganisms in the oceans plummeted, including the vast multitudes of free-floating plankton that are at the base of the food chain of many aquatic animal species. As a result, mollusks were decimated. Ammonites, the tightly coiled, swimming predator that had hunted the seas since the Devonian Period, went extinct. So did all of the large marine reptiles, including the dolphin-like ichthyosaur and the long-necked and large-finned plesiosaur.

But there were survivors, especially on land. Those animals that lived in watery environments, such as frogs and fish, alligators and turtles, passed through the mass extinction without much reduction in the number of their species.

But the flying reptiles, the pterosaurs, were wiped out completely. So were most of the birds. All birds that dove into the sea to find food went extinct. So did all of the birds that had teeth or claws or long bony tails. More than ten thousand species of birds exist today. The majestic eagles and the fluttering finches, the twittering sparrows and the deep-diving murres and penguins, the hummingbirds, the ostriches, all are descended from a few species that survived the mass extinction at the end of the Mesozoic Era. Perhaps as few as five species survived. And those few survivors were flightless ground-dwellers, scampering after insects or feeding on marine life washed up on shore or on carcasses found on land. Their closest living relatives are modern ducks and chickens.

And then there were the mammals. Mammals had originated at the same time as the dinosaurs during the Triassic Period, and yet most of them remained small and nocturnal throughout the Mesozoic Era and most disappeared at the end of it. About three-fourths became extinct. Those that did survive were small. And they were generalists that could live and thrive in many different types of environments and had a varied diet. There was also a major change in the type of mammal. The Mesozoic Era was the heyday of marsupials. After the mass extinction, placental mammals began to dominate the planet as they do today.

The recovery of animal life was slow. It took several million years for the diversity of life to reach the complexity it had during the Cretaceous Period. In the oceans there was an explosion in the types of spiny-finned fish, which today represent nearly one-third of all of the vertebrates living today. Flounders, pufferfish, seahorses, eels, and anglerfish made their first appearance. Fish that are targets today of commercial fishermen, such as cod, bass, mackerel, and tuna, also first appeared after the extinction.

Mass extinctions create new evolutionary opportunities, and so many new animal forms appeared on land. Among the new lizards were the iguanas, and among the great diversity of snakes that came after the extinction were several giant ones including the boa.

The rate of recovery was not the same across the planet. The recovery was faster in the southern hemisphere than in the northern hemisphere, probably because both the Chicxulub impact and the eruption of the Deccan Traps occurred in the northern hemisphere, and so the disruption of ecosystems was more dramatic in the northern half of the planet.

But recovery did happen. After ten million years, the diversity of life was almost the same as it had been during the Cretaceous Period. And one can find this renewed diversity of life in the hillsides of southwestern Wyoming.

◆

To those who are unfamiliar with Fossil Butte, they may think this is yet another badlands. The land is sparsely covered by vegetation. The lines of hills have steep ravines. The summers are hot and dry and the winters are unbearably cold and windy, adding to what looks to be a challenging environment. But examine the rocks. They tell of a time when the climate and the terrain were quite different from what they are today. Within the rocks one finds fossils of palm fronds. There are bones of crocodiles and of turtles, two animals that thrive in warm climates and in regions that are flat and where there is stagnant or slow-flowing water.

The first geologic period of the Cenozoic Era is the Paleogene. The Paleogene Period is divided into three epochs: the similarly sounding Paleocene, the Eocene, and the Oligocene. The rocks at the base of Fossil Butte are gray and purple and various shades of red and brown and are of the Paleocene. These are sandstones and siltstones that were washed down from the newly formed Rocky Mountains and deposited by rivers on floodplains. Above them are rocks of the Eocene Epoch. These are buff-colored clays that accumulated at the bottom of a lake and that form the thousand-foot-high cliff of Fossil Butte today.

The lake had no outlet. For six million years the climate was remarkably hot and stable and allowed sediments to accumulate and remain undisturbed on the bottom of the lake. Most of these thin sedimentary layers, which number in the millions, formed when torrential rainstorms swept over the area, sending slurries of limy mud down the hillsides and into the lake. The mud then settled slowly to the bottom into the paper-thin layers seen today.

The hot climate made the surface water unusually warm. This set up a stable system within the lake where there was a layer of warm, well-oxygenated water lying over cold water that had little oxygen. When something died, the dead body settled slowly down into the cold water and into the muddy bottom where there was no oxygen for bacteria to operate and decompose and disrupt the body. The result is an abundance of exquisitely well-preserved fossils. There are fossils of leaves and of insects and of birds and of mammals. But, most of all, there are fossils of fish.

There are gars and pike and large-eyed, fork-tailed fish known as mooneyes that resemble the modern-day shad. Large numbers of fossils found in small areas indicate mass mortalities, and that the fossilized fish swam closely in schools, such as the spiny *Priscacara*. There are stingrays and bowfins. There are paddlefish up to four feet in length that are often preserved with fish in the stomach, indicating this species was a predator.

But the most common type of fish are *Knightia*, a type of herring. It is one of the most common vertebrate fossils in the world. It is also the Wyoming state fossil.

Millions of specimens of fossilized fish have now been recovered in Wyoming. And this is not due solely to the lake deposits exposed at Fossil Butte. During the Eocene Epoch there was a system of much larger lakes, a so-called Great Lakes of Wyoming, that covered large parts of south-western Wyoming, northwestern Colorado, and eastern Utah.

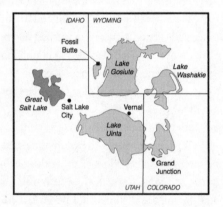

The Great Eocene Lakes of Wyoming and Utah.

These large lakes existed during the Laramide Orogeny and formed in the basins between the masses of rock that were rising and forming the Rocky Mountains. Though the water levels in these lakes changed continu-ously, causing the lakes to change shape and to merge or to become distinct bodies of water, in total, this was a system with no outlet to the oceans. These lakes were also stratified in the same manner as the one at Fossil Butte. And so an abundance of Eocene fossils, especially of fish, has been found not only in Wyoming but also in Colorado and in Utah.

For that reason, the geologic deposits that were left behind by this system of isolated lakes are famous in paleontology. But an interest in them extends far beyond a pursuit for highly detailed fossils. They also contain one of the largest potential reserves of oil on the planet. And they record an episode in the Earth's history when a sudden outpouring of carbon into the atmosphere drastically heated the planet.

◆

The rocks exposed at Fossil Butte are those of the Green River Formation. It is a geologic formation that covers a substantial part of the area where Utah, Colorado, and Wyoming meet. Find rocks of that formation, and you will have found deposits that accumulated over millions of years during the Eocene Epoch at the bottom of one of those ancient lakes.

A favored place to see the Green River Formation in Utah is along the state highway that runs through Indian Canyon southwest of the small town of Duchesne. It is possible to stop along the road and get a close-up look at the many thin layers of the formation. In addition to the paper-thin ones so evident at Fossil Buttes, these also contain an occasional layer of coal. In Colorado a prime place to see the formation is along a section of I-70 west of Rifle, Colorado. Here the interstate runs for more than twenty miles along the base of the Roan Cliffs. The upper thinly banded section of the cliffs is the Green River Formation.

And then there are the breathtaking views of the formation at its namesake along the Green River in Green River, Wyoming. This is where John Wesley Powell and nine other men climbed into four round-bottomed wooden boats and set out in 1869 to challenge the river and eventually to make the first recorded passage through the Grand Canyon. Their takeoff point was at a place now known in their commemoration as Expedition Park. The rock they stood on when they departed was the clay stone of the Green River Formation.

A short distance up the river from that point is a low cliff where a visitor's center has been erected by the local chamber of commerce. From there, one can sit quietly and look across a landscape with multiple spires and mesas. Each one is part of the Green River Formation. For those who want a rush when viewing this geologic formation, one can drive along the section of I-80 that runs along the eastern side of the river just skirting the limits of the town of Green River. The interstate passes close to both Castle Rock and Tollgate Rock. Both are spires of the Green River Formation. There are also twin tunnels along I-80 that pass through the formation. Two miles west of the tunnels is a high wall of rock that runs close to I-80. The buff-colored clays, as well as the greenish brown sandstones and the brown oil shales, are all part of the Green River Formation.

Now focus on the oil shale. The millions of fossils recovered from the Green River Formation attest to the multitude of animal and plant life that lived in and around these ancient lakes during the Eocene Epoch. But the main component of organic material within the formation comes from a different source. At times there was little oxygen anywhere in the lakes. When that happened, algae thrived. It was the remains of these microorganisms that produced the bands of oil shale.

The amount of organic remains in the Green River Formation is astounding: It exceeds that of all of the hydrocarbons in all of the known oil reserves in the world. But there is a problem in extracting these organic remains and converting them to oil. The remains are trapped in solid rock.

Oil shale is a misnomer. It is composed of neither oil nor shale. What seems to be shale is actually a type of limestone formed from the accumulation of tiny algal shells. The organic remains are in the form of a waxy hydrocarbon known as *kerogen*. To extract the kerogen, the oil shale must be drilled and blasted. The blasted rock must be loaded and hauled to a processing plant where it is crushed, and then heated. Kerogen is an early stage in the conversion of organic material to oil. To become oil, it must be heated in what is essentially a large pressure cooker where the kerogen vaporizes. The vapor condenses to form a thick oil that can be refined into a variety of oil products.

And so the entire process requires not only a great amount of digging and disruption of the landscape, but also large amounts of water must be used for mining and processing and large amounts of waste are produced that must be disposed of. And no one has developed an economical way to extract the kerogen and produce the oil. But the hydrocarbons are there in the ground in the Green River Formation. And there is enough to supply the world with its oil-producing energy needs for centuries.

But the burning of so much oil has consequences: It adds directly to the amount of carbon dioxide in the atmosphere. That in turn heats the planet. And so, when looking at the Green River Formation, there is geologic irony: Within this formation is one of the world's largest potential reserves of oil. There is also a record of what happens when a large amount of carbon dioxide is added suddenly to the atmosphere and the planet is quickly heated.

The climate cooled a bit after the end of the Cretaceous Period. After ten million years, there was a spike in the amount of carbon dioxide in the atmosphere and a heating of the entire planet. This shift from cooling to heating marks the transition from the Paleocene to the Eocene epochs. It is the warmest period in the Earth's history since the Cretaceous Period. And it is known as the Paleocene-Eocene Thermal Maximum (PETM).

Evidence of the PETM was first found in the early 1990s from rapid shifts in carbon and oxygen isotopes in carbonate rocks collected from deep drill holes made into the seafloor near Antarctica. Additional work showed that the same rapid shifts were seen in Eocene rocks collected anywhere in the world. Look at Fossil Butte. The same rapid shifts in carbon and oxygen isotopes have been found there. Drive through the twin tunnels along I-80 at Green River, Wyoming. These are rocks of the PETM.

In less than ten thousand years the amount of carbon dioxide in the atmosphere at least doubled. It may have tripled. And the global temperature increased by as much as 15° Fahrenheit (8° Celsius) to an average temperature as high as 73° Fahrenheit (23° Celsius).[2] The oceans became more acidic. And the patterns and rates of rainfall changed. The disruption lasted about 150,000 years.

An obvious source for the sudden injection of a large amount of carbon dioxide into the atmosphere is a massive volcanic eruption. One did coincide with the beginning of the PETM. It formed the lavas of the North Atlantic Igneous Province that can be found today scattered across the northeastern Atlantic. That eruption led to the formation of the Faroe Islands that lie between Scotland, Norway, and Iceland. Many of the volcanic rocks of western Norway and eastern Greenland date from this event, as do those found in western Scotland and in Northern Ireland. The Giant's Causeway, the famous site where tens of thousands of interlocking basaltic columns are laid out like giant stepping stones, was also formed during this event.

But the outpouring of those lavas is not sufficient to account for the amount of carbon dioxide that was needed to trigger the spike in global temperatures during the PETM. Another source is needed. And it must have had a biological origin.

One such potential source lies on the seafloor in the sediments that fringe the continents. Here on the seafloor today, there are vast deposits of a peculiar form of ice known as *clathrate*. Clathrate, in essence, is a chemical compound in which one type of molecule has been trapped within the lattice of another. In particular, on the seafloor, the compound is methane clathrate; molecules of methane, produced by decomposing bacteria in the sediments, are trapped in lattices of water ice. Under the current temperature of the seafloor, methane clathrate is stable. But raise the temperature of the surrounding water and this icy compound melts and vast amounts of methane are released and move into the atmosphere. This may have been one of the factors that led to the PETM.

Methane is a highly efficient greenhouse gas, much more efficient than carbon dioxide in being able to trap heat in the atmosphere. And so when large amounts of methane are released, the global temperature rises rapidly. In today's world, methane clathrate is one of the largest reservoirs of biological carbon on the planet. And if most of it was released, the global temperature would rise several degrees.

And so it is *possible* that it was the release of methane clathrate from the sediments of Paleocene seas that contributed to the heating of the planet during the PETM. The release would have been triggered by the release of vast amounts of carbon dioxide during the eruption of the North Atlantic Igneous Province, which would have heated the atmosphere, and which in turn would have heated the oceans. But this is far from proven.

The debate about the cause of the PETM is ongoing. What is certain is that a huge amount of carbon dioxide, and possibly methane, entered the atmosphere and that the planet suddenly heated up. It was one of the most intense and abrupt intervals of global warming in the geological record. And one of the places where this is clearly written is in the rocks of the Green River Formation.

◆

Throughout the Cenozoic Period, the Atlantic Ocean has been opening and the North American Plate has been moving westward, as both continue to do today. For forty million years the Farallon Plate collided with the North American Plate all along its western boundary. But that changed when an edge of the North American Plate had moved far enough west to begin to override the oceanic spreading center that defined the boundary between the Farallon and the Pacific plates. At that moment, the San Andreas fault was born.

As the North American Plate continued its westward movement, it ran over more and more of the Farallon Plate, lengthening the San Andreas fault. Today the fault extends for several hundred miles along the coast of California from Cape Mendocino to the eastern edge of the Salton Sea.

The fault is responsible for the long narrow inlet of Tomales Bay north of San Francisco and for its namesake, the San Andreas Valley, which is an equally long and narrow valley that runs south of the city. Continuing southward, the trace of the fault can be located exactly where it runs along the base of a steep hillside next to the old Spanish mission at San Juan Bautista. An exact location of the fault can also be found near the small town of Parkfield where slow movement along the fault has produced a noticeable bend in the guardrails that line the sides of a bridge.

At Carrizo Plain National Monument, midway between the cities of Santa Maria and Bakersfield, the San Andreas fault runs along the base of a low line of hills along the eastern side of the monument. Here the sliding of crustal blocks on either side of the fault has disrupted the normal straight-line course of several streams and has given them a sudden offset. In particular, at Wallace Creek, the stream bed runs straight down the hillside until it reaches the base of the hill. At that point, the stream bed makes an abrupt turn to the right, continues for several hundred feet, then turns to the left. The intervening section between the two turns is the San Andreas Fault.

The San Andreas Fault is also responsible for the three main corridors that are land connections into and out of the Los Angeles Basin. All of the interstate highways, main railways, and most of the pipelines, water lines, and power lines that link the basin to the rest of the continent run through one of these narrow corridors, through what would otherwise be a long line of continuous high mountains. And it is at each of these points—Tejon Pass north of Los Angeles, Cajon Pass north of San Bernardino, and San Gorgonio Pass east of San Bernardino—that the San Andreas Fault cuts through the mountains.[3]

Rock layers contorted by movement along the San Andreas Fault. Road cut is
along Highway 14 south of Palmdale, California (photo courtesy of Ben+Sam).

The relative motion between the Farallon Plate and the North American
Plate was such that the two plates collided, forcing the Farallon Plate to
slide under the North American and, through a variety of mechanisms,
forming mountains. The relative motion between the Pacific Plate and the
North American Plate is decidedly different.

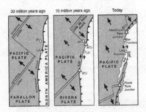

Origin and evolution of the San Andreas Fault.
(Adapted from a drawing by Kious and Tilling)

Both plates have a westward component to their motions, but the Pacific
Plate also has a significant component that moves it northward. As a result,
where the two plates meet, they slide against each other, the Pacific Plate
moving slightly faster to the northwest relative to the North American
Plate. That, in essence, is the nature of the San Andreas Fault. If you are

standing on the west side of the fault, as in Los Angeles, you are moving slowly northwest relative to the rest of North America.

The San Andreas Fault as it is today is the latest of a series of faults that have marked the boundary between the Pacific and the North American Plates. The first boundary fault was necessarily short and lies today somewhere off the coast of southern California, not yet clearly identified in the many oceanographic surveys that have been made of this area. One of the first long strands that ran along much of the coast of California is the San Gregorio Fault located off the coast of Northern California. The southern end of the fault is in southern Monterey Bay and the northern end is northwest of San Francisco near Bolinas Bay, where it intersects the San Andreas Fault. It crosses land between Moss Beach and Pillar Point north of Half Moon Bay and sections of it cross Big Sur between Point Sur and Piers Blancas.[4]

The San Andreas Fault runs remarkably straight for more than four hundred miles from Cape Mendocino to the southern end of Carrizo Plain. Then there is the first of two large bends, first to the left, then to the right, before it ends at the Salton Sea.

If one extends the long and straight northern section of the fault southward, it seems to continue along what is now called the Newport-Inglewood-Rose Canyon Fault that runs from Culver City to Long Beach to Laguna Beach to San Diego. This may have once been part of the boundary between the Pacific and the North American Plate. But then the boundary shifted twenty-five miles to the east and was along the Whittier-Elsinore-Laguna Salada Fault that runs from Whittier to Temecula across the Anza-Borrego Desert to Mexicali in Mexico. Then it shifted again to the present trace of the San Andreas Fault that runs a few miles east of the luxury resorts and celebrity mansions at Palm Springs and through Coachella Valley. The reason for this shift—and the evolution of the San Andreas fault—is that the two ends of the fault have been moving.

The northern end of the San Andreas Fault is at a feature known as the Mendocino Triple Junction, which is a point on the Earth where three tectonic plates meet. Two of the plates are the Pacific and the North American Plates. The third is the remnant of the Farallon Plate that has been renamed the Juan de Fuca Plate. This point, the Mendocino Triple Junction, has been migrating north as the boundary between the Pacific and the North

American Plates has continued to lengthen. Twelve million years ago the triple junction was near San Francisco. It is now near Eureka, California.[5]

The southern end of the San Andreas Fault was once also at a triple junction between the Pacific and the North American Plates and another remnant of the Farallon Plate known today as the Rivera Plate. But about six million years ago the position of the junction jumped a few hundred miles south and east to a point on the North American Plate. That caused a slab of North America to shear off and swing away from the continent. The slab is the peninsula of Baja California. The new seaway being formed and continuing to grow wider between Baja California and North America is the Gulf of California.

The southern end of the San Andreas Fault is still connected to the new triple junction through a series of short spreading centers that run along the length of the Gulf of California. And so, though much of South California and all of Baja California were once part of North America, both are now firmly attached to and are moving slowly in a northwest direction as part of the Pacific Plate.

The northern section of the San Andreas Fault is also undergoing an evolution. Both the Hayward Fault that runs close to Oakland and Berkeley and the Calaveras Fault that is several miles farther east run along trends that are nearly parallel to the San Andreas Fault. And so, in the future, it is quite reasonable to expect that the boundary between the Pacific and the North American plates might shift eastward, and that one of these faults will become the new plate boundary. But there is a more dramatic eastward shift that may be in the making.

Extend the trend of the southernmost section of the San Andreas Fault, the section that runs roughly from Palm Springs to the Salton Sea, northward. That extension runs along a region of earthquake activity known as the Walker Seismic Lane.

The Walker Seismic Lane is not a continuous fault like the San Andreas Fault. It consists of a series of hundreds of shorter faults that run for hundreds of miles from Owens Valley and Death Valley northward to Pyramid Lake north of Reno in Nevada. The nature of the earthquakes along the Walker Seismic Zone is the same as along the San Andreas Fault: a horizontal sliding such that the region west of the Walker Seismic Zone is slowly moving northwestward. And so it is conceivable, given the current state of earthquake activity, that the plate boundary could shift again and, in the distant future, essentially, the entire state of California will detach from the North American Plate, become part of the Pacific Plate, and drift slowly as a collection of large islands into the northern Pacific Ocean.

And there is yet another zone of earthquake activity that runs across the western United States. Known as the Intermountain Seismic Zone, it runs through western Montana and into the Yellowstone region and through western Wyoming and central Utah and into southern Nevada. It includes earthquakes that occur beneath the Wasatch Mountain Front that forms the backdrop of Salt Lake City. If you drive along I-15 from Las Vegas to Great Falls, Montana, you are driving along that zone.[6]

It is important to note that the Walker Seismic Zone and the Intermountain Seismic Zone run along opposite ends of the Basin and Range Province. And so it is reasonable to think that they must be related to the formation of the province. But before that can be pursued, two more important pieces of the puzzle must be considered. One is yet another extended zone of activity in the western United States. It is the Rio Grande Rift.

To stand anywhere along the I-25 corridor that bisects the state of New Mexico, say in Albuquerque or Socorro, is to stand within the Rio Grande Rift. In size and scale it is similar to the spreading centers that lie along the ocean floor. There are high walls that define the edges of the rift. The center has dropped down several hundred feet. And there is a broad flat floor. All of this indicates that this is a place where the rigid lithosphere has been stretched. And where the lithosphere has been stretched, it has also been thinned. That in turn has allowed volcanoes to form and erupt recently. Along the skyline west of Albuquerque is a short line of low, dark volcanic cones that were part of those eruptions. Fifty miles north of Albuquerque near Los Alamos is Valles Caldera, an explosive volcano that erupted huge volumes of ash and pumice about a million years ago. In nearby Bandelier National Monument prehistoric people built their pueblos from bricks that they carved out of the ash and pumice—the great thickness and heat of the deposit caused the ash and pumice to weld into a hard rock—and dug rooms into canyon walls in the same material.

The Rio Grande Rift marks an important boundary. Drive east of Albuquerque and one is soon on the Great Plains, a broad expanse of flat land that runs from hundreds of miles across the central part of the continent and where the land surface seldom rises more than a few thousand feet above sea level. Now drive west of Albuquerque. One soon reaches another region that is flat, much smaller in extent than the Great Plains, but here the land surface is more than a mile above sea level.

This region of flat elevated land is the Colorado Plateau.

◆

Never travel with more than one geologist at a time. Otherwise you will be hopelessly confused. This is especially true when traveling across the Colorado Plateau.

Geology is still very much a descriptive science. Though much progress has been made in using instruments that can reveal extremely small variations in isotopic compositions or that record the continuous slow shifting of positions on the Earth's surface or that capture the minuscule vibrations caused by the passage of seismic waves, much of geology still relies on what can be seen in the field with the eye. And that can lead to different interpretations—each interpretation is based on how each individual decides to rank the importance of different field observations. This is especially true on the Colorado Plateau because here there is such a diversity of spectacular rock forms that are exposed over a wide area, and those various forms have been studied by a multitude of investigators who have worked intently in so many remote corners of the plateau.

Notwithstanding the occasional objection that will inevitably be raised by some investigators who have traipsed over a particular section of the plateau and who maintain that an isolated rock unit should be interpreted in one way and not in another, there is a story to be told. And it is one that can bring together the many different geologic elements of the American Southwest into a coherent story.

The Colorado Plateau is centered on the Four Corners region and includes much of Arizona, Utah, Colorado, and New Mexico. Its eastern edge is the Rio Grande Rift. To the north are the Rocky Mountains and to the west and to the south is the Basin and Range Province. More than thirty national parks and monuments are within the confines of the Colorado Plateau, including Petrified Forest National Park, Dinosaur National Monument, Zion National Park, Arches National Park, Mesa

Verde National Park, and Grand Staircase Escalante National Monument, which indicates how diverse and spectacular the rock forms actually are. And then there is Grand Canyon National Park.

Stand on the South Rim and look down the sequence of flat-lying layers. The lowermost layer is the Tapeats Sandstone that sits atop the Great Unconformity. Everything above it—that is, everything that formed later, with one exception[7]—formed either under the sea or near sea level on a floodplain or tidal flat. For most of the last five hundred million years this region of the continent was either below or near sea level. Then during the last few tens of millions of years, this region was raised more than a mile. And it was raised in such a way that the horizontal layers remained undisturbed.

Now look up into the sky. Until a few tens of millions of years ago, the rock that is now exposed on the South Rim, the Kaibab Limestone, was buried by a mile or two of layered deposits that formed later. They included the Moenkopi Formation, the Chinle Formation, probably the Navajo Sandstone, and many others. Return to the Grand Staircase north of the Grand Canyon and you will see the sequence of these later deposits, the rocks that once lay over the Kaibab Limestone on the South Rim and have since been eroded and transported away by fast-moving streams and rivers. Go look in Monument Valley and you will see where the same process removed much, though not all, of the overburden, leaving behind the many spires and mesas that have made Monument Valley so famous.

The erosion of a mile or so of rock from the region of the Grand Canyon and the broad region that surrounds it removed a great weight of rocks from the Colorado Plateau. That is one of the main reasons why it has been so difficult to understand the evolution of the Colorado Plateau: Much of the geologic record has simply been erased.

And in response to the removal of so much rock, the plateau rose. But it was not enough to explain why the Colorado Plateau rose more than a mile, why everything between the Tapeats Sandstone and the Kaibab Limestone that formed below or near sea level is now high above sea level.

The Laramide Orogeny might have contributed some of the uplift that is so evident on the Colorado Plateau today. But there are neither folds nor contortions in the sequence of rocks revealed at the Grand Canyon. And so whatever contribution might have been made by the Laramide Orogeny was undoubtedly slight. Another mechanism must have acted to raise the Colorado Plateau.

Seismic tomography shows that the rigid lithosphere beneath the Colorado Plateau is substantially thicker than beneath the Basin and Range Province. This greater thickness may have made the Colorado Plateau sufficiently strong to resist being stretched out as it was pulled away from the Great Plains and the rest of North America and becoming a series of basins and ranges. Seismic tomography also shows that a large slab has broken away from the base of the lithosphere, through delamination, and is now descending into the hot and pliable mantle. No such large slab is descending beneath the base of the lithosphere under the Colorado Plateau. But a slab of the Farallon Plate has been detected deep beneath the plateau. Its downward movement has allowed hot mantle material to rise beneath the plateau and is probably responsible for most of the rise, essentially lifting the entire plateau like a giant piston.

Furthermore, when studied closely, the images produced by seismic tomography suggest a small cell of hot mantle material is convecting beneath this region of the American Southwest, rising beneath the Colorado Plateau and Rio Grande Rift and overturning and descending beneath the Great Plains.

And that brings the story back to the San Andreas Fault.

Ever since the Pacific and the North American Plates first touched twenty-five million years ago, the influence of the Pacific Plate on the North American Plate has been increasing. The length of the San Andreas Fault has continued to grow longer as more and more of the North American Plate has been transferred to the Pacific Plate. And the northwestward motion of the Pacific Plate has been influencing the motion of a larger and larger part of the North American Plate.

The EarthScope program that operated hundreds of seismometers across the United States also operated a dense network of satellite receivers that made use of the Global Positioning System (GPS) to record slight movements of the Earth's surface.[8] From those receivers, relative positions across the continent could be determined to a small fraction of an inch. When operated over a period of months to years, speeds as slow as a tenth of an inch per year could be measured. It was this ability to make remarkably precise measurements that revealed how the Earth's surface is slowly shifting today.

Imagine yourself and three friends spaced out across the western half of the United States, each of you with a GPS receiver that can make precise measurements of your position. You are in Los Angeles standing on the Pacific Plate. Your most distant friend is standing in the center of the continent, say in Kansas or Iowa. You make your measurements, compare them, and discover that, relative to Kansas or Iowa, Los Angeles is moving northwestward at a rate of five inches per year. That is the relative movement between the Pacific and the North American Plates.

Now compare your measurements with those made by another friend who is standing in the Mojave Desert, east of the San Andreas Fault. After a few years, your measurements will show that, relative to the Mojave Desert, Los Angeles is moving northwestward at a rate of four inches per year. That is, movements along the San Andreas Fault and its many ancillary faults—the Hayward Fault, the Calaveras Fault, the Newport-Inglewood-Rose Canyon Fault, the San Jacinto and the Elsinore Faults, and hundreds of others—account for *most* of the relative motion between the Pacific and the North America Plates. But there is still a significant amount of motion east of the Mojave Desert.

Now compare the measurements made by your friend standing in the Mojave Desert and your fourth friend who is standing somewhere in the Basin and Range Province in Nevada. Your friend in the Mojave will be moving northwestward by about three-quarters of an inch per year relative to your friend in Nevada. A last comparison will show that Nevada is moving northwestward about one-quarter of an inch per year relative to Kansas or Iowa.

Now put out more than a thousand GPS receivers—which is what the EarthScope program did—record measurements for several years, and look at the results. They will show that the motion of four inches per year between the Los Angeles and the Mojave Desert is occurring along the San Andreas Fault and its associated system of faults. The motion of three-quarters of an inch per year is occurring across the Walker Seismic Zone that runs close to the California-Nevada border. And the very slow motion of a quarter of an inch per year is occurring across the Intermountain Seismic Zone that runs through Salt Lake City and across the Rio Grande Rift.

In short, a transition is underway in the geological evolution of the southwestern United States. For more than a hundred million years the Farallon

Plate controlled the geology of this region, including the Laramide Orogeny, the formation of the Basin and Range Province, the raising of the Colorado Plateau, and the development of the Rio Grande Rift. For the last twenty-five million years, the giant Pacific Plate, which is sliding beneath Alaska and eastern Asia and is transporting the Hawaiian Islands to the northwest, is gaining more and more control of the geology of the southwestern United States. It is this transfer of influence that formed the San Andreas Fault and its associated system of faults, as well as the Walker Seismic Zone and the Intermountain Seismic Zone.

It is also this transfer of influence from the Farallon Plate to the Pacific Plate that has given rise to one of the most spectacular features on the planet: the Grand Canyon.

◆

In northern California the faults generally run parallel to the San Andreas Fault. In southern California some faults do run parallel to the San Andreas Fault, notably the San Jacinto Fault and the Elsinore Fault, but many do not. And there is a simple explanation for this difference: Southern California is caught in a vise.

In southern California, the earthquake faults, especially those in the Los Angeles Basin, run in many different directions and are of many different lengths. The pattern seems to resemble what happens when a pane of glass is shattered.

The Hollywood Fault runs roughly parallel to and just north of Sunset Boulevard in Hollywood. The Raymond Fault, which may be an eastward extension of the Hollywood Fault, has been followed from Griffith Park north of Los Angeles to the Santa Anita Racetrack in Arcadia, a distance of about twenty miles. The Northridge Hill Fault, which ruptured in 1994 and caused more than $10 billion in damages and killed fifty-seven people and injured thousands, has a trend that runs northwest from Panorama City in the San Fernando Valley. The Charnock Fault, one of the shortest known faults in the Los Angeles Basin, runs for just five miles along a north-south line just east of Los Angeles International Airport. And the Puente Hills Fault, which is actually a system of several fault strands, of which some have been discovered only in the last few decades, runs for at least twenty-five miles from beneath downtown Los Angeles southward to the Chino Hills in Orange County.

This chaotic pattern of faults is owed to the sudden shift of the Rivera triple junction to a point beneath North America and the subsequent eastward migration of the San Andreas fault in southern California. That, in turn, caused a large slab, what is now the peninsula of Baja California, to shear off and separate from North America. That slab is now swinging out away from the continent as it also slides northwestward as part of the Pacific Plate.

The combined motion of Baja California—the swinging outward away from the continent and the northwestward movement—has placed southern California in a vise.[9] It is compressing the region. And from the compression, mountains have formed.

The mountains of the Transverse Range that run north of Los Angeles—the Santa Ynez, the San Gabriel, the San Bernardino, and the San Jacinto—all formed because the Rivera triple junction shifted its position and Baja California swung outward and moved northwestward. So did lesser ranges, including the Santa Monica Mountains and the Santa Ana Mountains. As did short ranges of hills, such as the Elysian Hills near Dodger Stadium, that may, in the future, rise upward and become part of major mountain ranges.

As far as the formation of the Grand Canyon is concerned, the key element is what has happened behind Baja California as it has swung out and away from the continent. That movement has stretched the crust, leading to a thinning and dropping of the crust. From that, a deep seaway, the Gulf of California, has formed. In fact, the seaway would extend all the way up the Coachella Valley to Indio, California, but it has been filled with sediment. The sediment that now fills Coachella Valley—and the long low region south of it known as the Salton Trough—and prevents the sea from intruding farther is material that was eroded away by the formation of the Grand Canyon.

Deep dissection of the Colorado Plateau occurred with the opening of the Gulf of California. The gradient of the land that the Colorado River flowed over had increased, and so the flowing water eroded material at a faster rate. The steepness had increased so much—the Colorado Plateau had recently been raised more than a mile and the Gulf of California was more than three miles deep—that erosion occurred at a phenomenal rate. And the Grand Canyon was born.

That is the simple explanation. There is a more complex, and intriguing, one based on more than a hundred years of geologic studies of the canyon. In a way, those studies and the history of ideas that came from them are somewhat reminiscent of the multitude of sculptures of David made by Renaissance artists. Michelangelo, Donatello, Bernini, Verrocchio, and many others all created sculptures of David. Each one felt compelled to offer a new insight about the biblical figure. The long train of geologists who have studied the Grand Canyon and proposed ideas have done the same: Each one has added what he regarded as a crucial observation and, from that, proposed something new about the way the canyon must have formed.

The story begins with John Strong Newberry. He was the first geologist to see the canyon. He was part of an expedition led by Joseph Ives of the United States Corps of Topographical Engineers in 1857.[10] In his report about the expedition, Ives stated that "the region . . . is altogether valueless. [The canyon] can be approached only from the south, and after entering it there is nothing to do but leave. Ours has been the first, and will doubtless be the last party of whites to visit this profitless locality." He then summed up his impression of the canyon by writing: "It seems intended by nature, that the Colorado River, along the greater portion of its lonely and majestic way, shall be forever unvisited and undisturbed." The eyes of a geologist, carried by Newberry, saw something distinctly different.

Newberry looked upon the same rugged and steep terrain as Ives and sensed a landscape with a different potential. To him the Grand Canyon and its surroundings were part of a "paradise" where one found the "most splendid exposure of stratified rocks that there is in the world." He made the first crucial observation as to how the canyon had formed. He recognized that the sequence of layered rocks in the walls of the canyon on either side of the river was not offset, as by a rift or a fault, and so he concluded, rightly, that the Grand Canyon had been carved "by the exclusive action of water."[11]

In 1900, William Morris Davis of Harvard University completed a month-long trip through northern Arizona and southern Utah that included considerable time examining the landforms of the Grand Canyon and those farther upriver in Marble Canyon and concluded that, originally, the Colorado River had flowed to the northeast, then reversed direction and flowed to the west as it does today.

In the 1930s, Chester Longwell of Yale University was studying the section of the Colorado River that was soon to be covered by the rising waters behind Hoover Dam and recognized a way to determine the age of the Grand Canyon. It rests on the age of the Muddy Creek Formation. The Colorado River exits the Grand Canyon through the Grand Wash Cliffs, an imposing, west-facing wall of rocks that runs close to the Nevada-Arizona border and marks the transition between the Colorado Plateau and the Basin and Range Province. The small unincorporated town of Meadville is one of the best places to see the cliffs. The northwestern section of town and farther along the road that eventually leads to the edge of Lake Mead and the Colorado are the rocks of the Muddy Creek Formation.[12]

These rocks are mostly marine mudstones and siltstones that were deposited in an estuary at the end of an embayment of the Gulf of California, long before the Coachella Valley and the Salton Trough were filled with sediments. What is most important is that there are *no* pebbles or cobbles within the Muddy Creek Formation that were swept down by the Colorado River from the interior of the continent. That is, when the Muddy Creek Formation formed, there was no Colorado River piercing the Grand Wash Cliffs. And there was no Grand Canyon.

The rocks of the Muddy Creek Formation were deposited six million years ago. That is a firm age. And so the Grand Canyon that we see today must be younger than that age. But there is a caveat. If one looks at the rocks along the entire length of the Grand Canyon and farther upriver into Marble Canyon, one finds evidence for a flowing river and for deeply incised canyons that are much older. These canyons formed early during the Cenozoic Era, about fifty million years ago, when rivers were flowing toward the northeast off what is now the region of the Colorado Plateau.

When one is looking at the Grand Canyon at its greatest extent, to the east and to the west, one is seeing the legacies of at least two major rivers. The older river flowed northeastward into the Western Interior Seaway. Then, about ten million years ago, the Colorado Plateau was raised. As it was raised, the land surface tilted slightly to the south.[13] A second major river developed, one that may have flowed initially to the northwest, but eventually flowed to the west. When the land surface dropped and the Gulf of California and the Salton Trough formed, the second river cut deeply into the Colorado Plateau. As it cut, its headwaters progressed eastward until the flow of the older river was captured and the modern Colorado River was formed. The capture point was probably near the confluence of the modern Colorado River and the Little Colorado River.[14]

There is one more important question to ask: When did the Grand Canyon take on the appearance that it has today, when did it reach its current depth and length? And that question does have an answer.

Ninety miles downriver from the popular viewpoints around Grand Canyon Village in Grand Canyon National Park on the North Rim of the canyon is the Uinkaret volcanic field. Lava flows from this field have cascaded down into the Grand Canyon, damming the Colorado River. Most of the lava has been swept away by the continued flow of the river, but there were at least thirteen times when the lava poured down in such great volumes that natural dams were created and huge lakes formed on the upriver side.

These dams ranged from several hundred to more than a thousand feet in height, reaching up from the river bottom. The largest dam created a large lake that extended more than two hundred miles upriver, all the way to Moab, Utah. It probably took a few decades to fill with water. At its highest point, the lake reached up to the level of the Redwall Limestone, halfway up the walls of the Grand Canyon.

These lakes were short-lived, probably not playing a major role in controlling the flow of river water for more than a few centuries. Powell encountered the remnant of one of these great lava dams in late August 1869, noticing black rocks that still hung on the canyon walls and a series of giant boulders in the river as the cause for the largest rapid he had yet to challenge. "What a conflict of water and fire there must have been here!" he wrote. "Just imagine a river of molten rock running down into a river of melted snow. What a seething and boiling of the waters; what clouds of steam rolled into the heavens." He named these rapids Lava Falls. Modern boatmen who ride the Colorado River have referred to these rapids as "the scariest ten seconds on the river." Powell chose to portage his boats and supplies around them.[15]

It is the age of these lava flows that is the key to understanding when the Grand Canyon reached its current appearance. Some of the lava still sits in the river. A large dark angular rock known as Vulcan's Anvil formed from one of the flows and is about a mile upriver from Lava Falls. Other remnants of lava flows and the dams they created extend as far as eighty miles downriver along the current level of the Colorado River. And so when these flows erupted and cascaded down the canyon walls, the Grand Canyon was about as deep at that time as it is today.

The age of these lava flows is 1.2 million years. By then, the Grand Canyon that we see today had formed. The volume of material that was moved is more than a thousand cubic miles. That immense amount of material is now sitting in the Salton Trough, a feature that would have been miles deep, and that would have allowed ocean water from the Gulf of California to reach far into southern California if the Grand Canyon had not formed.[16]

Because so much more of the geologic record contains information about recent events than ancient ones, as the story of the geological history of North America—and of the Earth—gets closer to the present time, it gains in complexity. For that reason, a pause is needed to briefly recall what has happened.

The first geologic eon, the Chaotian, though not yet officially recognized by the International Commission on Stratigraphy, was the time period when the solar system began to form. The Sun had condensed and planets, especially Jupiter and Saturn, were moving about in chaotic orbits. Hold a meteorite in your hands, and you are probably holding a rock that formed during the Chaotian Eon.

The Moon had formed and the Earth had achieved its final size and mass by the beginning of the next eon, the Hadean. During the Hadean Eon, life probably existed on the planet, though whether it originated from the physical and chemical environments of the early Earth or from an extraterrestrial source is still debated. The physical relics of this eon are the few hundred almost microscopic grains of zircons recovered from some of the planet's oldest rocks. Those grains indicate that an ocean did exist and that the Earth's surface was hot, but not the fireball that was once thought.

The beginning of the Archean Eon is the beginning of the rock record. The final adjustments to the orbits of Jupiter and Saturn sent debris into the inner solar system and, for a hundred million years or so, a barrage of large meteors hit the Moon and the Earth. The effects of the Late Heavy Bombardment are clearly seen as the great circular scars on the Moon's surface. The effects of the Late Heavy Bombardment have not yet been recognized on the Earth. After the bombardment, conditions on the Earth's surface stabilized. The 3.524-billion-year-old Morton Gneiss found in Minnesota formed. Life proliferated and photosynthesis began to add oxygen to the atmosphere. That led to the first episode of Snowball Earth and the end of the Archean Eon, recorded in the La Cloche Mountains in Killarney Provincial Park in Ontario.

The two supercontinents of Columbia and Rodinia are the bookends that define the beginning and the end of the Proterozoic Eon. Rocks of that eon are exposed in the Adirondack Mountains and as Mount Rushmore in the Black Hills of South Dakota. They also comprise almost all of the rocks of southeastern Canada. Within those rocks of southeastern Canada are the trace fossils of Ediacarans, the earliest indication of the proliferation of multicellular life. Whether Ediacarans are the ancestors of modern animal life is still unknown.

Rocks of the Proterozoic Eon are also exposed in the steep walls of the inner canyon of the Grand Canyon. The Vishnu Schist and the Zoroaster Granite were remnants of Columbia and Rodinia. Above them there is a second episode of Snowball Earth and the Great Unconformity. And above the Great Unconformity are rocks of the current geologic eon, the Phanerozoic.

The first part of the Phanerozoic Eon is the Paleozoic Era. Rocks of that era are the great sequence of layered rocks at the Grand Canyon. At the bottom is the Tapeats Sandstone and the Cambrian Explosion. Midway up the wall, in geologic time, plants and animals moved from the oceans and onto land. On the rim of the canyon is the Kaibab Limestone. Soon after this limestone formed, the Permian mass extinction happened and the Paleozoic Era ended.

The next geologic era is well recorded in the sequence of rocks of the Grand Staircase. Here is the beginning, the rise, the dominance, and the end of the dinosaurs. The Mesozoic Era ended with the Chicxulub impact.

The next 66.04 million years, that is, from the Chicxulub impact to the present time, is the Cenozoic Era.

Sequences of Cenozoic rocks can be found scattered across North America. A record of several million years is typical, such as at Fossil Butte. But there is an exceptional and easily accessible sequence that spans two-thirds of the Cenozoic Era. It is in north-central Oregon at John Day Fossil Beds National Monument.

Today this is a dry landscape of sagebrush and the occasional juniper. Fifty-four million years ago, when the rock sequence at John Day begins, the land was also dry and desolate. Plant and animal life had not yet recovered from the devastation of the mass extinction that had ended the dinosaurs. Life was also struggling with the prolonged period of extreme heat of the Paleocene-Eocene Thermal Maximum.

After ten million years, the planet had cooled sufficiently that a dense tropical jungle that included small bananas, palm trees, dates, and fig trees existed in what is now north-central Oregon. After another ten million years and more cooling, the climate was a temperate one with wooded areas similar to today's eastern United States with deciduous forests of alder and beech and elm and oak in the lowlands and coniferous forests, including the dawn redwood, growing at higher elevations.[17] This transition in climate can be seen in the Painted Hills Unit of the national monument near Mitchell, Oregon. The dark red bands in the hills are periods of warm tropical forests. The lighter bands are cooler, drier periods of deciduous and coniferous forests and of grasslands. Follow the bands upward. The dark red ones become less pronounced and the lighter ones eventually dominate, indicating a cooling of the planet.

Fossilized mammals are especially plentiful in the rocks at John Day. Thousands of skeletons have been unearthed. Many are from a small area known as the Hancock Mammal Quarry. This high concentration of fossils occurred along a tight bend in an ancient river channel. Seasonal flooding washed down an incredible variety of dead animals that accumulated along the bend where silt, sand, and gravel built up. There were various rodents, rhinoceroses, a three-toed horse, tapirs, sheeplike oreodonts, and saber-toothed cats. Elsewhere in the sediments at John Day are fossils of a four-toed, dog-sized horse and a bear-sized omnivorous pig.

The discoveries made in the fossil beds at John Day have revealed a truly remarkable and highly detailed history of mammalian evolution. The mammals did not immediately take over the Earth after the dinosaurs vacated the planet. Ten million years passed without much change in the degree of diversity of mammals. Then came the PETM. Those extreme environmental conditions did cause an extinction of some deep-sea microorganisms, especially foraminifera, but there was no major extinction of land plants or land animals, including mammals. Instead, there was a great diversity of plants and animals. That included the first appearance of three major types of mammals.

Two of the types were ungulates, animals with hoofs. The *Artiodactyl* were the even-toed ungulates, such as camels, deer and elk, bison, hippos, giraffes, alpacas and llamas, sheep, goats, and cattle. The *Perissodactyl* were the odd-toed ungulates, such as horses, zebras, tapirs, and rhinos. The third type were the *Primates*, a biological order that would lead to many different types of monkeys and to the great apes, which includes ourselves.

Close study of deposits that were laid down during the PETM, such as those at Fossil Butte in Wyoming, have indicated that the environmental change was not simply a smooth rise to a high temperature followed by an equally smooth recovery to a cooler one. Instead, there were several spikes and drops in temperature, which set up repeated and rapid changes in the environment that skewed genetic diversity through a bottleneck-and-flush mechanism. That was the nature of the physical forces that led to the appearance and rapid diversification of mammals. The genetic basis came from the Hox genes.

Recall that Hox genes are the part of an animal's genetic code that controls the development of an individual. For example, Hox genes determine whether a limb becomes an arm or a leg, and which type of vertebrate bone goes into which place along the spinal column.

Simple animals like sponges have a single Hox gene. Most invertebrates have twelve or fewer Hox genes, all clustered on a *single* chromosome. All vertebrates have a few dozen Hox genes clustered on *four* separate chromosomes.[18] That difference between invertebrates and vertebrates suggests the genetic code of an early shared ancestor was doubled twice—from one to two, then from two to four—to produce the vertebrates. Other studies of the genetic code of a wide variety of invertebrates and vertebrates support this idea of multiple duplications of the entire genome, or, as it is usually expressed: At least twice in the lineage that led to the vertebrates, there was a *whole genome duplication.*

According to molecular phylogenetics, a comparison of the chemical sequences in the Hox genes of invertebrates and vertebrates indicates that the two duplications that led to the vertebrates occurred about five hundred million years ago—that is, during the Cambrian Period. That is consistent with the fossil record that indicates the earliest vertebrates did appear at that time.

Whole genome duplication—the simultaneous duplication of an entire genetic code—is yet another way that life can evolve quickly. By having a duplicate set of genes, organisms within a population can vary more in form and individuals may be more resistant to environmental changes, such as those that occur during extinction events. Even so, there are few such duplications in the evolutionary history of animals—as indicated by the presence of the same four clusters of Hox genes in all vertebrates. In comparison, the genome of many plants show multiple duplications.[19] And the timing of the duplications has been related to major events in the evolutionary history of plants.

One such duplication occurred about 320 million years ago and may have led to the development of the first plants that produced seeds. Another occurred about 190 million years ago at the time when the first flowering plants, the angiosperms, appeared. A third occurred about fifty million years ago and the effects of it can be seen in the fossil beds of John Day and elsewhere in North America. This episode of whole genome duplication allowed a different type of plant to dominate the land surface: the grasses.

Grasses were the last major group of plants to evolve. They have existed since the middle of the Cretaceous Period. Evidence of their existence is found in coprolites (fossilized excrement) of dinosaurs. Some dinosaurs did graze on grass, but it was not a major part of their diet.

The dominance of grasses came about because a whole genome duplication produced a new class of grasses that had a new characteristic. The leaves of early grasses grew from the tip. The leaves of the new grasses grew from the base and they grew continuously. (This is why lawns need to be cut every week or so.) This gave the new grasses an evolutionary advantage over shrubs and trees in that they could recover and spread faster across a land surface that had been devastated by prolonged drought or by fire. They could also recover faster when heavily grazed upon by the growing number and diversity of *Artiodactyl* and *Perissodactyl*.[20]

There was at least one more major development in the evolution of grasses. The concentration of carbon dioxide in the atmosphere had been high for hundreds of millions of years, since the first appearance of animals and land plants. After the PETM, the concentration of carbon dioxide started to decrease, dropping nearly to current levels about twenty-five million years ago. When that happened, some grasses—and some other plants—modified the process of photosynthesis. For more than two billion years, photosynthesis had relied on the production of an intermediate molecule that had three carbon atoms. The modified process produced one with four carbon atoms. For that reason, plants, including grasses, that use the modified process are known as C4 plants or C4 grasses. Those that still process carbon dioxide and sunlight into oxygen and water using the older process are C3 plants or C3 grasses.[21]

The C4 photosynthetic process is more efficient than the C3 one. A C3 plant can capture and store a maximum of 4 percent of the energy received in sunlight. Some C4 plants, such as sugarcane, can capture and store as much as 7 percent. The C4 photosynthetic process also works better at lower concentrations of carbon dioxide. And so, as the concentration of carbon dioxide in the atmosphere continued to drop during the Cenozoic Era, a new ecosystem developed and soon dominated the landscape: the grassland.

From this, two questions must be asked: Why was the planet cooling? And what were the consequences of a switch from a continent dominated by forests to one dominated by C4 grasslands?

15

A Drowned River
at Poughkeepsie

Late Cenozoic Era: 2.58 Million Years Ago to the Present
Quaternary Period

Sitting at the edge of a sandy beach next to a well-maintained trail no more than two miles north of the border between British Columbia and Washington is a large boulder. Its weight is nearly five hundred tons. For years it was kept white by shellfish-eating seabirds whose guano covered the rock. Today it is kept white through monthly applications of white paint. Its prominence and its white color make it a familiar beacon for sailors who sail close to the coast.

If one scratches away the paint and whatever may remain of the guano, one will discover that this huge boulder is a granite. Then look around. There is no granite in the nearby exposures of rock. In fact, there is no exposure of granite for many miles. This rock is an *erratic*, a boulder that is found in isolation and differs from the types of rocks that are native to the surrounding area.

This particular erratic has a name, the Great White Rock, named long ago by sailors at a time when seabirds were adding their guano. The name has since been adapted by local citizens as the name for their community, White Rock, British Columbia.

This lone boulder that sits on a sandy beach at the western end of the continent is just one of hundreds of large erratics that are scattered across southern Canada and the northern part of the United States. And names have been given to them, too, in recognition of their large size, prominent locations, and the fact that they seem to be misplaced,

On the other side of the continent is Balance Rock, a famous landmark on the waterfront of Bar Harbor, Maine. Bubble Rock in Acadia National Park, also in Maine, sits on a steep slope and seems ready to roll off any second. In Rhode Island there is Cobble Rock in a forest between the two communities of Woonsocket and North Smithfield. In Massachusetts, there is Doane Rock, the largest boulder exposed along the Cape Cod National Seashore. Prayer Rock, which has life-sized hands etched on it by people who lived on the Great Plains long ago, has been moved from its original location and is now on the grounds of the public library in Ipswich, South Dakota. On Bainbridge Island in Washington, two erratics that sit at the intersection of major roads have been brightly painted and are known locally as Frog Rock and Ladybug Rock.[1]

The Brassfield Erratic is near Oregonia, Ohio, and was once more than a dozen feet thick and covered almost an acre. This giant block of limestone was chipped away for more than a hundred years, the material used in local construction. In nearby Sunbury, Ohio, a much smaller erratic of hard crystalline rock was used in a practical way as the base for a local hero, General William Rosecrans, who led the Union Armies of Ohio during the Civil War.[2]

The most famous erratic is Plymouth Rock near the site where the *Mayflower* landed in 1620. Two hundred years later, tourists who visited the site were provided hammers to chip off a piece of the rock as a souvenir. One large fragment was split off in 1834 and moved to the town square in Plymouth, Massachusetts. In 1880 it was returned and mortared back in a place and a memorial with a fence was erected around the famous stone. The smaller fragments taken by tourists are now scattered across the country and around the world. Some, undoubtedly, were discarded. Others, so the owners will say, are now in the form of paperweights, tie clips, and cufflinks.

Notwithstanding the scattering of fragments of Plymouth Rock by tourists, the challenge in geology is to determine where each of these erratics originated. The original location of Plymouth Rock was easy. It is part of the Dedham Granite that forms the basement rock south and west of

Boston. The giant Brassfield Erratic was moved more than seventy miles from where it was once firmly attached to others limestone of early Silurian age. An erratic known as the Wedgwood Rock in the Wedgwood neighborhood of Seattle has a mineral content that matches the exposure of rock at Mount Erie on Fidalgo Island in Washington, a distance of about fifty miles from where it lies today. As to the hundreds of other named erratics, and the thousands of smaller ones, though much effort has been made to identify the locations of their sources, little progress has been made, except to say that they have come from great distances.[3]

Look around where one finds an erratic or where fields of these mysterious rocks exist and one will see other unusual features. If an erratic sits on hard bedrock, that hard rock almost always has smooth contours and is covered by sets of closely spaced, parallel grooves. Several erratics lie atop the Hartland Formation in Central Park. Examine the surface of the Hartland Formation and one will see the smoothing and the grooves. Split Rock, another erratic, is in Pelham Bay Park, also in New York City, and it sits on the Hartland Formation. It seems undeniable that a connection must exist between the transport of these erratics and grooves. To be convinced—and thrilled—go to Kelleys Island in Lake Erie in Ohio. Erratics are plentiful here. So are grooves that have been dug into a limestone and, because of their size, are more properly described as troughs or as *megagrooves*, tens of feet deep, scores of feet wide, and hundreds of feet long.[4]

In addition to fields of erratics and sets of long parallel grooves in bedrock, there are also long low hills tens of feet high, a few miles wide, and tens to hundreds of miles long that run across southern Canada and the northern part of the United States. These are *moraines*. Dig into them. They consist of piles of unsorted sand and gravel. Cape Cod, Nantucket Island, Martha's Vineyard, and the Elizabeth Islands are parts of a series of moraines. Two exceptionally large moraines form the backbone of Long Island. The Harbor Hill Moraine runs along the North Shore and the Ronkonkoma Moraine runs through the center of the island through Prospect Park and through Nassau and Suffolk counties. Fort Wayne, Indiana, received the nickname Summit City because it lies at a high point along the Valparaiso Moraine. The highest point in Ohio is Campbell Hill, at 1,550 feet above sea level, along the Union Moraine.

Another form of low hill, which number in the hundreds in North America, can also be found scattered across the same part of the continent as moraines. These hills usually have a teardrop outline and when viewed from the ground, have the shape of an inverted spoon or half-buried egg and are known as *drumlins*. One of the most famous drumlins in North America, located near Manchester, New York, is the one called Cumorah. That is where Joseph Smith said he found a set of golden plates with writing on them that he translated into English and published as the Book of Mormon. So is Breed's Hill in the Charlestown section of Boston, where a monument has been erected to commemorate the Battle of Bunker Hill.[5]

One also finds long narrow valleys in the same part of the continent, such as the Minnesota River Valley and the Hudson River Valley, where the rivers contained within them seem much too small to have created the valley, no matter how much time is allowed.

There are also great drifts of material, *tills*, that are composed of a wide mixture of sedimentary material. There are silt, clay, sand, gravel, and boulders that are jumbled together. These great drifts cover most of the land surface of Minnesota, Iowa, Wisconsin, Michigan, Illinois, Indiana, and Ohio. In places, they are hundreds of feet thick. They are often difficult to distinguish and so they are given general names, such as the Des Moines Lobe, or the Southern Iowa Drift Plain, or the Illinoian Lobe.

But what could account for such unusual and diverse landforms strung across a limited region of the continent and reaching from coast to coast?

◆

By the eighteenth century, after the first cursory examinations of the geology of North America had been completed, most people accepted the fact that these various landforms were related, that they had been formed simultaneously. Moreover, most people were convinced that they knew the cause: It was the Great Biblical Flood—the Noachian Deluge, as those who were seeking a rational understanding of the Earth and its history called it—an event in which the entire planet had been flooded with water and only the highest mountain peaks had stood as dry land.

This explanation remained unchanged and unchallenged until the beginning of the nineteenth century. By then a sufficient area of the Earth's land surface had been explored and described so that it was clear that such features did not occur everywhere, that they could not have been the products of a global event. Erratics and moraines were in the northern part of the United States, but not in the southern part. Furthermore, they were also found in the British Isles and across much of Europe, especially in the Scandinavian regions and around and within the Alps, but were absent in Africa, southern Asia, and South America.

The beginning of a scientific explanation came from field observations made in the Alps. Here, for a long time, naturalists had been puzzled by the presence of large boulders of granite and schist that were lying atop hard bedrock surfaces of limestone. Jean de Charpentier, a director of copper mines and salt mines in the Jura region of the Alps, was one of those who noticed and was curious about the unusual association. One day when he was on his way to Lucerne, he met a local farmer and the two men began a discussion about the mysterious rocks. The farmer was very clear "there are many stones of that kind around here" and "they had come from far away" because "the mountains of this vicinity are not made" of the same stone. Charpentier asked him how he thought the boulders had been moved and the farmer did not hesitate in giving his answer: They had been moved by glaciers, long ago, when the glaciers were much larger than they are today.

This was quickly accepted by Charpentier and others who were familiar with the Alps as the explanation. But how large had the glaciers once been? Given this new insight, Charpentier and others soon found evidence that glacial ice had once filled entire valleys. In 1837, German naturalist Karl Schimper made the startling claim that all of Switzerland and most of Europe, northern Asia, and the northern half of North America had once been covered by thick sheets of ice. He also suggested that the ice sheets had formed multiple times. Each time that a sheet had advanced was a period when extreme cold gripped the planet, a period of time that led him to propose the term *Eiszeit*, or "ice age."

Many naturalists came to the Alps to conduct field studies and contribute their own observations to the growing idea of ancient massive glaciers and ice ages, but their work was haphazard with no one understanding how the various pieces fit together or what the implication of past ice ages had in understanding the Earth's history. And then came Louis Agassiz.

Agassiz had begun his scientific work studying fish fossils under the guidance of George Cuvier in Paris. He soon became a well-recognized authority. He was also a childhood friend of Charpentier. It was Charpentier who took him to the Alps and showed him the evidence of past glaciations.

Agassiz took up the idea of past ice ages with a passion becoming, as one biographer has remarked, a "glacial evangelist." In 1837, he assembled the observations that he and others had made and put them into a coherent form and presented them at a meeting of the Swiss Society of Natural Sciences. He explained how giant boulders (erratics) had been carried long distances by the slow flow of ice within large glaciers. He described how the movement of glaciers had caused erratics and smaller particles of hard rock to be scraped against the underlying surface of bedrock leaving large grooves and, in many cases, the much smaller scratches known as striations. He also described how the advancing front of a glacier plowed up the land and formed moraines. How the melting and retreat of large glaciers had left behind the great thicknesses of unsorted sedimentary debris known as glacial till. And how ancient giant glaciers had once covered much of Europe.

It was the first time anyone had proposed a continental theory of glaciation, one that spanned geologic time. And the reception by members of the society was hostile.

In the minds of many, including the most learned, it was too outrageous for anyone to suggest that the Earth had been so different so recently, that huge sheets of ice had recently covered large regions of some of the continents. But Agassiz continued the pursuit. And many suggested, and not always politely, that he return to a study of fish fossils. Then an opportunity came from a friend who was trying to save Agassiz and his promising scientific career.

The friend was Alexander von Humboldt, the Prussian geographer and explorer who, in the early 1800s, had traveled extensively in the Americas. Humboldt mourned the direction that Agassiz's interest had taken him. He was convinced that Agassiz's fascination and wild ideas about glaciers and past ice ages would bring professional ruin to the much younger man. And so Humboldt obtained for him the financial support from the king of Prussia for a scientific mission to North America to study the faunas and fossils of that continent and compare them with those of Europe.

Agassiz left Europe in 1846. He settled in Cambridge at Harvard University, where he gave a series of lectures and where he eventually became a professor and remained to the end of his life. For the first few years after arriving, everything went according to Humboldt's plan. Agassiz returned to a study of fossils. He traveled along the East Coast collecting for Harvard's museums. But in 1850, he went to the region of the Great Lakes and saw the multitude of evidence—here more striking than in the Alps—for recent ice ages. He placed the idea of massive continental ice sheets firmly into the realm of scientific study.

◆

Walk out over the Hudson River on the old Poughkeepsie-Highland Railroad Bridge, rejuvenated for pedestrians and bicycles, in Poughkeepsie, New York. The bridge's boardwalk hangs more than two hundred feet above the river's surface. Far to the west are the Catskill Mountains, part of the immense wedge of sedimentary material washed down from the ancient Taconic Mountains. To the east, in the city of Poughkeepsie, is College Hill, a drumlin that has maintained a hint of its original teardrop shape, one of thousands of similar ice-age features scattered across North America. Now look down beneath the bridge. Here is a geologic oddity: a drowned river that runs in two directions.

When the great sheets of ice moved southward from the high-latitude regions of North America, as first clearly envisioned by Agassiz, they scoured out several large basins that are now filled with water: the Great Lakes. The rock that lay beneath the western half of Lake Superior was somewhat less durable—it was part of the lavas erupted from the Midcontinent Rift—than the rock elsewhere, and so the sheets of ice dug deeper here, making this area of the Great Lakes deeper than the rest. The sheets of ice also dug into the ground and formed the elongated depressions now filled by the waters of the Finger Lakes of western New York. And they dug down and carved out the 100-mile-long Hudson River Valley.

As the sheets of ice advanced, eventually covering most of the northern half of North America and much of northern Europe, they also grew in volume, the water for the ice coming from the oceans. That, of course, caused the level of the oceans to drop several hundred feet and the shorelines of the continents to move outward by many miles. When the ice sheets were at their greatest extent and their largest volumes, the Atlantic coastline was more than a hundred miles from where Sandy Hook in New Jersey or Rockaway Beach on Long Island are today.

The planet did warm again and the sheets of ice did retreat. As they retreated, ice dams formed at some of the natural gaps between rocky prominences, and large lakes of melted ice formed behind the dams. In northern New York a large ice dam formed at what is now the northern edge of Lake Champlain, a massive wall of ice preventing water from running eastward. And so a huge glacial lake formed. This ancient lake, known as Lake Iroquois, was positioned in the basin where Lake Ontario is today, though it grew to be more than three times the size of the modern lake. The ice dam did collapse near present-day Rome, New York. The ridges of sand west of the city probably formed at this time as water poured out, first surging down the valley of the Mohawk River, then down the Hudson River Valley.

The surge was stopped temporarily when the water reached what is now New York Harbor by a huge earthen dam formed by the Hill Harbor Moraine, which then extended from Staten Island across New York Harbor to Long Island.[6] But the stoppage was short-lived as water eroded through the earthen dam and again surged across the landscape for another hundred miles until it reached what was then the Atlantic coastline and poured into the ocean.

As the water poured out of Lake Iroquois and down the Hudson River Valley, the erosive power of the water deepened the valley. And so, when the ice sheet retreated far to the north and enough water returned to the oceans to raise sea level to its present position, the lower stretch of the Hudson River Valley was flooded by seawater, drowning the Hudson River.

Today the ocean's tidal pulse is felt all the way to Troy, New York, more than a hundred miles from New York Harbor. Put a stick in the water's edge anywhere between the City of New York and Troy and one can watch the twice-daily rise and fall of the river's surface. Look down from the pedestrian bridge at Poughkeepsie and one is likely to see a sinuous line that separates the freshwater that is flowing southward from the saltwater that is flowing northward. Follow the slow movement of a log floating on the river's surface. If the weather is calm, the movement of a log or debris will change direction from going downriver to upriver then downriver twice a day. The Hudson River is a river that can flow in two directions. The reversal of direction is a consequence of an ice age.[7]

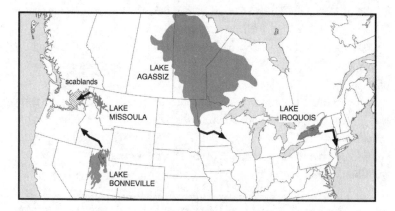

Four of the large lakes that formed during the retreat of the latest
large ice sheet. The scablands of eastern Washington were scoured
by repeated floodwaters released from Lake Missoula.

Several large meltwater lakes formed along the edge of the ice sheet as it retreated. The shapes of these lakes changed greatly, and so did their volumes. Lake Iroquois was, at most, modest in size comparable to a series of truly great glacial lakes that once existed in the region of the modern Great Lakes. Among these, Lake Algonquin was the largest. It blanketed all of the region that today is covered by both Lake Michigan and Lake Huron and much more. Most of the Upper Peninsula of Michigan was flooded by the waters of Lake Algonquin, as was much of the land where the cities of Green Bay, Milwaukee, and Chicago are today. And, yet, this was not the largest of the great glacial lakes that formed in North America. That record-breaking distinction belongs to the largest freshwater lake or inland sea that ever existed on the continent—Lake Agassiz.[8]

At its maximum, Lake Agassiz covered much of Manitoba, parts of eastern Saskatchewan, northern Quebec, western Ontario, a sliver of eastern North Dakota, and much of northern Minnesota. Along its western, northern, and eastern shorelines was a high wall of ice. The main outlet for much of its existence was along the southern shoreline where the massive waters of this ancient glacial lake were held back by a hundred-foot-high earthen dam formed by the Big Stone Moraine. The dam eventually failed and the flow of water out of the lake cut down through the earthen dam and a wide gap formed, Traverse Gap, that lies along the border between South Dakota and Minnesota at Browns Valley.

The volume and rate of flow of water through this gap were staggering. At the time the dam was breached, Lake Agassiz held almost twice the volume of water contained in all of the Great Lakes today. The sudden surge of water spread out across the countryside. It eroded deep into the loose glacial till and produced a long and wide spillway, the Minnesota River Valley. Eighty miles from the Big Stone Moraine and Traverse Gap, the fast-flowing water cut down to the hard ancient rock of the Morton Gneiss.

When it reached the Mississippi River, near present-day Fort Snelling, just south of Minneapolis, the rushing water deepened that river, and eroding along the steep banks of the river caused several waterfalls to form along tributary streams.[9] A colossal waterfall also formed along the Mississippi River near St. Paul, Minnesota. The width and the height of this waterfall were nearly identical to those of Niagara Falls, though up to ten times more water once cascaded over these falls on the upper Mississippi River than has ever been seen to fall at Niagara.

It took several years, perhaps a few decades, for Lake Agassiz to empty of water. During that time, the flow rate of water was comparable to the discharge of freshwater from all of the world's rivers today. It would have been a spectacular event to see. The impact that so much fast-flowing water had on the development of landforms across the central part of North America is still being evaluated. Nevertheless, the rate that water poured out of Lake Agassiz was relatively small compared to the great torrent of water—perhaps the *greatest* torrent of water to ever surge across the Earth's surface—that poured out of Lake Missoula.

The city of Missoula in western Montana lies on the floor of this ancient glacial lake. The lake water there was about a thousand feet deep. A series of horizontal lines etched into a west-facing hillside marks the strand lines and beaches made by wave action on the lake.[10]

The volume of water in Lake Missoula was never more than one-fifth the volume in Lake Agassiz. The reason for the difference in the intensity of the flooding is due to the nature of the dams that were breached. At Lake Agassiz, it was an earthen dam about a hundred feet high. The gap that eventually opened was about a thousand feet wide. At Lake Missoula, the dam was a lobe of ice that extended across what is now the valley of the Clark Fork River in northern Idaho. Along the eastern edge of the lobe where the water was being held back, the dam rose as a wall of ice more than two thousand feet high.

When the water reached a critical depth so that the eastern end of the ice dam became buoyant and started to rise, tunnels opened along the base of the dam. Water flowed into those tunnels, weakening the dam. When the dam eventually broke it was catastrophic.

The gap formed was several miles wide. Water surged through it. It raced southward down the Purcell Valley then made a right turn and crossed Rathdrum Prairie and headed for where the city of Spokane is today. From there, the water began to spread across eastern Washington, carving valleys, forming giant potholes where the water swirled, and leaving behind giant flood bars and giant ripple marks.[11]

The flow of water slowed as it reached and piled up into a new lake behind the first major constriction at Wallula Gap near the twin cities of Richland and Kennewick, Washington. No more than several hours had passed between the breaking of the ice dam and the buildup of water behind Wallula Gap. It took about two weeks for the water to drain through the gap.

Then a second constriction reduced the flow at another narrow point at The Dalles. From there, the water funneled through and cut a channel through the Columbia River Gorge, eroding the sides of the gorge and leaving hanging streams and forming high waterfalls. Six-hundred-foot-high Multnomah Falls east of Portland is the most famous of these.

The water crested over Crown Point, also east of Portland. It washed up against a high ridge on the north side of the gorge and flowed through a narrow notch and the churning of water created an elongated lake known as Lacamas Lake.

West of Portland, the flow of water encountered the third and last constriction, a mile-wide gap near Kalama, Washington. That caused the water to back up a second time, this time into the Willamette Valley as far south as Eugene. What is now downtown Portland was then four hundred feet under water. Many hills in the Portland area were flooded; only the tops of Rocky Butte and Mount Tabor remained dry.

Giant icebergs rafted on the fast-moving water as it flowed across eastern Washington and into and through the Columbia River Gorge. Within some of these icebergs were giant rocks that had been picked up in Canada and deposited in Montana. When the water finally receded and the icebergs settled, then melted, the large rocks were left sitting on the land surface. Several hundred can be seen scattered today across the well-maintained fields of the Willamette Valley. A fifteen-ton, iron-nickel meteorite—now known as the Willamette Meteorite—that was found near West Linn was once among them. It had fallen in Canada long ago, was transported to western Montana by the ice sheet, then, still encased in ice, was caught in one of the icebergs that floated down during one of the Missoula Floods, ending its long journey in the Willamette Valley.[12]

A few weeks were required for these floodwaters to recede. Then, over time, another lobe of ice closed the gap at the Clark Fork River in northern Idaho. And another glacial lake formed in western Montana. The refilling may have taken as little as sixty-five years. And then this ice dam broke and another great flood of water flowed across eastern Washington and into the Columbia Gorge and out to the Pacific Ocean. At least forty times this happened over a period of a few thousand years. The last great flood from Lake Missoula happened about twelve thousand years ago.

◆

So far the story has been about a recent age when a thick sheet of ice covered the northern half of North America. But a basic question needs to be addressed: What happened that the Earth became cool enough for ice to cover the poles and high mountains?

Throughout the Mesozoic and the early Cenozoic eras—that is, the last 250 million years—the planet was without polar ice caps. Two of the warmest periods in the planet's history occurred during the Cretaceous Period when *Tyrannosaurus rex* raged across the land and when a pulse of carbon dioxide entered the atmosphere and the great lakes of Wyoming formed and the extremely hot PETM period happened. Then, within a few tens of millions of years after the PETM peaked, two tectonic events started the planet on a long and slow Cenozoic cooling.

One of these events was the collision between India and the southern part of Asia. The collision forced up the Himalayan Mountains. And it was the rise of those mountains—and the newly exposed rock in the steep mountain cliffs—that increased the rate of physical and chemical weathering, which in turn drew down the amount of carbon dioxide in the atmosphere and caused the planet to cool.

The other event was the movement of Antarctica over the South Pole and its isolation from the other continental masses. Antarctica and Australia had been connected, but their separation caused a waterway, the Tasmanian Gateway, to form. Antarctica was also connected, though only by a thin strip of land, to South America. That, too, was severed and the Drake Passage formed. The isolation of Antarctica became more severe now that it was surrounded by a circumpolar sea. The warm-water currents of the Pacific and the Atlantic could no longer reach it, and Antarctica cooled faster than the rest of the planet. That was thirty-five million years ago.

But the collision of India and Asia and the isolation of Antarctica were not enough to cool the planet and allow an ice cap to form. Another element came into play: the rapid expansion of grasslands and the simultaneous development and spread of large mammals.

Thirty-five million years ago the concentration of carbon dioxide in the atmosphere was about 1,500 parts per million (ppm),[13] about the same as it had been throughout the Cretaceous Period. By twenty-five million years ago—after the collision of India and Asia and the isolation of Antarctica—the concentration had dropped to about 700 ppm. That is a threshold. Now C4 photosynthesis was favored over C3 photosynthesis. And that allowed an expansion of C4 grasses and of grasslands.

Forests were now in retreat and grasslands began to cover vast areas of the planet, as they do today. The African veldt, the Asian steppes, and the South American pampas developed. So did the North American prairie.

And as grasslands began to dominate, forest-dwelling mammals, which had been surviving on leafy trees, evolved into prairie-dwelling animals that grazed on grass. Now that the protection of a forest was gone, some of the mammals evaded predators by growing to enormous sizes, such as the rhinoceros, the hippopotamus, and the elephant. The largest land mammal ever to exist, *Megacerops*, appeared then and may have been twice as large as the modern elephant. But the main strategy to evade predators was speed. Fast-moving large mammals, such as bison, camels, cattle, horses, and zebras, evolved. In response, the mammalian predators grew larger and faster. This included *Arctodus*, the short-faced bear, the largest-ever carnivorous land mammal, which stood as high as twelve feet and weighed up to a ton, and *Amphicyon*, or "bear-dog," a giant wolf-like predator larger than any of today's bears. And there was a multitude of cats. Most famous among them were the saber-toothed cats with long, daggerlike canines.[14]

And so there was a parallel development in the evolution and expansion of grasslands and in the evolution and diversity and size of mammals. And, from that, came the ice cap that now covers Antarctica.[15]

But the northern hemisphere was still mostly clear of ice. Then a third tectonic event entered into the planet's history and caused the climate to cool quickly.

◆

The key event that led eventually to a series of ice ages was the creation of the Isthmus of Panama and the closure of a seaway between the Pacific and the Atlantic. This happened about 2.5 million years ago. Warm tropical water could no longer be exchanged between the two oceans. Instead, warm water was transported northward, activating the Gulf Stream, which brought higher amounts of moisture to higher latitudes and produced more snowfall.

And so ice ages came to the northern hemisphere. From that, great sheets of ice grew southward across North America and retreated. A record of this is found in the annual accumulation of sediments in ponds and on the seafloor and in the slow growth of stalagmites in caves. It is also found in the pattern of snow and ice layers in the polar regions themselves.

Holes have been drilled into the ice caps that cover Antarctica and Greenland and long cores of ice have been retrieved. These cores have been subjected to much examination. Hundreds of thousands of individual layers have been identified. Each one corresponds to an annual accumulation of snow and ice.

Tiny bubbles trapped in the layers are actual samples of the air from past ages. And from those bubbles it has been possible to reconstruct a history of the atmosphere, in particular, a history of the past concentration of carbon dioxide. In the water molecules themselves, within the snow and ice, is a history of past global temperatures determined from slight variations in the amounts of hydrogen and oxygen isotopes within the water molecules. In all, from the pond sediments and the stalagmites and especially from the ice cores, a remarkably precise and highly detailed history of climate has been constructed.

This history of climate shows that the rate of Cenozoic Era cooling did increase 2.5 million years ago at the time the seaway between the Pacific and the Atlantic closed and the thin strip of land known as the Isthmus of Panama finally connected North America and South America. That marks the beginning of the current geologic period, the Quaternary. And as the accelerated cooling commenced, an ice cap did form over the Arctic and grew and retreated and, as it did, there were wild swings in the concentration of carbon dioxide in the atmosphere and in global temperature. Each swing corresponds to an individual ice age. And each swing came with nearly clockwork regularity, suggesting that some force was at work that triggered an ice age, then caused the ice to retreat.

At first attempts were made to explain the regularity by invoking familiar geologic processes, such as suggesting a regularity in the eruption of volcanoes, but these were unsuccessful. Eventually, an answer was found, not in Earth's internal processes but in its orbit in space.

As the Earth moves in its orbit around the Sun, its orbit is constantly being perturbed by the slight gravitational tugs it receives from the Moon and Jupiter and the other planets. In particular, the slight amount of elongation to the Earth's orbit—that is, the *eccentricity*—is always changing. So, too, is the direction of its spin axis, which has a slight wobbling motion, or a *precession*, similar to the wobbling motion of a spinning top. These two slight motions, changes in eccentricity and a constant precession, mean that all regions of the Earth do not receive exactly the same amount of solar radiation. Instead, there is a periodicity to the amount of solar radiation and, hence, to the amount of solar heating different parts of the Earth receive.

All of this can be calculated. The first to do so was Serbian astronomer Milutin Milankovitch, who began a series of laborious hand calculations in 1911 and completed, at least this first series, in 1920. These calculations showed the amount of solar radiation by latitude and by season over the entire Earth over the last several hundred thousand years. He improved on his initial calculations and completed a second series in 1941 and published them in a now famous book, *Canon of Insolation of the Earth and Its Application to the Problem of the Ice Ages*. In that book he introduced a third type of orbital variation, a variation in *obliquity*, the angle of tilt of the Earth's axis.

His efforts showed that there were three periodicities—now known as Milankovitch cycles—in the amount of solar radiation that the Earth receives. The shortest is due to precession and has a periodicity of 23,000 years. Obliquity has a periodicity of 41,000 years. And due to the fact that eccentricity is highly variable, it has a periodicity of about 100,000 to 400,000 years.

The recent state of the planet—that is, the most recent series of ice ages, detailed in the ice cores—is a complex interaction of all three periods. And so whether there has been twenty to forty or more ice-age cycles is in the eye of the beholder. What is clear is that a hundred-thousand-year cycle has been the dominant one for the last million years or so. And it is due to slight variations in the eccentricity of the Earth's orbit. And so that is what is seen in the three most recent glacial lobes—Des Moines, Southern Iowa, Illinoian—a hundred-thousand-year cycle in the ice ages.

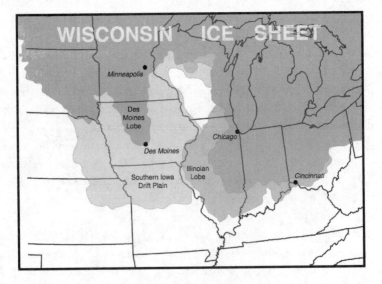

Lobes of ice created by successive ice ages.

The most recent glacial maximum—that is, the farthest extent of the ice sheet, the sheet that formed the Des Moines Lobe, dropped the Great White Rock and the Plymouth Rock erratics, produced glacial moraines on Long Island and the drumlins of Cumorah and Breed's Hill—occurred eighteen thousand years ago. That was followed by a general warming of the planet, one occasionally interrupted by sudden periods of cold and brief advances of the ice sheet. Then, according to the ice cores, there was a sudden increase in global temperature, an increase that can be timed to have occurred, within a hundred years or so, to an age of 11,700 calendar years before the year 2000.

It marks the end of the last cold episode, and so it has been given special significance: It marks the division of the Quaternary Period into two epochs. The earlier one, the Pleistocene Epoch, runs from 2.58 million years to 11,700 years ago. It is the epoch of the ice ages. After it is the Holocene Epoch, identified as a location in a borehole, NGRIP2, at a depth of 4,896 feet 5.9 inches (1492.45 meters) on the central Greenland ice sheet.[16]

That is the beginning of the current geologic epoch, the beginning of "recent time."

◆

It is often misunderstood that the Earth was more geologically active—with more cataclysmic earthquakes and more colossal volcanic eruptions—in the distant past than today. That is not true. Earthquakes and eruptions comparable to the largest of the past have happened recently and will continue to happen in the future. Consider the largest earthquake to strike North America in the last several hundred years.

Cascadia is that region of the continent where the Juan de Fuca Plate, a remnant of the Farallon Plate, is sliding down into the Earth. The land area of Cascadia is generally regarded to be the southern part of British Columbia, all of Washington and Oregon, and northern California. At either end of Cascadia, the Pacific Plate makes direct contact with the North American Plate and two major fault systems have developed.

The San Andreas Fault system stretches from Cape Mendocino on the northern California coast to the Salton Sea in southern California. Major earthquakes occurred along this system in 1857 and 1906. Both were magnitude 8 events. The other fault system, the Queen Charlotte Fault system, runs from Dellwood Knolls off the coast of Vancouver Island northward and into Alaska. The most recent major earthquakes along this system occurred in 1949 and 2012. And both were magnitude 8 events.

A magnitude 8 earthquake is a major event. Damage can be severe. The 1906 earthquake demolished much of San Francisco and led to a fire that engulfed most of the city. But the impact of these recent events pales when compared to the severity of the ground shaking and the level of destruction caused by a magnitude 9 earthquake.

The most recent magnitude 9 earthquake occurred on March 11, 2011, in the Tohoku region of western Honshu, Japan. It produced sea waves that reached heights of more than a hundred feet. It killed nearly thirty thousand people and damaged nuclear reactors. The previous magnitude 9 earthquake occurred in 2004 off the western coast of Sumatra. It, too, produced large sea waves. The death toll from that earthquake was over two hundred thousand. And previous to that a magnitude 9 earthquake struck in 1964 in Alaska, where the ground shaking lasted five minutes, and in 1960 in Chile. Both events produced large sea waves that spread across the Pacific Ocean and left wide areas devastated from the intense ground shaking. Early in the year 1700 an earthquake of similar magnitude to these struck Cascadia.[17]

Though there are no written records of the earth shaking, the date is known because there are written records in Japan of a giant sea wave rushing up along the Japanese coastline at that time and flooding hundreds of miles of coastline, sweeping away small boats and devastating villages. This giant wave was known as the "orphan" tsunami because no earthquake could be linked to it as a source of the wave.[18] But a connection was made with a Cascadia earthquake when the time of death of "ghost" forests along the coastlines of Washington and Oregon was finally determined.

These ghost forests are stands of dead trees that died suddenly by being flooded by seawater.[19] The growth rings of these trees showed that they died during the winter months of 1700. In support of that, stories told by Indigenous people whose ancestors lived in the region for many generations tell of a time when "the land shook (and) a big wave smashed into the beach" and when "there was a great earthquake and all of the houses . . . collapsed." Put this together with the date of the orphan tsunami and the latest magnitude 9 earthquake in Cascadia occurred on January 26, 1700.

When might the next one occur? To be clear, earthquake prediction is not possible. But there are curious events—discovered only recently and repeating like clockwork—that happen beneath Cascadia and might lead to one.

The motion of the Juan de Fuca Plate is not head-on with the edge of North America. It is angled a bit to the northeast. And that slight angle is causing Cascadia to rotate, ever so slowly clockwise.

The pivot point is in northeastern Oregon. All of Oregon and most of Washington and the southwestern part of Idaho are rotating around this point as a rigid block, sweeping through one angular degree every million years. Such rotational motion is slow, but Astoria, Oregon, is moving a fraction of an inch farther north each year than Bend, Oregon.

And, at the same time, there is another and faster type of motion across Cascadia. Imagine living in Seattle or Portland or Baker City or Spokane or anywhere in Cascadia. Because the Juan de Fuca Plate is moving ever so slowly under the North American Plate, the ground beneath your feet is moving ever so slowly eastward about two or three inches every year. But, every fourteen months, there is a reversal. And the ground in Seattle and Portland and elsewhere in Cascadia lunges back westward about an inch. The lunge takes a few days, maybe a week or two, then the ground surface resumes a slow eastward movement at a rate of two or three inches a year.

This reversal is the surprising part. When it happens, there is also an increase in seismic rumbling deep beneath Cascadia—or as seismologists say, an increase in "seismic tremor." It is the regularity of the reversal that surprises and bewilders seismologists. Nature usually does not behave this regularly. And that has opened up the door on whether this periodic movement *might* be related to a buildup to the next catastrophic magnitude 9 earthquake beneath Cascadia. Such a pattern was seen before the 2011 Tohoku earthquake in Japan.

◆

Earthquakes are part of the Quaternary Period history of Cascadia, and so are volcanic eruptions. And if one was to choose a "volcano city" in North America, it would be Portland, Oregon.[20]

The oldest volcanic rocks in the Portland area are exposed on Elk Rock Island in the Willamette River near Milwaukee, Oregon—thirteen small acres of black basalt that are part of the Siletzia Terrane that collided with and accreted to North America about fifty-five million years ago. Jump forward forty million years and the massive lava flows of the Columbia River Basalt are surging through the Columbia River Gorge and ponding and congealing in the Portland Basin. The remnants of several thick lava flows can be found in the Tualatin Hills west of Portland and in the steep cliff face at Cape Horn east of Portland.[21]

The series of eruptions that gives Portland the status as a volcano city began 2.6 million years ago. The Boring Volcanic Field formed from these eruptions, the name derived from a group of hills capped by volcanic rocks near the community of Boring southeast of Portland. Within this field are more than eighty small volcanoes scattered throughout the Portland-Vancouver metropolitan area of northwestern Oregon and southwestern Washington. That most people are unaware that the greater Portland area was recently a vast field of volcanic cinders and cones is due to the great erosive power of the Missoula Floods that swept much of the evidence away.[22]

The eruptions began near Oregon City and the lava that was erupted forms the Highland Butte. That was followed by a hiatus of almost a million years. When eruptions resumed, the renewed activity occurred over a wide area north and east of Oregon City.

The volcanic rocks that form the prominent features of Rocky Butte and Mount Tabor erupted at this time. These were short eruptions that lasted a few weeks to a few years.[23] Eruptions that lasted years to decades formed Mount Scott (where the rim of a thousand-foot-wide crater can still be seen), Mount Sylvania, and Larch Mountain in Oregon, and Prune Hill and Mount Norway in Washington.

The youngest feature in the Boring Volcanic Field is the rocky monolith known as Beacon Rock that towers above the northern shoreline of the Columbia River in the Columbia River Gorge. It was named in 1805 by members of the Lewis and Clark Expedition and was one of the landmarks anxiously anticipated by travelers along the Oregon Trail in the 1840s because it meant that their journey to Portland and the Willamette Valley was near an end.

Geologically, Beacon Rock is the hard rock that solidified beneath a large cinder cone that erupted fifty-seven thousand years ago in the Boring Volcanic Field. Most of the cinders and other loose material were eroded away about fifteen thousand years ago by the Missoula Floods. Cinders in the tree roots near the summit are evidence that Beacon Rock was not completely covered by floodwater.

People who live in the greater Portland area live among and commute around the scores of volcanic relics that are the Boring Volcanic Field. If questioned about volcanic rocks, they will invariably ignore what is in their immediate vicinity and point to the great volcanic peaks that dot the horizon. Directly east of Portland is Mount Hood. Visible to the north are Mount St. Helens and Mount Adams. Journey into central and northern Washington and one finds Mount Rainier and Mount Baker. Far south of Portland in California is Mount Shasta. And between Mount Shasta and Portland is the missing volcanic peak of Mount Mazama, today's Crater Lake. These are the Cascade Volcanoes.

In broad terms, there is a link between the Cascade Volcanoes and the scattered volcanic cones and craters of the Boring Volcanic Field. Both are, ultimately, the consequence of the sliding of the Juan de Fuca Plate beneath North America. As the plate slides downward—in fact, as any oceanic tectonic plate slides into the Earth—it carries not only oceanic crust, but also water-rich sediments. Here water plays a crucial role in lowering the melting temperature of the hot mantle. At a depth of about sixty miles the interaction of the cold oceanic plate and the hot mantle and the effect of the water causes molten rock to form. The molten rock is less dense than the surrounding mantle and so it rises, feeding molten rock to the Cascade Volcanoes and to the Boring Volcanic Field.

What type of molten rock is produced, whether a fluid basalt or a viscous, granitelike rock, determines what type of volcanic feature will be formed and how violent an eruption will be. Eruptions of fluid basalt are relatively quiet and produce cinder cones and lava flows. Molten rock that is granitelike in composition may explode violently and produce hot ash or may ooze out slowly as a volcanic dome.[24]

The Boring Lava Field is mostly basalt. The tall volcanic peaks of the Cascades grew mainly by the buildup of volcanic ash and the coalescence of volcanic domes. The larger ones, Rainer, Adams, and Shasta, are the oldest:

Those volcanic peaks began to grow about four hundred thousand years ago. Baker and Hood are somewhat smaller and significantly younger, about two hundred thousand years old. And Mount St. Helens, the smallest and the youngest, has an eruptive history that goes back a mere thirty thousand years.

And all of these volcanic peaks are active. Mount St. Helens famously exploded in 1980, and eruptions continue intermittently. Early settlers in the Portland area reported fire, smoke, flying rocks, and voluminous steaming from Mount Hood in 1859 and 1865. Mount Rainier and Mount Shasta have each erupted several times in the last few thousand years. The most recent eruption of Shasta was in 1786 and was witnessed by French explorer Jean-François de la Pérouse from a ship off the California coast.

The most recent major eruption in the Cascades was several thousand years ago. It came from a volcano that in modern times is known as Mount Mazama, the name derived from an American Indian word meaning "mountain goat." Before the eruption, the summit peak probably rose to higher than 12,000 feet, but following a catastrophic eruption this was reduced to the much more modest 8,157 feet that is the elevation of the highest point today.

Mount Mazama did not explode outward, but dropped downward. Molten material from a shallow chamber erupted suddenly from the base of the volcano. That reduced structural support for the overlying mass of the volcanic peak, and the entire volcano collapsed catastrophically downward. What remained was a deep crater, today's Crater Lake, the deepest lake in North America.

Just as the intensity of earthquakes is measured by an earthquake magnitude scale, the intensity of volcanic eruptions is measured by an explosivity index. The explosivity index takes into account a variety of parameters, such as the volume of material erupted and how fast it was erupted and the volume of ash produced. And just as an increase of one value on the earthquake magnitude scale means the intensity has increased many times—it is a logarithmic scale—an increase of one value on the volcano explosivity scale means the intensity of an eruption has increased many times.

The slow and steady outpourings of lava in the Hawaiian Islands have an explosivity index of 1 to 0. An eruption that produced a feature like Mount Scott in the Portland Basin probably had an index of 3. The explosion of Mount St. Helens in 1980, which blew down hundreds of square miles of forest and caused volcanic ash to drift down out of the sky as far east as Minnesota, had an explosivity index of 5. The eruption that destroyed Mount Mazama and resulted in the formation of Crater Lake and the spreading by air of ash across half of the continent had an index of 7.

And yet, many larger eruptions, much more explosive and more destructive, have occurred on North America. They had an index of 8. They are described, in more of a Hollywood fashion than poetically, as "megacolossal." And, at least in recent geologic time, they originated from a so-called "supervolcano"—Yellowstone.

◆

The immense size of Yellowstone is difficult to comprehend. It does not seem to be a single volcano like Mount Hood or Mount Rainier, or the volcano Vesuvius in Italy or Mount Fuji in Japan. And yet it is.

The many fields of geysers and boiling mud pots are one measure of the size of the Yellowstone volcanic system. And beyond them are the boundaries of a national park that is larger in area than the state of Connecticut. And Yellowstone extends far beyond those boundaries.

Look at the mountains of the Wind River Range that are southeast of Yellowstone National Park. Those mountains were lifted upward during the Laramide Orogeny. Follow the crests of those mountains. They slope down to the southeast away from Yellowstone, a suggestion that a second, more recent episode of uplift has added to the height of the Wind River Range and an uplift that was centered somewhere near Yellowstone.

Now look at the mountains of the Absaroka Range along the eastern boundary of Yellowstone National Park. These mountains were also lifted upward during the Laramide Orogeny. But that was long ago and erosion should have smoothed those mountains down. Instead, some of the steepest and fastest eroding hillsides anywhere in the Rocky Mountains are found in the Absaroka Range, another hint of a more recent and localized uplift.

Now drive around Yellowstone and follow the rivers. The Yellowstone, the Boulder, the Stillwater, the Shoshone, and the Wind Rivers all flow radially *away* from Yellowstone National Park. That is actually easy to explain: A great bulge has risen and rain and melting snow collect into streams, then rivers flow downward off the bulge in radial directions. But how wide is this bulge? Look at a map of the pattern of rivers. The bulge is more than two hundred miles across, greater than the distance between Portland and Seattle, or between Dallas and Houston, or between New York and Boston.

Now run off to the west of Yellowstone and into Idaho to the long, wide plain known as the Snake River Plain. This is where the famous Idaho potatoes are grown. Examine the mountains north of the plain and pay attention to the rivers that run through them. The Salmon, the Lemhi, and the Pahsimeroi Rivers are all flowing northward. But look at some of the deposits in the rivers. Those deposits indicate that, long ago, each of these rivers flowed southward.[25]

This tilting of mountain ranges and the reversal of river directions has been sorted out: The wide bulge at Yellowstone has been creeping across North America, moving from southwest to northeast. But there is one more crucial element that is needed to understand it.

From Ashton, Idaho, which is just west of Yellowstone, take the long drive that runs southwestward through the cities of Idaho Falls and Pocatello to Twin Falls. Beneath the fields of potatoes and the thick soil are volcanic rocks, erupted from a series of Yellowstone-like volcanic centers. Continue southeast from Twin Falls and climb up onto the Owyhee Plateau. More volcanic rocks, this time exposed in grand style on the surface. Now measure the ages of these various volcanic rocks.

There is a clear age progression. On the Owyhee Plateau they are sixteen million years old. Moving back toward Yellowstone, beneath Idaho Falls the volcanic rocks are ten million years old. Then, in succession, eight million years old at Pocatello, five million years old at Idaho Falls, and two million years old at Ashton.

The bulge and the progression of volcanic ages associated with Yellowstone are reminiscent of another major feature on the planet: the Hawaiian Islands.

The main Hawaiian Islands extend for more than three hundred miles. The oldest is Kauai. Then, in order from the next oldest to the youngest, is Oahu, Molokai and Lanai, Maui, and the island of Hawaii.[26] The island of Hawaii is much larger than the other islands for two reasons: This is where volcanic activity is happening, and so the island continues to grow, and this island lies atop a great bulge in the Earth, a bulge that, like Yellowstone, is hundreds of miles across and at its highest point rises about a mile above normal seafloor.

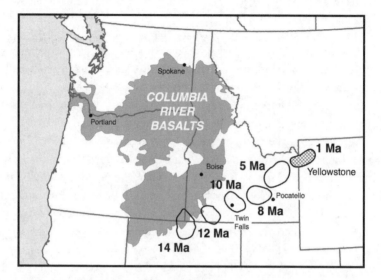

The Columbia River Basalts and the migration of the Yellowstone hotspot.

The age progression has a simple explanation: The Hawaiian Islands are on the Pacific Plate, which is moving slowly to the northwest. As the plate moves, it slides over a hot plume that is rising up through the Earth's mantle. The plume is continuously supplying magma, which, because the Pacific Plate is moving, erupts and forms a succession of islands. The same is happening at Yellowstone—but with an important difference.

At Hawaii, the magma passes up through oceanic crust, which is basaltic in composition, and so there is little change in the composition of the rising magma, leading to eruptions with an explosivity level of 1 or 2. At Yellowstone, the magma passes up through a continental crust, which is close to granite in composition, and so the rising magma mixes, becomes more granitic, and eruptions are highly explosive, level 7 or 8—the most explosive eruptions on the planet.

Three such eruptions have occurred at Yellowstone in the last few million years. Each one sent clouds of hot ash rolling across the landscape at high speed and shot vast amounts of ash into the air.

The first of the three occurred 2.1 million years ago and the most recent 640,000 years ago. Each time a massive amount of ash erupted, the ground surface above a huge magma chamber that was feeding the eruption collapsed, and so three large volcanic calderas formed. They are nested together and the high walls that were formed are easy to see. The trail from Canyon Village to Observation Peak climbs one of the walls. The eighty-foot-high cascade of water known as Gibbon Falls drops down another section of caldera wall.

These eruptions not only sent massive amounts of hot volcanic ash surging and churning across the landscape, but also blasted huge volumes of ash high into the air that settled down over large sections of North America. In southern Idaho the accumulation of Yellowstone volcanic ash is more than fifteen feet. Several feet accumulated across the middle section of the continent in Colorado, Nebraska, Kansas, Oklahoma, and northern Texas. In Elko, Nevada, there is as much as ten feet of Yellowstone ash. And in parts of southern California, several hundred miles from Yellowstone, more than two feet of volcanic ash fell.[27]

The geyser field where Old Faithful is located is a product of much later volcanic activity. About 160,000 years ago a huge block of hot rock began to rise in what is now the southeastern part of the national park and the Mallard Lake Dome began to form.[28] The arching profile of this dome is the backdrop to the scene at Old Faithful. As one watches the periodic displays of Old Faithful, one should be aware that the broad volcanic dome behind it is also rising by minuscule amounts. In fact, the entire region of Yellowstone is doing a complex dance. Some areas will rise by several inches over a few years, then they will fall and will rise again. It is all part of the fascination that is Yellowstone.

16

A World Bequeathed and the Great Acceleration

The Anthropocene

Of the many specialists in the field of geology, the paleontologist is the most interesting to talk to, capable of describing fantastic previous worlds inhabited by barely believable creatures. The seismologist seems to be the most self-assured, always armed with vast arrays of numbers that are used to discover the many unseen mysteries that lie within the Earth. The volcanologist tells the most adventurous and wildest tales, of which some might be true. But it is the person who seeks mineral ores who is the most passionate, convinced that in another year—or perhaps tomorrow—unbelievable riches may be found beneath one's feet.

I had a pleasant dinner one evening with such a person. He described how he was leaving the next morning and would be traveling to a place that was far away. Other seekers of mineral wealth had clamored over the ground before him, he said, but they had missed the obvious. He was confident that he would find what others had been unable to see. In fact, there were probably riches lying right in plain sight and scattered across the ground surface and one needed to do no more than pick up what one wanted. I was intrigued. I got excited. I was ready to jump up on the table and say, "Why wait until morning? Let's leave tonight!" But I did not. We retired to our respective places and I went to sleep. I woke up the next morning in a more sober mood. My friend had departed. I never heard from him again.

When one mentions mineral riches, one immediately thinks of gold. In ancient Egypt, it was the "flesh of the gods." To the Incas of South America it was derived from the "sweat of the sun." And to the Aztecs of Mexico this most malleable of all metals was the "excrement of the gods."

Gold is easy to reshape, does not corrode, and is fairly rare. That makes it ideal as a medium for money. It has been used in the treatment of syphilis, heart disease, smallpox, and alcoholism without positive results. It has been used successfully in the treatment of arthritis. Gold is often regarded as a symbol of high attainment, given as Olympic medals, Nobel Prizes, Oscars, and Emmys. It is pursued, often with reckless abandon. And multiple times in history societies of people have stopped what they were doing and changed course to follow a gold rush.

The first successful rush for gold in North America followed the discovery of a seventeen-pound gold nugget in North Carolina in 1799. The discovery was made by a twelve-year-old boy who showed the heavy yellow rock to his parents and his seven brothers and sisters. None of them knew what it was, but they saved it and used it as a doorstop. Years later, the father sold it to a jeweler. Word was now out and people started coming to North Carolina to dig and pan for gold. Mines were established and North Carolina became the center of gold production in the United States and it remained so until 1849 when word arrived of a greater discovery of gold in California.

More than a hundred thousand people went to California to search for gold along the western slope of the Sierra Nevada. Another rush came in 1898 when gold was reported along the Klondike River in the Yukon Territory in Canada. And a fourth major gold rush, known as the Porcupine Gold Rush, began in 1909 when fortune-seekers poured into Northern Ontario into what is now known as the Abitibi Gold Belt centered on Timmins, Ontario. As with almost any story about gold, there was irony: Those people who headed to California by hurrying across the continent traveled through regions where there was more gold to be found than has ever been recovered in California. This includes the Comstock Lode that lies beneath Virginia City and the "invisible" gold along the Carlin Trend near Elko, both in Nevada.

But why is gold found in places that have had such different geologic histories as North Carolina, California, the Klondike, northern Ontario, and Nevada, as well as in the Black Hills of South Dakota, near Pikes Peak in Colorado, and near Butte, Montana? And what does it mean in the quest to understand the geological history of North America?

An old adage is still true: You discover gold wherever you find it. Gold is rarer than other metals, and so it takes considerable more effort to find it.[1] And it is found in many different geologic settings because there are different ways to concentrate gold. But there is one factor that is common to all: The formation of a gold deposit requires the circulation through rock of very hot water that picks up and carries gold atoms from a very large region and, by the sudden cooling of the water, drops the atoms and concentrates them in a small area.

The formation of gold deposits is not evenly distributed in geologic time. A major episode occurred 2.7 billion years ago during the Super Event of the Archean Eon when much of the continental mass that is seen today was created. The gold deposits of the Abitibi Gold Belt formed at that time. The richest periods of lode, or orogenic, gold—that is, gold deposits that formed during mountain-building events—occurred 1.8 billion years ago during the formation of the first supercontinent, Columbia, and since 0.8 billion years ago—that is, during the Phanerozoic Eon during the breakup of the second supercontinent, Rodinia, and during the formation and breakup of the third, Pangea. These deposits include those of the Mother Lode that sparked the California Gold Rush and those that led to the Klondike Gold Rush in the Yukon.

Gold can also be concentrated on the seafloor where hot water is circulating through the recently erupted lavas along the crest of the Mid-Ocean Ridge. Such gold is associated with sulfur and includes the gold of North Carolina. And then there are microscopic particles of gold, the so-called "invisible" gold, that can only be found through chemical analysis and is being recovered along the Carlin Trend in Nevada. Here huge blocks of sedimentary rocks have slid against each other along long faults and hot water has circulated along the fault lines. Where the hot gold-rich water came from is still debated. But its circulation through these faults produced one of the richest gold mining districts in the world. Eighty percent of all of the gold now mined in the United States comes from the Carlin Trend.[2]

The gold in Colorado also concentrated along fault lines, but in a very different geologic setting. It and many other valuable mineral ores are found along the Colorado Mineral Belt that runs from the Four Corners region in Arizona to the Front Range of the Rocky Mountains just west of Denver. The origin of this mineral-rich belt has been a long-standing problem because its trend cuts across the geologic grain of Colorado; that is, it runs across the range of mountains produced by the Laramide Orogeny and the trends of much older features that formed as far back in time as the Archean Eon. But a solution might now be at hand given what has been revealed recently by seismic tomography and by an unraveling of the history of the Farallon Plate.

The details are, admittedly, still being worked out. But it seems that the alignment and the great enrichment in mineral ores—billions of dollars of gold, silver, lead, zinc, copper, and molybdenum have already been extracted—might be related to a tearing apart of the Farallon Plate as part of the Laramide Orogeny. The tear would have allowed fluids and hot mantle material to rise up between the two pieces of the plate, carrying gold and other metallic elements close to the surface, where they were then concentrated by the circulation of hot water. It remains to be seen whether this idea gathers additional support. But it does illustrate that something as fundamental to geology as the search for mineral ores—which was one of the reasons that geology initially developed as a science—is still evolving.

◆

The pursuit of gemstones has also undergone a dramatic change. Jade, which like gold forms by the circulation of hot water, is now seen as a product of the subduction of oceanic plates. Find an ancient subduction zone along the western edge of North America and jade might be present. Rubies and sapphires grow under slightly different conditions of pressure and heat than jade—though also by the circulation of hot water—but the exact conditions required for rubies and sapphires to form have been unusual in the history of North America, and compared to other regions of the world, such in Southeast Asia, few have been found on North America.[3] And then there are diamonds.

As mentioned in an earlier chapter, a diamond is a high-pressure form of carbon that cannot form at or near the surface of the Earth. Diamonds can form only at pressures that correspond to depths greater than about ninety miles. They are literally blasted out of the Earth from these great depths at high speeds, rising up to the surface in less than a day, by a rare type volcanic explosion known as a *kimberlite eruption*, named after one of the world's most famous and productive diamond mines in Kimberley, South Africa. Hundreds of such eruptions have occurred in North America, though fewer than 1 percent of them have had the right eruptive conditions to bring up diamonds. One of these diamond-rich eruptions did occur about a hundred million years ago along the southern edge of the Ouachita Mountains in Arkansas.

For a modest fee one can spend an entire day sifting through barren and leached soil in search of diamonds at a seventy-three-acre site at Crater of Diamonds State Park near Murfreesboro, Arkansas. Diamonds have been found here since 1906 when a local farmer found two on the surface. Several attempts have been made to mine diamonds, but each one failed. The quality of the diamonds is low. Yet, hundreds of people come each day to search. And one or two diamonds are found each week, usually about the size of the head of a pin.[4]

There is no obvious sign of a volcanic feature here. There is no crater or raised rim. The evidence is in the rock fragments that one finds while sifting for diamonds. These include a dark basalt-looking rock known as *lamproite* that melted directly from the mantle at depths that were many times deeper than the magma supplied to most volcanoes.

The diamond-rich crater in Arkansas is at the southern end of a wide corridor of hundreds of kimberlite craters that run from Somerset Island at the northern edge of Canada, into Saskatchewan, through Montana and Wyoming and Kansas, and finally into Arkansas. This includes the recently discovered fields of diamonds, and the multiple mines that have been excavated, in the North Slave Region of the Northwest Territories. Because this corridor roughly aligns with the position of the Western Interior Seaway and because most of these kimberlite eruptions occurred during the Cretaceous Period, one feels compelled to associate the origin of these kimberlite eruptions—and diamonds in Arkansas—with the history of the Farallon Plate. But it is not quite that simple.

One of the most mysterious things about diamonds and kimberlite eruptions is that almost all diamonds were formed deep inside the Earth more than a billion years ago, and, yet, almost all of the diamonds that have been and are being mined were erupted and brought up to the surface during a specific episode of the Earth's history, during the Cretaceous Period between about 140 and 66 million years ago. Without this brief episode of kimberlite eruptions, the surface of the Earth would be almost devoid of diamonds. And that illustrates a major point that has been made repeatedly in this book.

Much of our plant's history has been punctuated by colossal events, of which some have been catastrophic. The diamonds in an earring or a ring, a sapphire or a ruby that highlights a necklace, the gold in a bracelet are all indicators of the Earth's dynamic past. Each one is as clear an indicator of that dynamic past as are the many thick ash deposits that surround Yellowstone or the great blocks of granite that have been shoved along the San Andreas system of faults from the Sierra Nevada to the San Francisco Bay Area. Stand on the pink granite at the top of Pikes Peak in Colorado and one can quickly imagine how an event more than a billion years ago caused rock deep in the Earth to melt, then to form the granite, and how some tens of millions of years ago the slow motion of the Farallon Plate produced the Laramide Orogeny and caused the Front Range to rise. Or follow the Niagara Escarpment from Green Bay, Wisconsin, to Rochester, New York, and stop at Niagara Falls and see how it is yet another indicator of the dynamism. Or consider that a slab of cold material is peeling away and dropping into the hot mantle and, thus both the Sierra Nevada and the modern topography of the Appalachian Mountains has formed. Or stand in front of the great collection of dinosaur bones exhumed at Dinosaur National Monument and compare them with the mammalian skeletons unearthed at John Day. Then rub your finger along the thin band of gray clay at Trinidad Lake in southern Colorado and understand how for days, perhaps weeks, fire and ash rained down from the sky after a giant asteroid stuck the planet along the northern coastline of the Yucatán Peninsula.

Here the story of North America and of the great forces and the great events that have formed and shaped the continent could end. And this is how the story would have ended if this book had been written a decade or so ago. But the science of geology continues to evolve. Ideas continue to be challenged and to be fine-tuned. From this, one gains a better understanding of how the continent—and the Earth, at large—came to its current state. And that has introduced a new question: What role have the activities of human beings played in the recent geologic evolution of the planet?

◆

Here is the history of the Cenozoic Era so far: After the demise of the dinosaurs and the pulse of heat indicated by the PETM, the Earth began to cool slowly. The cooling was probably a response to the end of land bridges that connected Antarctica to South America and Australia, causing Antarctica to be surrounded by a circumpolar sea that isolated it from warm ocean currents, thereby setting the conditions for the growth of a massive ice sheet, and to the great uplift of mountains from the Alps to the Himalayas by the collision of Africa with Europe and India with southern Asia. The exposure of fresh rock faces in those mountains fostered an increase in the rate of weathering that lowered the amount of carbon dioxide in the atmosphere and contributed to the cooling.

In response to the cooling, grasslands expanded rapidly across the continents, including North America. Mammals that had adapted for life in forests now evolved into ones that could survive in those grasslands. In part, unable to hide in forests to escape predators, those mammals became larger and were able to run faster than their predecessors. The predators also grew larger and could run faster. And so a diverse megafauna developed. Within the megafauna were three major types of mammals: even-toed and odd-toed ungulates and primates. It is from those primates that the giant apes evolved. And one type of those giant apes, which first appeared as a distinct species about three hundred thousand years ago, was *Homo sapiens*, modern human beings.

Great sheets of ice began to develop over the northern regions about 2.5 million years ago. And so when human beings first appeared, they were largely prevented from moving into North America from their origin in Africa and their expansion across Europe and Asia until the retreat of the most recent ice sheet. The oldest evidence found of human beings living across most of North America dates back twelve thousand years ago. It was also about then that many of the mammals of the megafauna that lived in North America disappeared in a very short time, most of them within just a thousand years. The list of those lost includes the woolly mammoth, the mastodon, two species of buffalo, several kinds of giant ground sloths, a large bear, and several kinds of horses, as well as at least one species each of moose, antelope, and deer.

This coincidence in time and space of the first major dispersal of human beings across North America and the rapid extinction of much of the megafauna naturally leads to the proposal that it was the actions of human beings that caused the extinction. It is estimated that just before the extinction, about a hundred million large mammals lived and thrived in North America. And so there should be ample evidence of mass butchery. A search has been done, but few sites have been found.

There are scores of Pleistocene- and Holocene-age sites of human occupation in North America. And, yet, only about a dozen sites contain evidence that early humans were directly involved in the deaths of large mammals that are now extinct. Big Bone Lick in Kentucky is one of the most famous and most accessible of the Pleistocene-age sites. Here there is evidence that mastodons and humans overlapped in time; however, other than a single bone fossil that has a possible cut mark, there is no clear proof that humans hunted or butchered animals at the site.

At Kimmswick, Missouri, near the confluence of the Mississippi, Missouri, and Illinois Rivers, there is ample evidence for this activity. Stone tools have been recovered that were used to hunt and butcher a mastodon cow and calf, a ground sloth, and white-tail deer, as well as a wide spectrum of amphibians, birds, fish, reptiles, and small mammals including rabbits and rodents. Near Wally's Beach in Alberta there is a site where camels were butchered. At Lange-Ferguson in South Dakota, ancient hunters left evidence of the killing and butchering of two mammoths, one adult and one juvenile. And near Murray Springs in Arizona there is evidence that hunters killed at least one, but perhaps two mammoths, at least eleven bison, and possibly one horse. But such sites are few in number. And so it is likely that there was another cause for the megafauna extinction.[5]

A possible one is rapid climate change. As the most recent ice sheet was retreating and as the planet was warming, there was an abrupt reversal in the climate and the ice sheet again started to advance and the planet began again to cool rapidly. Known as the Younger Dryas Event,[6] this cooling period lasted about a thousand years and occurred at about the same time as the megafauna extinction. The trigger was probably one, or perhaps several, of the great floods that flowed from Lake Agassiz down the St. Lawrence Seaway and into the Atlantic Ocean, where the cold freshwater mixed with salty seawater. This mixing may have shut down the Gulf Stream that was carrying warm tropical waters to the North Atlantic. Whatever the trigger, the Younger Dryas Event is a sudden cooling of the Earth and may have been responsible for the megafauna extinction. But there is also a problem with this explanation. Though the megafauna extinction was almost global—it did not seem to affect greatly large mammals in Africa or southern Asia—the extinction did not occur simultaneously in North America and in Europe and northern Asia.

So there is still a debate as to the cause of the sudden disappearance of such a wide variety large mammals in North America and why there were some— bison, caribou, deer, elk, moose, pronghorn, mountain goat, mountain sheep, and musk ox—that survived. Human beings may have played a role, for example, by destroying habitats, but there is not yet clear evidence for this.

Human beings have had an impact on what followed. After the Younger Dryas Event abruptly ended 11,700 years ago, there was a period of a few thousand years of global warming, followed by a remarkably long period, one that leads up to the present time, of climate stability. This long-term stability is almost certainly due to agriculture.

Agriculture originated independently in several regions. The two earliest developments were about twelve thousand years ago in the Fertile Crescent region of Mesopotamia at the eastern end of the Mediterranean and in the Yellow River Valley in north China. By eight thousand years ago, agriculture had also started in Central America, in the Peruvian Andes, and along the tropical belt of Africa. It is from these regions that agriculture expanded.

The expansion of agriculture required the clearing of additional land, usually by the cutting and burning of forests and grasslands. Irrigation systems were constructed. Animals were domesticated and tended. All of this activity led to an increase in the emission of carbon dioxide, an increase that is clearly seen in ice cores.

Eight thousand years ago the concentration of carbon dioxide in the atmosphere was 260 parts per million. By five thousand years ago, it was 270 ppm, a small but significant increase. By a thousand years ago, it was 280 ppm.

The variations in the Earth's orbit around the Sun that have been linked to the advance and retreat of ice sheets during the last few million years—the Milankovitch cycles—were still happening. Calculations show that these variations could have been such that a new ice age should have started about five thousand years ago, and that a sheet of ice should be covering part of Canada today. How large the sheet should be is a matter of debate. But no such sheet of ice exists. Instead, northern Canada is barren of permanent ice. Also the climate of the planet has been stable during the last several thousand years.

The stable climate reflects a counterbalance between a natural slow cooling of the planet that has been going on for most of the Cenozoic Era and the recent slight warming caused by humans and agriculture. In other words, it seems that the slow production of carbon dioxide through human activity during the last several thousand years has stopped a glaciation!

A stable climate means little or no change in sea level. It also means repetitive seasonal weather. And because weather patterns are repetitive, there is little change in the type of vegetation that covers a region—including the type of plants that can be grown successfully through agriculture—and the type and number of animals that can be hunted or that can be raised and tended through farming. In short, the stable climate of the last several thousand years was one of the factors that allowed the rise and the development of civilizations.

Then, in the mid-eighteenth century, human beings started a new activity—industrialization.

◆

Twenty miles southwest of Salt Lake City is the Bingham Canyon Mine. It is the deepest open-pit mine in the world. It is currently more than three-quarters of a mile deep and more than two and a half miles wide. It has been in operation for more than a century. There, large amounts of copper have been extracted as well as appreciable amounts of gold, silver, molybdenum, platinum, and palladium, each one essential to the consumer-driven desire for ample electricity and for electronic devices. But for now, focus on the size of the hole.

Look at the Grand Canyon one more time. It took a few million years for the waters of the Colorado River to erode and move material to form the Grand Canyon. Apply the amount and type of equipment now being used at the Bingham Canyon Mine and the number of people and, using this method of digging, the Grand Canyon could be excavated in a few hundred thousand years. That is still a long time period when compared to a human lifetime, but it is barely the blink of an eye in geologic time.

Now consider the level of human activity that is happening on a global scale.

Human beings are now moving ten times more rock and soil and sediment, both intentionally through mining and construction and unintentionally by accelerating erosion through agriculture and urbanization, than is moved by all of the Earth's rivers combined. Ninety percent of all of the concrete that has ever been used to construct buildings, roads, dams, tunnels, and docks has been used in the last seventy-five years, the length of a single human lifetime. Ninety percent of all of the fertilizer that has ever been used has also been used in the same time period, doubling the production of food and greatly increasing land, water, and air pollution. Almost every major river system has been redesigned by reservoirs, levees, and dams. Human activity has been re-forming the Earth's surface. It is now a major geologic force.[7]

Human activity has also changed the content of the biological world. In 1900, the combined weight of all domesticated animals was about three times as large as that of all wild mammals. Today it is twenty-five times larger. The total weight of chickens is now three times the total weight of wild birds. Three animals—cattle, humans, and chickens—are now the planet's dominant forms of large animal life.

As the surface of the Earth has been re-formed and as the dominant form of large animal life has shifted, there has also been a dramatic increase in the production of salt, aluminum, and iron, in water usage, and in the production of paper. The size and needs of cities have increased greatly. So has the ability of people to transport themselves and to communicate. This episode of rapid change that has occurred geologically, biologically, economically, and culturally, much of it happening in the last seventy-five years—the "Great Acceleration," as it is often called—came about because human beings were able to harness and utilize more energy.

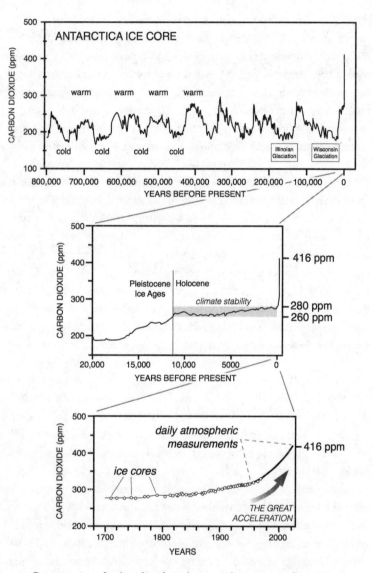

Concentration of carbon dioxide in the atmosphere, measured in parts
per million (ppm). Daily measurements of the concentration of carbon
dioxide have been made since 1958. Earlier measurements are of air
bubbles trapped in ice caps in the Antarctic and in Greenland. (Data
from the US National Oceanic and Atmospheric Administration).

Most of the energy that has driven the Great Acceleration has come from two sources: coal and oil. This is energy that was captured by plants and photosynthesizing microorganisms from sunlight, then stored within those plants and microorganisms, which were then buried, encasing the captured energy within the Earth for eons. The release of this energy has had untold, and seldom recognized, benefits, such as leading to the development of medicines and medical treatments that have greatly reduced human suffering. It has also had a detrimental side. Most of the mercury in the air and in lakes, rivers, and oceans came from the burning of coal. Essentially all of the lead that is in the air and that is inside our bodies came from the burning of oil. The great increase in the amount of waste and pollutants produced during the Great Acceleration was owed ultimately to the great amount of energy released by the burning of coal and oil. But the greatest consequence seemed, at first, to be rather benign: It was the burning of huge amounts of fossils fuels that has dramatically increased the concentration of carbon dioxide in the atmosphere.

The burning of coal and oil occurs by the combining of oxygen in the air with the hydrocarbon molecules contained within the fossilized plants and microorganism and results in the release of energy from the molecules and the production of carbon dioxide. Carbon dioxide is, of course, essential for photosynthesis. All animal food is derived ultimately from the products of photosynthesis. And it plays another role: Carbon dioxide keeps the planet from being frigid and encased in ice.

Carbon dioxide is a greenhouse gas. Its presence in the atmosphere acts like the panes of glass in a greenhouse in that it allows the short-wavelength visible sunlight to shine through, but blocks the longer-wavelength infrared light given off by the planet's surface from radiating back into space. The effect is to trap the Sun's energy and insulate the planet.

The greenhouse effect was first described in the 1820s by French physicist Joseph Fourier, who noted that the planet's temperature was a balance between the energy received from the Sun and infrared radiation emitted back into space. In 1859, Irish-born physicist John Tyndall identified carbon dioxide as a greenhouse gas. And in 1896, Swedish chemist Svante Arrhenius brought the two ideas together and showed through a series of calculations that an increase in atmospheric carbon dioxide caused by the burning of fossil fuels would be enough to cause global warming.

The first measurement of the increase of carbon dioxide in the atmosphere was made in the late 1950s at a new global weather station established on the high slopes of the volcano Mauna Loa in the Hawaiian Islands. But what was the cause of the increase? Was it due solely to the burning of fossil fuels or might it be a natural fluctuation in the concentration of carbon dioxide in the atmosphere and the increase due to the emission of the gas from natural sources?

Those questions were clearly answered in the 1980s when it was determined that the increase in carbon dioxide was mimicked by an increase in light carbon atoms—that is, isotopes of carbon-12—in the atmosphere. This was the human-caused fingerprint because photosynthesis in plants and microorganisms takes up more carbon-12 than carbon-13 atoms, and so, when the molecules of fossilized plants and microorganisms are released by the burning of fossil fuels, the atmosphere becomes slightly enriched in carbon-12 atoms. But was there a corresponding increase in global temperature as predicted by Arrhenius? It required another two decades to decide the question.

In 1999, one of the most important—and most maligned—diagrams ever made in the history of science was first presented. It showed the history of surface temperature in the Northern Hemisphere for the last millennium. From a thousand years ago to about 1900, there had been a slow cooling. Since 1900, there had been two large steplike increases, the first lasting from about 1900 to about 1930 and the second beginning in about 1970 and continuing today.[8]

When it was first presented, this diagram was dismissed by both scientists and nonscientists who questioned whether there was sufficient data and data of adequate quality so that an accurate diagram could be made. Since then, additional research has challenged the original diagram. It has been refined, but mostly it has been supported by reams and reams of additional data that confirm the original conclusion: During the last millennium, a general cooling of the planet occurred until the twentieth century. During the twentieth century and continuing into the twenty-first century, there has been a dramatic increase in global temperature.

It also took decades to confirm that this increase in global temperature was due to an increase in the concentration of carbon dioxide in the atmosphere, and, hence, caused by the burning of fossil fuels and other human activities. In 1990, the first report issued by the Intergovernmental Panel on Climate Change, a panel of scientists and policymakers charged with evaluating the causes and degree of climate change, concluded that human activities were "substantially increasing" the concentration of greenhouse gases. In the second report, issued in 1995, the panel suggested there was a "discernible human influence on global climate." In the third report, issued in 2001, there was "stronger evidence" of human-caused climate change. In the fourth report, in 2007, it was "likely" that global warming was caused by human activities. And in the fifth and most recent report, released in 2014, it is definitive: The activity of human beings, mainly through the burning of fossil fuels, is "the dominant cause of the observed warming since the mid-twentieth century."[9]

A rise in global temperature is not the only evidence that the climate is now changing. There has been a dramatic reduction in the area covered by the Arctic ice sheet and a dramatic retreat of mountain glaciers. All of the glaciers in the North Cascades of Washington State are now in retreat. The terminal end of the Columbia Glacier near Valdez in Prince William Sound in Alaska has retreated more than nine miles in the last twenty-five years. Grinnell Glacier in the heart of Glacier National Park in Montana has receded so much that it now covers only half the acreage it did just fifty years ago.

Sea level has also been rising due in part to the melting of the polar ice sheets—both the Arctic and the Antarctic—and to the thermal expansion of warming seawater. Except for the most recent decades, the rate of sea-level rise during most of the last several thousand years has been less than a hundredth of an inch per year. But it has accelerated significantly during the most recent twenty-five years. Sea level is now rising globally at more than a tenth of an inch per year, a seemingly small rate until it is multiplied over time periods of several decades. Then it becomes significant, increasing the impact of ocean storm surges and the occurrence of high, or king, tides and, due to the intrusion of saltwater, decreasing the quality of freshwater taken from coastal plains. This has been true on the Sacramento-San Joaquin Delta, the Louisiana Delta in southern Florida, and especially along the broad low-lying areas that line Chesapeake Bay.

Global warming has also caused a significant change in weather patterns, in both extremely wet and extremely dry weather. California's rapid shift from severe drought to great amounts of precipitation and wide-spread flooding during the winter of 2016 and 2107 is just one compelling example. The severity of a drought now occurring across the Missouri River basin is not only unprecedented over the last two centuries but, as indicated by an examination of recent lake and river sedimentary deposits, has not occurred over this region of the planet in the last millennium.

The tendency is to ascribe such changes to natural variations in weather patterns, variations that first occur across one region of the planet, then another. Such shifts have been quite common and have been documented throughout most of the historical period.

The most famous such shift is the so-called Little Ice Age. It began as a prolonged period of unusually cold weather over the central and eastern Pacific Ocean during the fifteenth century as indicated by a slow growth of coral and of trees, the latter indicated by the close spacing of tree rings. By the sixteenth century the central and eastern Pacific had recovered and a prolonged period of extreme cold began to grip Europe. Its occurrence is amply documented in diary entries and newspaper accounts of severely cold weather. Government reports tell of massive crop failures due to the cold. And numerous paintings of landscapes and cityscapes indicate that this was a period of heavy ice and massive snowfall.[10] Then, by the late eighteenth and early nineteenth centuries, a period of cold weather had settled over North America.

A period of shifting warm weather has also been documented, first in the Pacific in the tenth century, then over North America in the eleventh century, and finally across central South America in the thirteenth century. The main point is that these periods of unusual cold or warmth shifted and occurred regionally, not simultaneously across the planet. That is not what is happening today.

The warmest period of the last few thousand years has occurred during the last several decades. And it is occurring in all regions. What appear to be regional variations in weather patterns are actually part of a global shift in weather patterns—that is, a change in climate.

Much has been made—at times, in an alarming and exaggerated way—about the possible effects of the warming. The current warming of the planet is a direct consequence of the dramatic increase of carbon dioxide in the atmosphere. And the increase is caused mainly by the burning of fossil fuels, which means it is one of the outcomes of the Great Acceleration.

But how to determine the future? One measure can be gained by looking into the past and considering how the Earth responded when large amounts of carbon dioxide were suddenly injected into the atmosphere.

◆

Such sudden injections of carbon dioxide have happened during the eruption of flood basalts. The release of massive amounts of carbon dioxide during the eruption of the Wrangell flood basalts in the Yukon and British Columbia led a prolonged period of wet climate and the Carnian Pluvial Event. The evidence for this is seen in the petrified forest in Arizona. The eruption of the Siberian Traps caused such a major change in climate that it led to the Permian extinction and to a planet that was barren for tens of millions of years. The eruption of the Deccan Traps and the climate change it produced led to another mass extinction and was followed by another period when the Earth was covered by a barren landscape. The effect of the eruption of those flood basalts is complicated by the formation of Chicxulub Crater, which was produced by a large meteor striking the Earth in a shallow sea and from the energy of the impact vaporizing a large amount of limestone and injecting a large amount of carbon dioxide into the atmosphere. The impact also injected large amounts of fine debris that provided a temporary reflective blanket over the Earth causing the temperature of the planet to cool.

Because there are such complications and because flood basalts are erupted over time periods that are measured in tens of thousands of years, such sudden events are not good analogies for the current rapid rise in the concentration of atmospheric carbon dioxide. But there is a past event that might be.

To recall, the PETM was caused by a sudden emission of a huge amount of carbon dioxide—the source of the carbon may have been the sudden melting of a form of ice known as methane clathrate that lies in large quantities on the ocean floor—that lasted a few thousand years, perhaps as little as a few hundred years. What makes it a possible analogy to today's buildup of carbon dioxide is that the average rate that carbon was being released fifty-six million years ago during the beginning of the PETM was about the same rate as carbon is being released today.[11]

As this pulse of carbon dioxide was being released, average global temperatures increased by about 8° Fahrenheit (about 4° Celsius) and remained high for the next 150,000 years.[12] The equatorial region of the Earth became too hot. There was, however, no mass extinction. The rise in temperature was slow enough and there were apparently insufficient impediments to stop the migration of animals. Instead, major northward and southward migrations took place as the planet warmed. Mangrove thickets and rainforests occurred as far north as Wyoming and as far south as Tasmania. In what is now Canada arctic turtles, hippopotami, and palm trees were common.

The increase of carbon dioxide in the atmosphere caused an increase of carbon dioxide to dissolve in the oceans. And that led to an acidification of seawater. So much so that there was a major diversification of many major marine groups, such as echinoids, bivalves, gastropods, sharks, and bony fish. On land, as already told, the extreme heat of the PETM eventually led to one of the great diversifications of mammals—that is, the origin of the odd-toed and even-toed ungulates and the primates.

At the time when the global temperature was at its highest during the PETM, the concentration of carbon dioxide in the atmosphere was about 2,000 parts per million (ppm); that is, out of a million molecules in the atmosphere—mostly nitrogen and oxygen—about 2,000 of them were carbon dioxide. As the Cenozoic Era cooling resumed, by thirty-five million years ago, when a permanent ice cap started to formed over Antarctica, the concentration was about 1,000. When about 2.5 million years ago, when the Isthmus of Panama formed and ocean water was no longer circulated between the Atlantic and Pacific Oceans and when an ice cap started to form over the Arctic, the concentration was about 400 ppm. And it continued to drop.

During the last period when ice sheets were at their maximum extent over North America, eighteen thousand years ago, the concentration of carbon dioxide in the atmosphere was 180 ppm. Then came the Holocene warming and the stabilization of the atmosphere by agriculture. Between about eight thousand years ago and up to the beginning of the Industrial Age in the late eighteenth century, the concentration fluctuated between 260 and 280 ppm.

Consider, one last time, the four men portrayed at Mount Rushmore. What was the condition of the planet when each of these men was alive? When Washington and Jefferson were living, the concentration of carbon dioxide in the atmosphere was about 280 ppm, at the high end of what it had been for previous eight thousand years. Aristotle, Cleopatra, and Genghis Khan all breathed the same air as Washington and Jefferson. By the time of Lincoln, the concentration of carbon dioxide had increased slightly to 285 ppm. By the time of Theodore Roosevelt, it was at 295 ppm.

By the middle of the twentieth century it had reached 310 ppm and by the end of the century it was at 370 ppm. As of 2020, it stands at 410 ppm, a 50 percent increase since the start of industrialization. There is more carbon dioxide in the atmosphere than there has been for the last three million years. And most of the increase has occurred during the last seventy-five years, during the length of a human lifetime, during the Great Acceleration.

If the emission of carbon dioxide continues at the current rate, the concentration will reach about 900 ppm by the end of the twenty-first century. The global temperature will increase by about 8° Fahrenheit (about 4° Celsius), similar to the temperature rise during the PETM. The global climate is beyond the point where the Arctic ice cap will no longer be perennial. By the end of the twenty-first century, the ice cap that covers Antarctica will be on the verge of permanently disappearing.

Typical temperatures in the equatorial regions will then be above 140° Fahrenheit (60° Celsius), too hot for most plants to grow and too hot to be habitable by most animals. The equatorial regions will be barren. But will this lead to a major mass extinction?

The extinction rate has increased during the Great Acceleration, mainly due to a loss of habitat. The extinction rates of birds and of mammals are up. So are the extinction rates of many reptiles and amphibians. But there is a bias here. These extinctions have been mostly of large animals, which are not able to adapt as quickly to environmental changes as some small animals, and of animals that are isolated on islands where there is less opportunity to migrate. When considered as a whole, the extinction of modern animals is far less than the 75 percent threshold that was used originally to define a major mass extinction.

That is not to say that the current rates of extinction are not a concern. Major changes have occurred in the chemistry of ocean water, mainly from the pouring in of great amounts of fertilizers and a rise in acidity caused by an increase in the amount of dissolved carbon dioxide, which is a direct result of the increased concentration of carbon dioxide in the atmosphere. These changes in chemistry mean it is more difficult for corals, plankton, and creatures with shells, like clams and oysters, to extract carbonate and build their skeletons.

And there has not only been a great loss of habitat on land, but also restrictions on the ability of animals to migrate in response to climate change. Such restrictions have been created by the expansion of cities and urban areas and the construction of dense networks of roadways and water systems. The poleward migration of animals has essentially been stopped.

Whether the planet is now on the edge of another major mass extinction is still undecided. But this does illustrate just how dramatic the changes must have been to produce the major mass extinctions seen in the fossil records.

What has happened is that the many rapid changes to the environment during the Great Acceleration have taken the climate out of the long period of stability that had occurred over the previous several thousand years, a stability that was, in part, a consequence of the development of agriculture.

In short, the global environment that we are confronting and that our descendants will have to deal with will be dramatically different from the one that allowed the rise of civilization. Rates of precipitation have already increased and patterns of precipitation have already changed.

But have the changes that have already occurred been sufficiently dramatic and will they have an impact far enough into the future to declare the beginning of a new geologic epoch?

◆

In 2000, Paul Crutzen, an atmospheric chemist of considerable renown, was attending a scientific conference and was becoming increasingly impatient with the speaker. Five years earlier Crutzen had been awarded a Nobel Prize for showing that the emission of industrial gases was causing the ozone layer to thin, thus increasing the amount of ultraviolet radiation that was reaching the Earth's surface and threatening life.

The speaker was repeatedly referring to the present time as the Holocene Epoch, which, as formally approved by members of the International

Commission on Stratigraphy, was the correct designation. But Crutzen lost his temper. He stood up and interrupted the speaker and said, "Stop saying the Holocene! We are not in the Holocene anymore." Then he paused. And with another burst of temper said, "We are living in the . . . the . . . the Anthropocene!" There was silence in the room. Then as the meeting continued, and ever since, scientists have been discussing exactly what Crutzen meant.

The Anthropocene is literally the "Age of Humans." Crutzen later tried to clarify what he meant by saying the Anthropocene is that time when "what is happening on the Earth is strongly determined by what humans do." When pressed for a starting time, he suggested that the beginning of the industrial age in the mid-eighteenth century might be appropriate, though he admitted that the effects of industrialization where, at first, slow to spread and did not have a global beginning.

Other beginnings have been proposed. Perhaps the beginning of the Anthropocene should be marked by the beginning of agriculture several thousand years ago. That human activity seemingly left a global mark by delaying the onset of the next ice age. Or, perhaps the beginning should be placed at the completion of Columbus's first voyage across the Atlantic Ocean. That event was quickly followed by an exchange of animal and plant life between two hemispheres and by a global migration of people who introduced and applied new technologies and who, to the misfortune of many, caused the collapse and rise of new civilizations by the introduction of disease. Or, maybe the beginning should be marked by a more recent event. The Great Acceleration is a possible recent event because it has resulted in so much human-caused modification of the planet.

Whatever event might be chosen for the beginning of the "Age of Humans" it must have left a clear stratigraphic horizon in the rock record, one that is readily detectable if the Anthropocene is to be included in the Geologic Time Scale and take its place as a continuation of the recent series of geologic epochs already established—that is, Miocene to Pliocene and Pleistocene to Holocene. In that, the beginning of agriculture and Columbus's first voyage both fail. So does the original proposal made by Crutzen to mark the beginning of the Anthropocene at the beginning of the industrial age. It is the Great Acceleration that offers the most abrupt and greatest change in the rock record, one that could be recognized tens or hundreds of millions of years from now. And so it is worthwhile to explore a bit more how abrupt and widespread this change has been.

Consider how the fossil record will look to future geologists who examine strata from the twentieth century and that are being laid down today. They will notice not only an abrupt change in the type of animal fossils but also where they are found.

Humans and their domesticated animals now dominate the planet. The total weight of humans and their animals is at least twenty-five times greater than the total weight of all wild animals. In terms of numbers, chickens by far dominate the land surface. There are about eighteen billion of them, compared to almost eight billion humans and about 1.4 billion cattle. When looking at what is strolling across the land surface today, the Earth has truly become a planet of chickens, people, and cows.[13]

Considering chickens alone, their fossils may be a marker for the Anthropocene. The body size and body shape of today's "broiler chickens" are distinct from that of wild fowl and of their early-twentieth-century cousins. The worldwide distribution and massive population size will probably ensure that they will be well represented in the future fossil record. Future geologists might use a "chicken signal" to indicate that the stratum they are examining is close to the start of the Anthropocene.

Because the cause of death and the subsequent burial of humans and domesticated animals are different from that of wild animals, the types of geologic fossils that will be formed from them will also be different. Most fossils of wild animals are where sedimentary rocks have accumulated, which are usually areas associated with water—lakes, wetlands, tidal basins, and rivers. The majority of dead livestock and human remains are disposed by burial in landfills, mass graves, and cemeteries. The locations of landfills, mass graves, and cemeteries are specially chosen not to be in areas where water is accumulating or flowing. And so this, too, could be used as a sign of the Anthropocene by future geologists.

When examining the strata that are being laid down today, future geologists will also notice an abrupt modification in the pattern of sediments that has been caused by the construction of large dams and by the modification of the flow of rivers and tidal water by levees and other diversions. A study of the geochemistry of those strata will also reveal unusually high amounts of trace metals such as silver, chromium, copper, nickel, and lead. There will be complex chemicals that have never before appeared in the geologic records, such as polycyclic aromatic hydrocarbons and polychlorinated biphenyls—more commonly known today as PAHs and PCBs—that have come from the use of pesticides and can ultimately be tied back to the consumption of oil and its many byproducts.

There will be unusual trace fossils. Large areas of the ocean floor have been churned repeatedly by bottom trawling of the deep sea, a favored method used today by commercial fishing companies to catch large quantities of fish. There will also be an unusual number of boreholes that begin in strata of the Anthropocene and penetrate through a thick series of much earlier strata. In North America there are more than three million such holes, drilled in the search and for the production of oil. The average length of these holes is about a mile and a half. And so the total length of all holes drilled in North America for the search and production of oil is about five million miles, about the length of all roadways on the continent—a disruption that will probably be seen by those who will examine these strata far into the future.[14]

But what might be used to define this transition; that is, what will future geologists see or detect and say, with complete certainty, that, yes, this spot—this line—in the geologic record marks the beginning of the Anthropocene?

One choice is the rapid appearance of a human-made rock that can be found almost everywhere: concrete. This artificially produced aggregate of sand, gravel and crushed rock, cement, and water is now the most consumed substance in the world behind water. Each year two cubic miles of it are poured to form the superstructure of our buildings, roads, and dams. That is twice the volume of lava that is erupted from all of the volcanoes in the world.[15] Broken-up fragments are now common in the turned-over ground underneath towns and cities. As Jan Zalasiewicz of the University of Leicester in England has observed, concrete is the signature rock of the Anthropocene, together with human-made bricks and ceramics.

Another possible choice is plastic. Plastics were developed around 1900 and surged in production and in use after 1950. All of the plastic that has ever been produced weighs twice as much as all of the animals, marine and terrestrial, on the planet. Plastic litter is now common. It is found as both macroscopic fragments and as microscopic particles. Many sea animals inadvertently eat plastics that end up in the muds of the seafloor. Plastic fibers have been found in almost every part of the seafloor. And so it, too, is a potential marker.

But there is a much better one, one that is certainly global and more distinct than either concrete or plastics. It is the radioactive fallout that came during the atmospheric testing of nuclear weapons.

It has been suggested that the Anthropocene epoch be formally started on July 16, 1945, the day when the first atomic device—called Trinity—was detonated at Alamogordo, New Mexico. This would have a parallel with the K-T boundary that marks the exact beginning of the Cenozoic Era with the moment a large asteroid touched the surface of the Earth along the northern coastline of the Yucatàn Peninsula. But this is a poor choice because the fallout was local. It is only since 1952 that nuclear explosions have left a clear and global signature.

The peak in atmospheric nuclear testing came in the early 1960s. These tests released radioactive nuclei that fell across most of the planet. A key geochemical signature that has already been identified in soils is that of plutonium-239, which has a half-life of 24,100 years. The long half-life means appreciable amounts of it will remain in strata formed in the 1960s for hundreds of thousands of years.

And where might a horizon be defined for the beginning of the Anthropocene Epoch—that is, where might members of the International Commission on Stratigraphy agree on the location of a GSSP, the "golden spike" that defines where in the geologic record the Anthropocene began? It should come as no surprise that several have been proposed. One proposal is to define it in an Antarctic ice core, in a similar way that the beginning of the Holocene Epoch has been defined. Another is a peat bog in Poland where a buildup of microscopic plastics, complete chemical pollutants, and radioactive fallout has been recorded.[16]

One of the more promising sites for a golden spike is Lake Crawford, east of Toronto, Ontario. Though located in a protected conservation area, this small lake is close to industrial pollution sources. It is unusually deep and the bottom lacks oxygen and so there are no critters down there to disturb and turn over the sediments. Furthermore, chalky debris from the Niagara Escarpment enters the lake as silt, which drops to the bottom and forms a white layer. That white layer is interrupted annually by a dark layer formed of microorganisms that grow in the summer, then sink to the bottom.

Whether Crawford Lake or another site will be chosen as a GSSP is yet to be decided. In fact, whether members of the International Commission of Stratigraphy will decide to recognize and define the Anthropocene Epoch as a new and the current geologic epoch is unknown.

Some claim that it is too early to define a new geologic epoch, that not enough time has passed to know what might be the outcome of the Great Acceleration. Might it turn out that this period of high consumption and the production of odd chemicals and dispersal of rare metals and radioactivity is just a blip in the geologic record? It has also been noted that each of the geologic epochs before the Holocene—the Miocene, the Pliocene, and the Pleistocene—each lasted millions of years. The Holocene began 11,700 years ago. And so, it has been argued, it is too early to define a new geologic epoch. That may be true. But there is a moral importance to its consideration.

◆

The climate of the Holocene Epoch was favorable to human beings. It enabled them to switch from hunting and gathering to farming for a living. An unexpected consequence of farming was that agriculture ensured a stable climate. And from farming came cities. And from cities came a division of labor, which led to more productivity and more stability, this time of society. Accordingly, specialized groups of artisans and a commercial class grew. And so did priesthoods and politicians. In short, civilization and all of its various trappings and implications owes its existence to climate stability. And that stability is now gone.

Look at the recent geologic record. For thousands of years global temperatures remained within a small range. So did sea level and the temperature and the chemistry of the oceans. Weather patterns were regular. And so was the distribution of animals and plants.

Look at the more distant geologic record. The Carboniferous Rainforest Collapse. The Carnian Pluvial Event. The Paleocene-Eocene Thermal Maximum. Each one was a sudden pulse of carbon dioxide into the atmosphere and a sudden heating of the planet. The patterns and amounts of rainfall changed. If there were polar ice caps, sea level rose dramatically. The chemistry of the oceans changed. And there were increased rates of extinction. The difference between now and then was in the cause of the disturbance.

For the first time in history, humanity, as a consequence of the Industrial Revolution, has grown into a driver and disrupter of climate. We have pushed the climate into a state that has not existed for tens of millions of years. It stands to reason that if the Holocene climate gave rise to human civilization, then a profound disruption of the Holocene climate could lead to the collapse of that civilization.

Human beings will survive. We are a resourceful and resilient species. But what will be the condition of our survival? Some have argued that no action should be taken because it would be too costly. Others argue the opposite point that no action will produce economic distress.

Look again at the geologic record. The Earth has passed through tremendous swings in environment. Only recently has it been suitable for our occupation. And so the reason to take action is obvious: It is a moral imperative.

ABOVE: Mount Rushmore. The portraits are carved into a light-gray granite. Beneath George Washington is a dark metamorphic rock known as a schist. BELOW: Geology of the Black Hills, South Dakota. The central granite is surrounded by a dark schist which, in turn, is surrounded by concentric rings of a yellow limestone, a red shale, and a white sandstone. *Drawing courtesy of Alan H. Strahler, Boston University.*

ABOVE: Morton Gneiss, west of Morton, Minnesota, is the oldest rock in the United States, 3.524 billion years. LEFT: Morton Gneiss used as a facing stone along the base of the Exchange Building in Seattle, Washington. *Photo courtesy of David B. Williams.* BELOW: Layers of thin lake sediments deformed by a dropstone.

ABOVE: The Grand Canyon from the South Rim. BELOW: Geologic formations at the Grand Canyon. The oldest rocks are the Vishnu Schist and the Zoroaster Granite in the Inner Canyon, 1,750 billion years. The youngest rock is the Kaibab Limestone at the top of the South Rim, 270 million years.

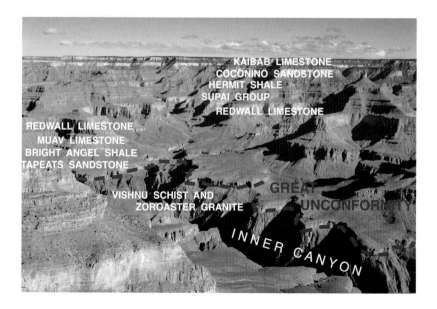

ABOVE: Massive vertical walls of the Navajo Sandstone at Zion National Park, Utah, which are about 190 million years old. The Navajo Sandstone was deposited when the climate was extremely dry and North America was part of the supercontinent of Pangea. BELOW: The convolutions in the Carmel Formation in Arches National Park may have been caused by shock waves that rippled through the region during the meteor impact that formed Upheaval Dome.

ABOVE: Monument Valley, Utah, Late Permian Period, 250 to 270 million years ago. Changes in the vertical outline of the butte indicates climate change. The broad apron around the base of the butte was a time of a wet climate. The steep section with vertical walls was a time of windblown sand during an extremely dry climate. The thin veneer of rocks on the top of the butte was a return to a wet climate. BELOW: Monument Rocks, Kansas, Late Cretaceous Period, 85 million years ago. Monument Rocks of western Kansas formed when a shallow sea—the Western Interior Seaway—extended down the center of North America from the Arctic to the Caribbean. The chalk in these cliffs formed at the same time as the chalk of the White Cliffs of Dover.

ABOVE: Meteor Crater, Arizona, a simple, bowl-shaped impact crater. Diameter is 0.8 mi (1.2 km). *Photo courtesy of Steven Jurvetson.* BELOW: Upheaval Dome, Utah, a complex, multi-ringed impact crater. About a mile of material of the upper part of this impact crater has eroded away. Diameter is 6 mi (10 km). *Photo courtesy of Doc Searls of Santa Barbara, California.*

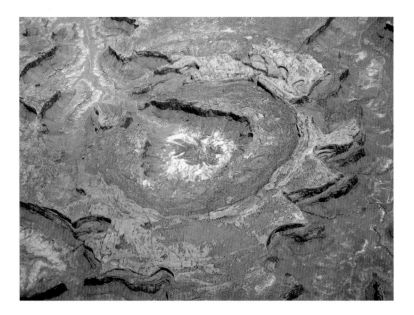

ABOVE: End of the dinosaurs. Cretaceous-Paleogene (Tertiary) boundary near Trinidad, Colorado, 66.043 million years old. The boundary is a thin light-gray layer of clay immediately beneath the massive layer of light-brown sandstone. BELOW: The light-gray layer of clay along the top of the photo is material that was thrown up then slowly drifted down out of the atmosphere during the giant impact that ended the Cretaceous Period 66.043 million years ago. (The quarter is for scale.)

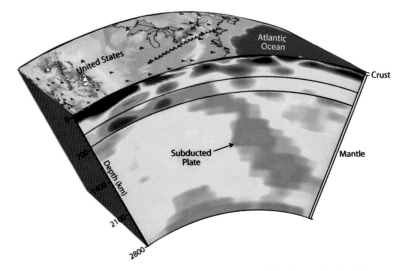

Tomographic image of the Earth beneath North America. The former Farallon Plate (now the Farallon Slab, indicated in green) is sliding from west-to-east beneath North America then slides at a steep angle into the Earth under the Atlantic Ocean. *Image courtesy of Suzan van der Lee, Northwestern University.*

EPILOGUE
Rugby, North Dakota

The Future

Soon after an obscure government publication made the announcement, local authorities worked quickly. A rock obelisk, fifteen feet high, was erected. The rocks are an assortment of fist-sized boulders of a glacial till that was deposited about twelve thousand years ago. But that is not the point. The obelisk marks, so the people of Rugby, North Dakota, believe, the geographical center of North America.

Stand next to the obelisk. The Rancho Grande Mexican restaurant is nearby. So are three poles that fly the flags of the United States, Canada, and the state of North Dakota. Face any direction. There is as much of North America in front of you as behind you. That is the point. This is the center of the continent—almost.

The announcement was made in 1930 by a government topographer who had made a cutout of North America and had pinned it to a wall several times in several different orientations. For each orientation, he hung a string from the pin and drew a line. After several lines were drawn, he had an intersection. He checked a map of North America and decided that the intersection was within a few miles of Rugby, North Dakota. Then, in 2015, an announcement was made that pointed to a new location.

Using digitized maps, a computer, and mathematical algorithms, a new geographical center of North America was determined. It was more than a hundred miles southwest of Rugby in the conveniently named community of Center, North Dakota. The community was named long before the mathematics was done. It was named "Center" because it was located near the center of the county it was in. But the few hundred people who live in Center know of the new determination. If you ask one of them what they think about the people of Rugby filing a trademark for the phrase "Geographical Center of North America," you will be told "Center is the center."

And so Rugby has its obelisk and Center has its mantra. But there is something much more important happening here. The people who live in and around Center, North Dakota, are attempting to change the future history of the Earth. They are attempting to capture and sequester carbon dioxide.

The effort that is still being tested and evaluated is to capture the carbon dioxide being released by a 1970s-era coal-fired plant before it leaves the plant chimney and transport it through a pipeline to a site where it would be injected deep underground for safe, permanent storage in a suitable layer of rock. Whether this will be successful is yet to be seen. The cost of making such an attempt is high, measured, even for a small coal-fired power plant that emits only three million tons of carbon dioxide annually—equivalent to the emission of six hundred thousand cars, the number of cars in South Dakota and North Dakota—in the billions of dollars.

The carbon dioxide would be forced down about a mile underground, where it will combine chemically with the surrounding porous rocks, a thick bed of Permian-age beach sand, to form a solid carbonate compound. Overlying the ancient beach sand is a layer of impervious shale that will act as a seal and prevent any carbon dioxide gas from rising back up to the surface.

If the effort in North Dakota is successful and meets the engineering and financial goals, making it possible to remove carbon dioxide from the atmosphere in this manner, there is still the challenge of scaling up such an effort. Just to maintain the current concentration of carbon dioxide in the atmosphere, the effort would have to be scaled up at least by a factor of ten thousand.

Another method is to use a biological force to remove carbon dioxide from the atmosphere. The most obvious choice is trees. Today there are about three trillion trees on the plant, about half as many as when human civilization arose. Add another trillion trees and more than half of the carbon released by humans since the industrial age began will be removed. But there is the obvious problem of finding the land. The land where most of these trees could be planted is now being used for agriculture. There is also the challenge of planting and growing the trees fast enough before deserts expand and regions of the planet become too hot for plants to survive.

The only certainty is that a carbon-based industrial society will eventually end. How and when it will end is unknown. But it will end. In the long aftermath after such a carbon-based society has ended, carbon dioxide will be removed slowly from the atmosphere by the chemical weathering of rocks. This process will take a hundred thousand years or so to return the concentration of carbon dioxide in the atmosphere to a preindustrial level and return to the Earth to the relative warmth—in human terms—of the preindustrial age.

If, in this far future, carbon emissions by humans end completely, then the climate of the Earth will resume the Cenozoic cooling. There will be a renewal of ice ages. And another series of giant ice sheets will advance and retreat and cover the northern regions of North America.

◆

If you want to ensure that your daily life gets recorded in the geologic record, move to New Orleans. There the Mississippi River is adding more and more sediments to the Mississippi Delta and the additional weight is causing New Orleans to sink. Eventually, the city will sink too deep to be sustainable. How far the modern city will drop and how deep will be the overlying layers of river sediments will not be known until the far future. And interleaved between those layers of river sediments will be layers of debris that were swept in by what are yet-to-happen hurricanes, adding interest to the entire section of deposits if a geologist in the far future decides to study the site.

Neither Seattle nor Chicago will contribute much to the geologic record, nor will New York or Toronto. In a few hundred thousand years, the series of thick sheets of ice that will be coming down from the north will scrape these sites clean. The cities will be bulldozed and the many accoutrements of society will be ground into powder by the ice. Whatever might survive might get caught up in the ice and get released when a massive flood of water from a future glacial lake bursts and sends such bergs across the landscape. The bergs will settle and the ice will melt and some of the relics of modern life might lie scattered across the landscape, much in the manner that glacial erratics are found today.

Miami is almost gone today. The continued rise of sea level will eventually force the abandonment of much of this city. The intruding seawater will corrode the iron and steel frameworks of modern buildings and weaken the concrete. Then, as a case in irony, the level of the sea will drop dramatically as the new ice sheets form, leaving Miami exposed to weathering and it will erode away.

Erosion is the fate of most cities in America. Atlanta and Dallas will be subjected to weathering and will be eroded away by water or by wind. So will San Antonio and Tulsa. Denver and Kansas City might have a different fate and be encased in the geologic record if they are suddenly blanketed by a thick layer of ash erupted from Yellowstone or from another major volcanic center. The fate of Portland, Oregon, is more problematic. It might be erased by ice if the sheets extend that far south or it might disappear by erosion. A major eruption of the Boring Lava Field could eject a thick layer of cinders that could entomb sections of the city and ensure that those sections are put into the geologic record, provided that no one tries to rescue the city and dig it out.

San Francisco and the other cities of the Bay Area are destined to be carved up and slide away from each other along earthquake faults. Two house sites that share the same street today might be hundreds of miles apart in the far future.

Los Angeles is a special case. It, too, could get carved up, but it is also a large sedimentary basin, and so it has the potential for a record of it to be held between rock layers. And it has the complication that mountain ranges are rising in the basin as the Baja California Peninsula continues to swing out into the Pacific. The Los Angeles Basin has already been transferred to the Pacific Plate, and so this section of North America is already headed north. One expects that major parts of Los Angeles will endure and appear somewhere on the planet in the future geologic record.

The faces on Mount Rushmore will last until the next ice sheet creeps in and grinds down and erases them. Hoover Dam will stand until the climate has changed sufficiently that there is not a great body of water behind it to provide essential structural support. But what will happen to the Grand Canyon?

Everything in geology is ephemeral. The Grand Canyon as seen today will persist—probably for millions of years—until a dramatic change in climate brings persistent and heavy rains to this part of North America. Then the canyon will not fill. Instead, its walls will be eroded back and the erosion will continue downward to the hard rock beneath the Great Unconformity. One wonders whether the few mesas and towers of rock that are left behind—similar to those in Monument Valley—will be enough for geologists in the far future to realize that a great canyon was once here.

Jump forward from today to twenty-five million years in the future. The southwestern corner of the United States is being transferred to the Pacific Plate today. Twenty-five million years in the future much of southern California will have slid toward the northwest. There may also be significant breaks and disruptions along the Walker Seismic Zone that follows the California-Nevada border and along the Intermountain Seismic Zone that runs from western Montana and into southern Nevada. Much of this land may have separated from North America in twenty-five million years. How much the Rio Grande Rift will widen is more problematic. But, in twenty-five million years, much of the southwestern part of the United States will probably be out in the Pacific Ocean as a close assembly of large islands. Add another hundred million years and they will be scattered across the northern part of the ocean.

The next supercontinent has already begun to form. Eurasia will be its core. The subcontinent of India was added to it thirty-five million years ago. Africa is closing in on Europe and the Mediterranean Sea will eventually disappear. Then Africa, too, will be added to this supercontinent.

Australia will come next, in about fifty million years. The large scattered islands that were formed by California, Nevada, Utah, Arizona, and New Mexico will probably slide up and collide against and hold tight against the new supercontinent in two hundred million years. How much of what remains of North America will be incorporated depends on the continued widening of the Atlantic Ocean. If that ocean continues to widen, then all of the land from Texas to Florida to Maine and northward through Quebec will lie along the coast of the new supercontinent. This new supercontinent has already been named: It is called Amasia.

Amasia will reach its maximum growth in about three hundred million years and be centered close to the North Pole. After another few hundred million years, it, too, will break apart and the continental pieces will be scattered. What we know today as North America will probably be fragments found across an entire hemisphere.

Amasia will probably be the last of the supercontinents. That is, in a billion years, the Earth will have cooled sufficiently that the sliding of oceanic plates into the mantle will no longer be efficient. And so, in its long history, Earth will have four supercontinents: Columbia, Rodinia, Pangea, and Amasia.

And because the sliding of oceanic plates will be retarded, if not ended, the movements of the plates will slow and finally stop. A new tectonic regime will mark this part of the Earth's history. Also the oceans will have disappeared by then. Estimates suggest that today the transport of water into the Earth by oceanic plates is five times greater than the return of water through volcanic eruptions. And so the oceans are disappearing. In a billion years, the surface of the Earth will be dry. That will end photosynthesis. And that will be the end of flourishing life.

What will follow is a geologic eon without visible life, though bacteria and archaea will probably survive. That will mark the end of the Phanerozoic Eon. And a new geologic eon will begin. Perhaps, the new eon should be called the Telozoic, a combination of the Greek words *telos* and *zoe*, the "end of life."[1]

But that is far into the future.

◆

One of the most wondrous moments of my life happened one night in Death Valley. I was alone and driving the highway that runs most of the length of the valley. The sky was clear and the landscape was flooded with light from a full Moon. I stopped the car and got out and walked away from the road.

I walked far enough so that I could no longer see the outline of my car. Beneath me was a rocky field of salty white. Above were stars in a brightened night sky. Running the length of both sides of the valley were two massive walls of mountains, the Panamint Range to the west and the Amargosa Range to the east. There was an initial feeling of awe. Then those feelings turned to thoughts about my moment in time.

There are more than four billion years of the Earth's history behind me and an untold number in the future. And so much has happened. But for a moment in that history I stood alone at night in Death Valley.

Then, as if to reassure me that I was not alone on the planet, the headlights of another car appeared on the highway at a great distance. The headlights were turned off at what I knew to be a high point in the road. And so I have often wondered: Was someone else out there that night and did they stop next to the road and what might they have thought about their moment in time?

ACKNOWLEDGEMENTS

F ifty years of travel and four years of writing were required to complete this book. Many, but not all of the sites mentioned in the book have been visited by the author. Over those few decades, ideas and interpretations have changed about the geologic history of some of the sites. Every effort has been made to provide the latest ideas as published in the scientific literature and to note where controversies still exist.

My quest to understand the Earth began, indirectly, with John Bollard, who instructed me in the fundamentals of aeronautics and astronautics. That led to my first crisscrossing of the western United States—at both high and low altitudes.

Seymour Rabinovitch taught me the value of quantitative analysis. Michael Schick changed my perception of nature with his relentless classroom lectures about the mathematics of quantum mechanics. From him, I formed the opinion that the Earth can be understood in the same mathematically precise way. Both men gave yearlong courses, consisting of nearly a hundred lectures—Rabinovitch in chemistry and Schick in physics. At the end of the final lecture of each course, I remember an auditorium filled with students applauding and giving each man a standing ovation.

Eugene Shoemaker took me to Meteor Crater, the Grand Canyon, and Frenchman Mountain. Robert Sharp took me to the beaches and to the deserts of California, showing me how the dynamic side of geology is told by the rocks. Heinz Lowenstam taught me that geology and biology were the same thing. Leon Silver hustled me through Baja California and, through his grand gestures, illustrated continental evolution. Don Helmberger showed me, through exquisite mathematics, how evidence of the Earth's interior is contained in the wild wave trains of earthquakes. Clarence Allen introduced me to the San Gorgonio Pass and the mud volcanoes near the Salton Sea. Though he never accepted the reality of mantle plumes, Don Anderson inspired me to think about how the Earth had formed and how it had evolved. Robert Decker opened a door that led to volcanoes. Paolo Gasparini showed me how to navigate Italian institutions. Legi in Indonesia reminded me how interesting—and pleasant—it was to live on planet Earth. And Roger Phillips, through his indomitable spirit, kept me focused until I understood the importance of impact craters and impact cratering. This book is a product of the years I spent with these people.

This book has benefited from discussions with Scott Baldridge, Kent Condie, Richard Hazlett, and Keith Meldahl. Alan Strahler, David Williams, and Suzan van der Lee have kindly allowed me to use their photographs and drawings. I have personally benefited from conversations, over many years, with Tom Peek, Richard Terrile, James Dieterich, Bruce Bills, Dan Dzurisin, and Tom Casadevall.

Sarah Dvorak was essential to my travels and Joyce Dvorak provided expertise on the correct use of language. Alex Sendegeya provided the means to complete the research.

I want to thank my agent Laura Wood for keeping me on a straight track. The editorial talent of Peter Kranitz greatly improved the text. And a special thank you to Jessica Case for her patience during the long process that it took to complete this book.

SOURCES

GENERAL REFERENCES

Abbot, Lon, and Terri Cook. *Geology Underfoot along Colorado's Front Range*. Missoula, Montana: Mountain Press Publishing Company, 2012.

Armstrong, Joseph E. *How the Earth Turned Green*. Chicago: University of Chicago Press, 2014.

Bartlett, Richard A. *Great Surveys of the American West*. Norman, Oklahoma: University of Oklahoma Press, 1962.

Beerling, David. *The Emerald Planet*. Oxford: Oxford University Press, 2007.

Bjornerud, Marcia. *Timefulness*. Princeton, New Jersey: Princeton University Press, 2018.

Blakey, Ronald C., and Wayne D. Ranney. *Ancient Landscapes of Western North America*. London: Springer, 2018.

Boenigk, Jens, Sabina Wodniok, and Edvard Glückstag. *Biodiversity and Earth History*. London: Springer, 2015.

Gradstein, Felix M., James G. Ogg, Mark D. Schmitz, and Gabi M. Ogg. *The Geologic Time Scale 2012*. Two volumes. Amsterdam: Elsevier, 2012.

Gradstein, Felix M., et al. *Geologic Time Scale 2016*. Second edition. Amsterdam: Elsevier. 2016.

McPhee, John. *Annals of the Former World*. New York: Farrar, Straus and Giroux, 1998.

Morton, Mary Caperton. *Aerial Geology*. Portland, Oregon: Timber Press, 2017.

Nudds, John R., and Paul A. Selden. *Fossil Ecosystems of North America*. Chicago: University of Chicago Press, 2008.

Oldroyd, David. *Thinking About the Earth*. Cambridge, Massachusetts: Harvard University Press, 1996.

Ogg, James G., Gabi Ogg, and F. M. Gradstein. *A Concise Geologic Time Scale: 2016*. Amsterdam: Elsevier, 2016.

Palmer, Douglas, et al. *Prehistoric Life*. New York: DK Publishing, 2012.

Prothero, Donald R. *The Story of the Earth in 25 Rocks*. New York: Columbia University Press, 2018.

Prothero, Donald R. *Fantastic Fossils*. New York: Columbia University Press, 2020.

Stanley, Steven M., and John A. Luczaj. *Earth System History*. Fourth edition. New York: W. H. Freeman, 2015.

Ward, Peter, and Joe Kirschvink. *A New History of Life*. New York: Bloomsbury, 2015.

SELECTED REFERENCES

Prologue: Mount Rushmore, South Dakota

Anderson, Sam. "Why Does Mount Rushmore Exist?" *New York Times Magazine*, March 22, 2017.

Casey, Robert Joseph, and Mary Borglum. *Give the Man Room*. Indianapolis: Bobbs-Merrill, 1952.

Graham, J. *Mount Rushmore National Memorial Geologic Resource Evaluation Report. Natural Resource Report NPS/NRPC/GRD/NRR—2008/038*. Denver: National Park Service, 2008.

Lisenbee, Alvis L. "Tectonic Map of the Black Hills Uplift, Montana, Wyoming, and South Dakota." *Geological Survey of Wyoming, Map Series 13*, scale 1:250,000, 1985.

Redden, Jack A. "Geology of the Berne Quadrangle Black Hills South Dakota." *US Geological Survey Professional Paper 297-F*. Washington, DC: US Government Printing Office, 1968.

Smith, Rex Alan. *The Carving of Mount Rushmore*. New York: Abbeville Press, 1985

1. The Relics of Hell

Morton Gneiss and the Radiogenic Age-Dating of Rocks

Bickford, M. E., J. L. Wooden, and R. L. Bauer. "SHRIMP Study of Zircons from Early Archean Rocks in the Minnesota River Valley: Implications for the Tectonic History of the Superior Province." *Geological Society of America Bulletin*, vol. 118, no. 1–2, pp. 94–108. 2006.

Moorbath, Stephen. "The Discovery of the Earth's Oldest Rocks." *Notes & Records of the Royal Society*, vol. 63, pp. 381–392. 2009.

Stern, Richard A., and Wouter Bleeker. "Age of the World's Oldest Rocks Refined Using Canada's SHRIMP: The Acasta Gneiss Complex, Northwest Territories, Canada." *Geoscience Canada*, vol. 25, no. 1, pp. 27–31. 1998.

Williams, David B. *Stories in Stone*. New York: Walker & Company, 2009.

Van Kranendonk, Martin J., R. Hugh Smithies, and Vickie C. Bennett (eds.). *Earth's Oldest Rocks*. Volume 15. Amsterdam: Elsevier, 2007.

Formation and Age of the Solar System

Batygin, Konstantin, and Greg Laughlin. "Jupiter's Decisive Role in the Inner Solar System's Early Evolution." *Proceedings of the National Academy of Sciences*, vol. 112, no. 14, pp. 4214–4217. 2015.

Batygin, Konstantin, et al. "Born of Chaos." *Scientific American*, vol. 315, no. 5, pp. 29–37. 2016.

Bouvier, Audrey, and Meenakshi Wadhwa. "The Age of the Solar System Redefined by the Oldest Pb-Pb Age of a Meteoritic Inclusion." *Nature Geoscience*, vol. 3, pp. 637–641. 2010.

Carlson, Richard W., et al. "How Did Early Earth become Our Modern World?" *Annual Reviews of Earth and Planetary Science*, vol. 42, pp. 151–178. 2014.

Planck Collaboration. "Planck 2015 Results. XIII. Cosmological Parameters." *Astronomy & Astrophysics*, Special Feature, vol. 594, p. A13. 2016.

The Grand Tack and the Age of the Earth

Barboni, Melanie, et al. "Early Formation of the Moon 4.51 Billion Years Ago." *Science Advances*, vol. 3. 2017.

Gáspár, András, and George H. Rieke. "New HST Data and Modeling Reveal a Massive Planetesimal Collision around Fomalhaut." *Proceedings of the National Academy of Sciences*, vol. 117, no. 18, pp. 9712–9722. 2020.

Rumble, D., et al. "The Oxygen Isotope Composition of Earth's Oldest Rocks and Evidence of a Terrestrial Magma Ocean." *Geochemistry Geophysics Geosystems*, vol. 14, no. 6, pp. 1929–1939. 2013.

Sallum, S., et al. "Accreting Protoplanets in the LkCa 15 Transition Disk." *Nature*, vol. 527, no. 7578, pp. 342–344. 2015.

Thompson, Maggie, et al. "Studying the Evolution of Warm Dust Encircling BD+20 307 Using SOFIA." *Astrophysical Journal*, vol. 875, no. 1. 2019.

Zahnle, Kevin, et al. "Emergence of a Habitable Planet," *Space Science Reviews*, vol. 129, pp. 35–78. 2007.

Earth's Oldest Rocks on the Moon

Armstrong, John C., Lloyd E. Wells, and Guillermo Gonzalez. "Rummaging through Earth's attic for remains of ancient life." *Icarus*, vol. 160, pp. 183–196. 2002.

Bellucci, J. J., et al. "Terrestrial-Like Zircon in a Class from an Apollo 14 Breccia." *Earth and Planetary Science Letters*, vol. 150, pp. 173–185. 2019.

Jack Hills and an Early Global Ocean

Mojzsis, Stephen J., T. Mark Harrison, and Robert T. Pidgeon. "Oxygen-Isotope Evidence from Ancient Zircons for Liquid Water at the Earth's Surface 4,300 Myr Ago." *Nature*, vol. 409, no. 6817, pp. 178–181. 2001.

Sarafian, Adam R., et al. "Early Accretion of Water and Volatile Elements to the Inner Solar System: Evidence from Angrites." *Philosophical Transactions, Royal Society*, series A, vol. 375. 2017.

Valley, John W. "A Cool Early Earth?" *Scientific American*, vol. 293, no. 4, pp. 58–65. 2005.

Origin of the Oceans

Jewitt, David, and Edward D. Young. "Oceans from the Skies." *Scientific American*, vol. 312, no. 3, pp. 36–43. 2015.

The Chaotian Eon

Goldblatt, C., et al. "The Eons of Chaos and Hades." *Solid Earth*, vol. 1, pp. 1–3. 2010.

2. Bombardment and Bottleneck

Seeing Deep Time

Darwin, Charles. *On the Origin of Species by Means of Natural Selection, or the Preservation of Favoured Races in the Struggle for Life.* First edition. London: John Murray. 1859.

Late Heavy Bombardment

Addison, William D., et al. "Discovery of Distal Ejecta from the 1850 Ma Sudbury Impact Event." *Geology*, vol. 33, no. 3, pp. 193–196. 2005.

Byerly, Gary R., et al. "An Archean Impact Layer from the Pilbara and Kaapvaal Cratons." *Science*, vol. 297, no. 5585, pp. 1325–1327. 2002.

Garde, Adam A., et al. "Searching for Giant, Ancient Impact Structures on Earth d." *Earth and Planetary Science Letters*, vol. 337–338, pp. 197–210. 2012.

Glikson, Andrew Y. *The Archean*. London: Springer. 2014.

Glikson, Andrew, et al. "A New ~3.46 Asteroid Impact Ejecta Unit at Marble Bar, Pilbara Craton, Western Australia: A Petrological, Microprobe and Laser Ablation ICPMS Study." *Precambrian Research*, vol. 279, pp. 103–122. 2016.

Lowe, Donald R., and Gary R. Byerly. "The Terrestrial Record of Late Heavy Bombardment." *New Astronomy Reviews*, vol. 81, pp. 39–61. 2018.

Taylor, G. Jeffrey. "The Scientific Legacy of Apollo." *Scientific American*, vol. 271, no. 1, pp. 40–47. 1994.

Tera, Fouad, et al. "Isotopic Evidence for a Terminal Lunar Cataclysm." *Earth and Planetary Science Letters*, vol. 22, pp. 1–21. 1974.

A Bottleneck and the Hyperthermophiles

Dodd, Matthew, et al. "Evidence for Early Life in Earth's Oldest Hydrothermal Vent Precipitates." *Nature*, vol. 543, no. 7643, pp. 60–64. 2017.

Gogarten-Boekels, Maria, et al. "The Effects of Heavy Meteorite Bombardment on the Early Evolution—The Emergence of the three Domain of Life." *Origin of Life and Evolution of the Biosphere*, vol. 25, no. 1–3, pp. 251–264. 1995.

Maher, K. A., and D. J. Stevenson. "Impact Frustration of the Origin of Life." *Nature*, vol. 331, no. 6157, pp. 612–614. 1988.

Mesler, Bill, and H. James Cleaves II. *A Brief History of Creation*. New York: W. W. Norton. 2016.

Tashiro, Takayuki, et al. "Early trace of life from 3.95 Ga sedimentary rocks in Labrador, Canada." *Nature*, vol. 549, no. 7673, pp. 516–518. 2017.

Birthplace of Life

Chan, Queenie H. S., et al. "Organic Material in Extraterrestrial Water-Bearing Salt Crystals." *Science Advances*, vol. 4, no. 1, p. 10. 2018.

Grotzinger, J. P., et al. "A Habitable Flucio-Lacustrine Environment at Yellowknife Bay, Gale Crater, Mars." *Science*, vol. 343, no. 6169, p. 48. 2014.

Hazen, Robert M., and John M. Ferry. "Mineral Evolution: Mineralogy in the Fourth Dimension." *Elements*, vol. 6, pp. 9–12. 2010.

Early Life: Cyanobacteria and Stromatolites

O'Brien, Charles F. "Eozoön Canadense 'The Dawn Animal of Canada.'" *Isis*, vol. 61, no. 2, pp. 206–223. 1970.

Schopf, J. William. "Solution to Darwin's Dilemma: Discovery of the Missing Precambrian Record of Life." *Proceedings of the National Academy of Sciences*, vol. 97, no. 13, pp. 6947–6953. 2000.

Schopf, J. William. "Fossil Evidence of Archaean Life." *Philosophical Transactions: Biological Sciences*, vol. 361, no. 1470, pp. 869–885. 2006.

Steele, John H. "A Description of the Oolitic Formation Lately Discovered in the County of Saratoga, and State of New-York." *American Journal of Science*, vol. 9, pp. 16–19. 1825.

Tyler, Stanley A.; and Elso S. Barghoorn. "Occurrence of Structurally Preserved Plants in Pre-Cambrian Rocks of the Canadian Shield." *Science*, vol. 119, no. 3096, pp. 606–608. 1954.

Van Kranendonk, Martin J., et al. "Geological Setting of Earth's Oldest Fossils in the Ca. 3.5 Ga Dresser Formation, Pilbara Craton, Western Australia." *Precambrian Research*, vol. 167, pp. 93–124. 2008.

Banded Iron Formations

Konhauser, K. O., et al. "Iron Formations: A Global Record of Neoarchaean to Palaeoproterozoic Environmental History." *Earth-Science Reviews*, vol. 172, pp. 140–177. 2017.

A First Snowball Earth

Bennett, Gerald. *The Huronian Supergroup between Sault Ste Marie and Elliot Lake, Field Trip Guidebook*, vol. 52, part 4, Institute on Lake Superior Geology, Sault Ste Marie, Ontario. 2006.

Coleman, A. P. "The Lower Huronian Ice Age." *Journal of Geology*, vol. 16, no. 2, pp. 149–158. 1908.

Holland, Heinrich D. "Volcanic Gases, Black Smokers, and the Great Oxidation Event." *Geochimica et Cosmochimica Acta*, vol. 66, no. 21, pp. 3811–3826. 2002.

3. The Children of Ur

Staten Island, New York

Merguerian, Charles. *Geology of Staten Island and Vicinity, New York. Field Guide for Hofstra University Freshman Seminar, Geology 014F*, 109 pages. 2008.

Alfred Wegener, Drifting Continents, and the Theory of Plate Tectonics

Herring, T. A., et al. "Geodesy by Radio Interferometry: Evidence for Contemporary Plate Motion." *Journal of Geophysical Research*, vol. 91, issue B8, pp. 8341–8347. 1986.

Hess, H. H. "History of Ocean Basins." In Engel, A. E. J., H. L. James, and B. F. Leonard (eds.). *Petrologic Studies: A Volume to Honor A.F. Buddington: Geological Society of America*. pp. 599–620. 1962.

Morgan, J. W. "Rise, Trenches, Great Faults, and Crustal Blocks." *Journal of Geophysical Research*, vol. 73, no. 6, pp. 1959–1982. 1968.

Oreskes, Naomi (ed.). *Plate Tectonics: An Insider's History of the Modern Theory of the Earth.* Boulder, Colorado: Westview Press. 2001.

Wegener, Alfred. *Die Entstehung der Kontinente und Ozeane.* Braunschweig, Germany: Viewer & Sons. 1915.

Wilson, J. Tuzo. "A New Class of Faults and their Bearing on Continental Drift." *Nature,* vol. 207, pp. 343–347. 1965.

When Did Plate Tectonics Begin on the Earth?

Condie, Kent C., et al. "Granitoid Events in Space and Time: Constraints from Igneous and Detrital Zircon Age Spectra." *Gondwana Research,* vol. 15, no. 3, pp. 228–242. 2009.

Korenaga, Jun. "Initiation and Evolution of Plate Tectonics on Earth: Theories and Observations." *Annual Review of Earth and Planetary Science,* vol. 41, pp. 117–151. 2013.

Diamonds and the Earth's Early History

Cartigny, Peter. "Stable Isotopes and the Origin of Diamond." *Elements,* vol. 1, pp. 79–84. 2005.

Condie, Kent C., and Craig O'Neill. "The Archean-Proterozoic Boundary: 500 My of Tectonic Transition in Earth History." *American Journal of Science,* vol. 310, pp. 775–709. 2010.

Ojakangas, Richard W., and Charles L. Matsch. *Minnesota's Geology.* Minneapolis: University of Minnesota Press, 2004.

Shirey, Steven B., and Stephen H. Richardson. "Start of the Wilson Cycle at 3 Ga Shown by Diamonds from Subcontinental Mantle." *Science,* vol. 333, no. 6041, pp. 434–436. 2011.

Cratons and the Oldest Mountain Ranges

Dhuime, Bruno, et al. "Continental Growth Seen through the Sedimentary Record." *Sedimentary Geology,* vol. 357, pp. 16–32. 2017.

Frisch, W., et al. "Chapter 12: Old Orogens," *Plate Tectonics.* London: Springer. 2011.

Early North America: Laurentia

Suess, Eduard. *Das Antlitz der Erde,* vol. 3, pt. 2, Prague: F. Tempsky. 1909.

Whitmeyer, Steven J., and Karl E. Karlstrom. "Tectonic Model for the Proterozoic Growth of North America." *Geosphere,* vol. 3, no. 4, pp. 220–259. 2007.

Columbia and Rodinia: The First Supercontinents and the Grenville Orogeny

Meert, Joseph G. "What's in a Name? The Columbia (Paleopangaea/Nuna) Supercontinent." *Gondwana Research,* vol. 21, no. 4, pp. 987–993. 2012.

Rogers, John J. W., and M. Santosh. "Configuration of Columbia, a Mesoproterozoic Supercontinent." *Gondwana Research,* vol. 5, no. 1, pp. 5–22. 2002.

Van Kranendonk, Martin J., and Christopher L. Kirkland. "Orogenic Climax of Earth: The 1.2–1.1 Ga Grenvillian Superevent." *Geology,* vol. 41, no. 7, pp. 735–738. 2013.

Voice, Peter J., et al. "Quantifying the Timing and Rate of Crustal Evolution: Global Compilation of Radiometrically Dated Detrital Zircon Grains." *Journal of Geology,* vol. 119, no. 2, pp. 109–126. 2011.

The Splitting of Laurentia: The Midcontinent Rift
Stein, Seth, et al. "Insights from America's Failed Midcontinent Rift into the Evolution of Continental Rifts and Passive Continental Margins." *Tectonophysics*, vol. 744, pp. 403–421. 2018.

The Boring Billion
Buick, Roger, et al. "Stable Isotopic Compositions of Carbonates from the Mesoproterozoic Bangemall Group, Northwestern Australia." *Chemical Geology*, vol. 123, pp. 153–171. 1995.

4. Gardens of Ediacaran

The Oldest Fossil: Grypania
Blankenship, Robert E. "Early Evolution of Photosynthesis." *Planet Physiology*, vol. 154, pp. 434–438. 2010.
Han, Tsu-Ming, and Bruce Runnegar. "Megascopic Eukaryotic Algae from the 2.1-Billion-Year-Old Negaunee Iron-Formation, Michigan." *Science*, vol. 257, no. 5067, pp. 232–235. 1992.

A Second Snowball Earth
Harland, W. Brian, and Martin J. S. Rudwick. "The Great Infra-Cambrian Ice Age." *Scientific American*, vol. 211, no. 2, pp. 28–39. 1964.
Hoffman, Paul F., et al. "A Neoproteerozoic Snowball Earth." *Science*, vol. 281, no. 5381, pp. 1342–1346. 1998.
Hoffman, Paul F., and Daniel P. Schrag. "Snowball Earth." *Scientific American*, vol. 282, no. 1, pp. 68–75. January 2000.
Hoffman, Paul F., et al. "Snowball Earth Climate Dynamics and Cyrogenian Geology-Geobiology." *Science Advances*, vol. 3, no. 11, p. 43. 2017.
Keeley, Joshua A., et al. "Pre- to Synglacial Rift-Related Volcanism in the Neoproterozoic (Cryogenian) Pocatello Formation, SE Idaho: New SHRIMP and CA-ID-TIMS Constraints." *Lithosphere*, vol. 5, no. 1, pp. 128–150. 2012.
Kirschvink, Joseph L., et al. "Paleoproterozoic Snowball Earth: Extreme Climatic and Geochemical Global Change and Its Biological Consequences." *Proceedings of the National Academy of Science*, vol. 97, no. 4, pp. 1400–1405. 2000.
Wilson, Samuel T., et al. "Kilauea Lava Fuels Phytoplankton Bloom in the North Pacific Ocean." *Science*, vol. 365, no. 6457, pp. 1040–1044. 2019.

The Discovery of Ediacarans
Boyce, W. D., and K. Reynolds. "The Ediacaran Fossil *Aspidella terranovica* Billings, 1872 from St. John's Convention Centre Test Pit CjAe-33." *Current Research, Newfoundland and Labrador Department of Natural Resources Geological Survey, Report 08-1*, pp. 55–61. 2008.
Ford, Trevor. "Precambrian Fossils from Charnwood Forest." *Proceedings of the Yorkshire Geological Society*, vol. 31, pp. 211–217. 1958.

Howe, Mike P.A., et al. "New Perspectives on the Globally Important Ediacaran Fossil Discoveries in Charnwood Forest, UK: Harley's 1848 Prequel to Ford (1958)." *Proceedings of the Yorkshire Geological Society*, vol. 59, pp. 137–144. 2012.

McCall, G. J. H. "The Vendian (Ediacaran) in the Geological Record: Enigmas in Geology's Prelude to the Cambrian Explosion." *Earth Science Reviews*, vol. 77, pp. 1–229. 2006.

Sprigg, Reginald G. "Early(?) Jellyfishes from the Flinders Ranges, South Australia." *Transactions of the Royal Society of South Australia*, vol. 71, pp. 212–224. 1947.

The Ediacarans at Mistaken Point, Newfoundland

Bobrovskiy, Ilya, et al. "Ancient Steroids Establish the Ediacaran Fossil *Dickinsonia* as One of the Earliest Animals." *Science*, vol. 361, no. 6048, pp. 1246–1249. 2018.

McMenamin, Mark A. S. *The Garden of Ediacara: Discovering the First Complex Life*. New York: Columbia University Press, 2000.

Misra, Shiva B. "Late Precambrian(?) Fossils from Southeastern Newfoundland." *Geological Society of America Bulletin*, vol. 80, pp. 2133–2140. 1969.

Pu, Judy P., et al. "Dodging Snowballs: Geochronology of the Gaskiers Glaciation and the First Apperance of the Ediacaran Biota." *Geology*, vol. 44, no. 11, pp. 955–958. 2016.

Three Forms of Life: Bacteria, Archaea, and Eukaryotes

Cohen, Phoebe A., and Francis A. Macdonald. "The Proterozoic Record of Eukaryotes." *Paleobiology*, vol. 41, no. 4, pp. 610632. 2015.

Margulis, Lynn. *Symbiosis in Cell Evolution*. San Francisco: W. H. Freeman, 1992.

Quammen, David. *The Tangled Tree: A Radical New History of Life*. New York: Simon & Schuster, 2018.

Whatley, John P. "*Paracoccus denitrificans* and the Evolutionary Origin of Mitochondria." *Nature*, vol. 254, no. 5500, pp. 495–498. 1975.

Woese, Carl R., and George E. Fox. "Phylogenetic Structure of the Prokaryotic Domain: The Primary Kingdoms." *Proceedings of the National Academy of Sciences*, vol. 74, no. 11, pp. 5088–5090. 1977.

Woese, Carl R., et al. "Towards a Natural System of Organisms: Proposal for the Domains Archaea, Bacteria, and Eucarya." *Proceedings of the National Academy of Sciences*, vol. 87, no. 6, pp. 4576–4579. 1990.

Bottleneck and Flush

Carson, Hampton L. "The Genetic System, the Deme, and the Origin of Species." *Annual Review of Genetics*, vol. 21, pp. 405–423. 1987.

Hedges, S. B. "The Origin and Evolution of Model Organisms." *Nature Review Genetics*, vol. 3, pp. 838–849. 2002.

Protohertzina: The First Terror with Teeth

Laflamme, Marc, et al. "The End of the Ediacara Biota: Extinction, Biotic Replacement, or Cheshire Cat?" *Gondwana Research*, vol. 23, no. 2, pp. 558–573. 2013.

5. The Great Unconformity

The Grand Canyon and Discovery of the Great Unconformity
Dutton, Clarence. *The Physical Geology of the Grand Cañon District*. Washington, DC: US Government Printing Office, 1882.
Dutton, Clarence. *Tertiary History of the Grand Cañon District*. Washington, DC: US Government Printing Office, 1882.
Powell, J. W. *Exploration of the Colorado River and its Tributaries*. Washington, DC: Government Printing Office, 1875.
Stegner, Wallace. *Beyond the Hundredth Meridian*. Boston: Houghton, Mifflin, 1953.

Finding the Great Unconformity: Frenchman Mountain, Nevada
DeLucia M. S., et al. "Thermochronology Links Denudation of the Great Unconformity Surface to the Supercontinent Cycle and Snowball Earth." *Geology*, vol. 46, no. 2, pp. 167–170. 2018.
Karlstrom, Karl E., and J. Michael Timmons. "Many Unconformities Make One 'Great Unconformity.'" *Grand Canyon Geology: Two Billion Years of Earth's History: Geological Society of America Special Paper 489*, pp. 73–79, 2012.
Keller, C. Brenhin, et al. "Neoproterozoic Glacial Origin of the Great Unconformity." *Proceedings of the National Academy of Sciences*, vol. 116, no. 4. pp. 1136–1145. 2019.
Miall, Andrew D. "The Valuation of Unconformities." *Earth Science Reviews*, vol. 163, pp. 22–71. 2016.

The Beginning of the Cambrian: Fortune Head, Newfoundland
Geyer, Gerd, and Ed Landing. "The Precambrian-Phanerozoic and Ediacaran-Cambrian Boundaries: A Historical Approach to a Dilemma." *Earth System Evolution and Early Life*, Geological Society, London, Special Publications, no. 448. pp. 311–349. 2017.
Gould, Stephen Jay. "The Evolution of Life on the Earth." *Scientific American*, vol. 271, no. 10, pp. 85–91. 1994.
Hou, Xianguang, and Jan Bergström. "The Chengjiang Fauna—the Oldest Preserved Animal Community." *Paleontological Research*, vol. 7, no. 1, pp. 55–70. 2003.

The Earliest Animals and the Burgess Shale
Conway-Morris, Simon. *The Crucible of Creation*. Oxford: Oxford University Press, 1998.
Gaines, Robert R., et al. "Mechanism for Burgess Shale-Type Preservation." *Proceedings of the National Academy of Sciences*, vol. 109, no. 14, pp. 5180–5184. 2012.
Gould, Stephen Jay. *Wonderful Life*. New York: W. W. Norton, 1990.
Losos, Jonathan B. *Improbable Destinies*. New York: Riverhead Books, 2017.

The Cambrian Explosion and True Polar Wandering
Kirschvink, Joseph L., et al. "Evidence for a Large-Scale Reorganization of Early Cambrian Continental Masses by Inertial Interchange True Polar Wander." *Science*, vol. 277, no. 5325, pp. 541–545. 1997.

Mitchell, Ross N., et al. "Sutton Hotspot: Resolving Ediacaran-Cambrian Tectonics and True Polar Wandering for Laurentia." *American Journal of Science*, vol. 311, pp. 651–663. 2011.

Mitchell, Ross N., et al. "Was the Cambrian Explosion Both an Effect and an Artifact of True Polar Wander?" *American Journal of Science*, vol. 315, pp. 945–957. 2015.

Peters, Shanan, and Robert R. Gaines. "Formation of the 'Great Unconformity' as a Trigger for the Cambrian Explosion." *Nature*, vol. 484, pp. 363–366. 2012.

Zhang, Xingliang, et al. "Triggers for the Cambrian Explosion: Hypotheses and Problems." *Gondwana Research*, vol. 25, pp. 896–909. 2014.

The Great Ordovician Biodiversification Event

Servais, Thomas, and David A. T. Harper. "The Great Ordovician Biodiversification Event (GOBE): Definition, Concept and Duration." *Lethaia*, vol. 51, pp. 151–164. 2018.

Hox Genes and Punctuated Equilibrium

Lappin, Terrence R. J., et al. "HOX Genes: Seductive Science, Mysterious Mechanisms." *Ulster Medical Journal*, vol. 75, no. 1, pp. 23–31. 2006.

Quinonez, Shane C., and Jeffrey W. Innis. "Human HOX Gene Disorders." *Molecular Genetics and Metabolism*, vol. 111, pp. 4–15. 2014.

Schwartz, Jeffrey H. "Homeobox Genes, Fossils, and the Origin of Species." *The Anatomical Record*, vol. 257, pp. 15–31. 1999.

Yanai, Itai, and Martin Lercher. *The Society of Genes*. Cambridge, Massachusetts: Harvard University Press, 2016.

6. An Ancient Forest at Gilboa

Cameron's Line and the Taconic Orogeny

Fuller, Tyrand, et al. "Tracing the St. Nicholas Thrust and Cameron's Line through the Bronx, NYC." *Sixth Annual Conference on Geology of Long Island and Metropolitan New York, 24 April 1999*, State University of New York at Stony Brook, NY, Long Island Geologists Program with Abstracts, p. 8. 1999.

Merguerian, Charles, and Charles A. Baskerville. "Geology of Manhattan Island and the Bronx, New York City, New York." *Geological Society of America Centennial Field Guide-Northeastern Section*, pp. 137–140. 1987.

Merguerian, Charles, and Cheryl Moss. "Newly Discovered Ophiolite Scrap in the Hartland Formation of Midtown Manhattan." *Twelfth Annual Conference on Geology of Long Island and Metropolitan New York, 16 April 2005*. State University of New York at Stony Brook, NY, Long Island Geologists Program with Abstracts, p. 7. 2005.

Niagara Falls and the Niagara Escarpment

Luczaj, John A. "Geology of the Niagara Escarpment in Wisconsin." *Geoscience Wisconsin*, vol. 22, no. 1, pp. 1–34. 2013.

Mikulic, Donald G., and Joanne Kulessendorf. *The Classic Silurian Reefs of the Chicago Area. Geological Field Trip 4: April 24, 1999. ISGS Guidebook 29.* North-Central Meeting, Geological Society of America, 33rd Annual Meeting, p. 41. 1999.

Raymo, Chet, and Maureen E. Raymo. *Written in Stone*. Delmar, New York: Black Dome Press, 2007.

Terror of the Devonian Seas: Dunkleosteus

Young, Gavin C. "Placoderms (Armored Fish): Dominant Vertebrates of the Devonian Period." *Annual Review of Earth and Planetary Science*, vol. 38, pp. 523–550. 2010.

The Green Invasion: Plants Move onto Land

Armstrong, Joseph E. *How the Earth Turned Green*. Chicago: University of Chicago Press, 2014.

Edwards, D., et al. "New Silurian Cooksonias from Dolostone of North-Eastern North America." *Botanical Journal of the Linnean Society*, vol. 146, pp. 399–413. 2004.

Leaves, Roots, and Seeds

Beerling, David J. "Leaf Evolution: Gases, Genes and Geochemistry." *Annals of Botany*, vol. 96, pp. 345–352. 2005.

Kenrick, Paul, and Christine Strullu-Derrien. "The Origin and Early Evolution of Roots." *Plant Physiology*, vol. 166, pp. 570–580. 2014.

Shubin, Neil. *Your Inner Fish*. New York: Vintage Books, 2009.

Suarez, Stephanie E., et al. "A U-Pb Zircon Age Constraint on the Oldest-Recorded Air-breathing Land Animal." *PLoS ONE*, vol. 12, no. 6, p. 10. 2017.

The First Great Forests: Archaeopteris and Eospermatoperis

Aldrich, Michele L., et al. "Winifred Goldring (1888–1971): New York Paleontologist." *Northeastern Geology and Environmental Sciences*, vol. 27, pp. 229–238. 2005.

Dawson, J. W. "On New Tree Ferns and Other Fossils from the Devonian." *Quarterly Journal of the Geological Society*, vol. 27, pp. 269–275. 1871.

Elick, Jennifer, et al. "Very Large Plant and Root Traces from the Early to Middle Devonian: Implications for Early Terrestrial Ecosystems and Atmospheric $p(CO2)$." *Geology*, vol. 26. no. 2, pp. 143–146. 1998.

Goldring, Winifred. "New Upper Devonian Plant Material," *Twenty-First Report of the Director of the State Museum and Science Department*, no. 267, pp. 85–88. 1926.

Stein, William E., et al. "Surprisingly Complex Community Discovered in the Mid-Devonian Fossil Forest at Gilboa." *Nature*, vol. 483, no. 7387, pp. 78–81. 2012.

Stein, William E., et al. "Mid-Devonian *Archaeopteris* Roots Signal Revolutionary Change in Earliest Fossil Forests." *Current Biology*, vol. 30, pp. 1–11. 2019.

Black Shale

Barth-Nafrilan, E., et al. "Methane in Groundwater Before, During and After Hydraulic Fracturing of the Marcellus Shale." *Proceedings of the National Academy of Sciences*, vol. 115, no. 27, pp. 6970–6975. 2018.

Harper, John A. "The Marcellus Shale—An Old 'New' Gas Reservoir in Pennsylvania." *Pennsylvania Geology*, vol. 38, no. 1, pp. 2–13. 2008.

7. Fires, Forests, and Coal

Crinoids and Mississippian Limestone

Gottfield, Richard. "The Burlington Limestone." *The Paleo Times. The Official Publication of the Eastern Missouri Society for Paleontology*, vol. 15, no. 10. 2016.

Hall, James. "On the Carboniferous Limestones of the Mississippi Valley." *American Journal of Science*, vol. 23, pp. 187–203. 1857.

The First Fires

Cressler, Walter L. "Plant Paleoecology of the Late Devonian Red Hill Locality, North-Central Pennsylvania, an Archaeopteris-Dominated Wetland Plant Community and Early Tetrapod Site." *Wetlands through Time, Geological Society of America, Special Paper 399*, pp. 79–102. 2006.

Falcon-Lang, Howard J. "Fire Ecology of the Carboniferous Tropical Zone," *Palaeogeography, Palaeoclimatology, Palaeoecology*, vol. 164, pp. 339–355. 2000.

Glasspool, I. J., et al. "Charcoal in the Silurian as Evidence for the Earliest Wildfire." *Geology*, vol. 32, no. 5, pp. 381–383. 2004.

Rimmer, Susan M., et al. "The Rise of Fire: Fossil Charcoal in Late Devonian Marine Shales as an Indicator of Expanding Territorial Ecosystems, Fire, and Atmosphere Change." *American Journal of Science*, vol. 315, no. 8, pp. 713–733. 2015.

Scott, Andrew C. "The Pre-Quaternary History of Fire." *Palaeogeography, Palaeoclimatology, Palaeoecology*, vol. 164, pp. 281–329. 2000.

Scott, Andrew C., and Ian J. Glasspool. "The Diversification of Paleozoic Fire Systems and Fluctuations in Atmospheric Oxygen Concentration." *Proceedings of the National Academy of Sciences*, vol. 103, no. 29, pp. 10861–10865. 2006.

Scott, Andrew C. *Burning Planet*. Oxford: Oxford University Press, 2018.

Lepidodendron and Pennsylvanian Coal

Spicer, Rachel, and Andrew Groover. "Evolution of Development of Vascular Cambia and Secondary Growth." *New Phytologist*, vol. 186, pp. 577–592. 2010.

Tudge, Colin. *The Tree*. New York: Crown Publishers, 2005.

Wanless, Harold R., and J. Marvin Weller. "Correlation and Extent of Pennsylvanian Cyclothems." *The Bulletin of the Geological Society of America*, vol. 43, pp. 1003–1016. 1932.

Wanless, Harold R., and Francis P. Shepard. "Sea Level and Climatic Changes Related to Late Paleozoic Cycles." *The Bulletin of the Geological Society of America*, vol. 47, no. 8, pp. 1177–1206. 1936.

Acadian Orogeny: The Final Assembly of New England

Hatcher, Robert D. "The Appalachian Orogen: A Brief Summary." *From Rodinia to Pangea: The Lithotectonic Record of the Appalachian Region. Geological Society of America Memoir 206*, pp. 1–19. 2010.

Alleghenian Orogeny: The Final Building of the Appalachian Mountains

Bedinger, M. S., et al. "The Waters of Hot Springs National Park, Arkansas—Their Nature and Origin." *US Geological Survey Professional Paper 1044-C*. 1979.

Poole, Forrest G., et al. "Tectonic Synthesis of the Ouachita-Marathon-Sonora Orogenic Margin of Southern Laurentia: Stratigraphic and Structural Implications for Timing of Deformational Events and Plate-Tectonic Model." *Geological Society of America Special Paper 393*, pp. 543–596. 2005.

Origin of Coal: Joggins Fossil Cliffs, Nova Scotia

Calder, John H. "'Coal Age Galapagos': Joggins and the Lions of Nineteenth Century Geology." *Atlantic Geology*, vol. 42, pp. 37–51. 2006.

Calder, John H., et al. "A Fossil Lycopsid Forest Succession in the Classic Joggins Section of Nova Scotia: Paleoecology of a Disturbance-Prone Pennsylvanian Wetland." *Geological Society of America Special Paper 399*, pp. 169–195. 2006.

Calder, John. *The Joggins Fossil Cliffs: Coal Age Galapagos*. Halifax: Formac Publishing Company, 2017.

Falcon-Lang, Howard J. "A History of Research at the Joggins Fossil Cliffs of Nova Scotia, Canada, the World's Finest Pennsylvanian Section." *Proceedings of the Geologist's Association*, vol. 117, pp. 377–392. 2006.

Lyell, Charles. *Travels in North America with Geological Observations on the United States, Canada, and Nova Scotia*. Two volumes. London: John Murray, 1845.

Scott, Andrew C. "The Legacy of Charles Lyell: Advances in Our Knowledge of Coal and Coal-Bearing Strata." *Lyell: The Past is the Key to the Present, Geological Society, London, Special Publications*, vol. 143, pp. 243–260. 1998.

Stevenson, John J. "The Formation of Coal Beds." *Proceedings of the American Philosophical Society*, vol. 50, pp. 1–116. 1911.

Lagerstätten

Clements, Thomas, et al. "The Mazon Creek Lagerstätte: A Diverse Late Paleozoic Ecosystem Entombed within Siderite Concretions." *Journal of the Geological Society*, vol. 176, no. 1, pp. 1–11. 2019.

Lyell, Charles, and J. W. Dawson. "On the Remains of a Reptile and of a Land Shell Discovered in the Interior of an Erect Fossil Tree in the Coal Measures of Nova Scotia." *Quarterly Journal of the Geological Society*, vol. 9, pp. 58–67. 1853.

Carboniferous Rainforest Collapse

Dunne, Emma M., et al. "Diversity Change during the Rise of Tetrapods and the Impact of the 'Carboniferous Rainforest Collapse.'" *Proceedings of the Royal Society B*, vol. 285, 20172730. 2018.

Feulner, Georg. "Formation of Most of Our Coal Brought Earth Close to Global Glaciation." *Proceedings of the National Academy of Sciences*, vol. 114, no. 43, pp. 11, 333, 337. 2017.

8. The Great Dying

Texas Oil, Part I: The Permian Story

Gaswirth, Stephanie, et al. "Assessment of Undiscovered Continuous Oil and Gas Resources in the Wolfcamp Shale and Bone Spring Formation of the Delaware Basin,

Permian Basin Province, New Mexico and Texas, 2018." *US Geological Survey Fact Sheet 2018-3073*. 2018.

Kilian, Lutz. "The Impact of the Shale Oil Revolution on US Oil and Gasoline Prices." *Review of Environmental Economics and Policy*, vol. 10, no. 2, pp. 185–205. 2016.

Permian Basin, West Texas

Mann, P., L. Gahagan, and M. B. Gordon. "Tectonic Setting of the World's Giant Oil and Gas Fields." *Giant Oil and Gas Fields of the Decade 1990–1999, AAPG Memoir 78*, pp. 15–105. 2003.

Speight, James G. *The Chemistry and Technology of Petroleum*. Fifth edition. Boca Raton, Florida: CRC Press, 2014.

Guadalupe Mountains and the Castile Formation

Alexander, J. I. D., and A. J. Watkinson. "Microfolding in the Permian Castile Formation: An Example of Geometric Systems in Multilayer folding, Texas, and New Mexico." *Geological Society of America Bulletin*, vol. 101, no. 5, pp. 742–750. 1989.

Anderson, Roger Y., et al. "Permian Castile Varved Evaporite Sequence, West Texas, and New Mexico." *Journal of Geophysical Research*, vol. 83, pp. 59–86. 1972.

Pray, L. C. "Geology of the Western Escarpment, Guadalupe Mountain, Texas." *Geologic Guide to the Western Escarpment, Guadalupe Mountains, Texas*. Midland, Texas: Society for Sedimentary Geology, Permian Basin Section, 1988.

Grand Canyon: Tapeats Sandstone to Kaibab Limestone

Abbott, Lon, and Terri Cook. *Hiking the Grand Canyon's Geology*. Seattle: Mountaineers Books, 2004.

Ancestral Rockies and Palo Duro Canyon

Matthews, William H. *The Geologic Story of Palo Duro Canyon*. Guidebook 8. Austin, Texas: Bureau of Economic Geology, University of Texas at Austin, 1969.

Nelson, W. John, and Spencer G. Lucas. "Carboniferous Geologic History of the Rocky Mountain Region." *New Mexico Museum of Natural History and Science*, Bulletin 53, pp. 115–142. 2011.

Ross, Marcus R., et al. "Garden of the Gods at Colorado Springs: Paleozoic and Mesozoic Sedimentation and Tectonics." *Geological Society of America*, Field Guide 18, pp. 77–93. 2010.

Permian Extinction

Black, Benjamin A., et al. "Systemic Swings in End-Permian Climate from Siberian Traps Carbon and Sulfur Outgassing." *Nature Geoscience*, vol. 11, pp. 949–954. 2018.

Bond, David P. G., et al. "The Middle Permian (Capitanian) Mass Extinction on Land and in the Oceans." *Earth-Science Reviews*, vol. 102, pp. 100–116. 2010.

Burgess, Seth D., et al. "High-Precision Timeline for Earth's Most Severe Extinction." *Proceedings of the National Academy of Sciences*, vol. 111, no. 13, pp. 3316–3321. 2014.

Cui, Ying. "Climate Swings in Extinction." *Nature Geoscience*, vol. 11, pp. 889–890. 2018.

Erwin, Douglas H. *Extinction: How Life on Earth Nearly Ended 250 Million Years Ago*.

Princeton, New Jersey: Princeton University Press, 2006.

Labandeira, Conrad C. "The Fossil Record of Insect Extinction: New Approaches and Future Directions." *American Entomologist*, vol. 51, no. 1, pp. 14–29. 2005.

Rampino, Michael R., and Shu-Zhong Shen. "The End-Guadalupian (259.8 Ma) Biodiversity Crisis: The Sixth Major Mass Extinction?" *Historical Biology*, vol. 33, issue 5, pp. 716–722. 2019.

Retallack, Gregory J., et al. "Middle-Late Permian Mass Extinction on Land." *Geological Society of America Bulletin*, vol. 118, no. 11/12, pp. 1398–1411. 2006.

Rothman, Daniel H., et al. "Methanogenic Burst in the End-Permian Carbon Cycle." *Proceedings of the National Academy of Sciences*, vol. 111, no. 15, pp. 5462–5467. 2014.

Sperling, Erik A., and James C. Ingle Jr. "A Permian-Triassic Boundary Section at Quinn River Crossing, Northwestern Nevada, and Implications for the Cause of the Early Triassic Chert Gap on the Western Pangean Margin." *Geological Society of America Bulletin*, vol. 118, no. 5/6, pp. 733–746. 2006.

Tabor, Neil J. "Wastelands of Tropical Pangea: High Heat in the Permian." *Geology*, vol. 41, no. 5, pp. 623–624. 2013.

Wignall, P. B. "Large Igneous Provinces and Mass Extinctions." *Earth-Science Reviews*, vol. 53, pp. 1–33. 2001.

End of the Paleozoic

Phillips, John. *Life on Earth: Origin and Succession*. Cambridge, London: Macmillan, 1860.

9. A Grand Staircase

Steps of the Grand Staircase

Chronic, Halka. *Pages of Stone: Grand Canyon and the Plateau Country*. Seattle: Mountaineers, 1988.

Dutton, Clarence. *Tertiary History of the Grand Canon District*. Washington, DC: US Government Printing Office, 1882.

Loope, David B., et al. "Wind Scour of Navajo Sandstone at the Wave (Central Colorado Plateau, USA)." *Journal of Geology*, vol. 116, pp. 173–183. 2008.

Megamonsoons

Dickinson, William R., and George E. Gehrels. "U-Pb Ages of Detrital Zircons in Jurassic Eolian and Associated Sandstones of the Colorado Plateau: Evidence for Transcontinental Dispersal and Intraregional Recycling of Sediment." *Geological Society of America Bulletin*, vol. 121, no. 3/4, pp. 408–433. 2009.

Dudbiel, Russell F., et al. "The Pangaean Megamonsoon—Evidence from the Upper Triassic Chinle Formation, Colorado Plateau." *PALAIOS*, vol. 6, no. 4, pp. 347–370. 1991.

Kutzbach, J. E., and R. G. Gallimore. "Pangaean Climates: Megamonsoons of the Megacontinent." *Journal of Geophysical Research*, vol. 94, no. D3, pp, 3341–3357. 1989.

Loope, David, et al. "Tropical Westerlies over Pangaean Sand Seas." *Sedimentology*, vol. 51, pp. 315–322. 2004.

Carnian Pluvial Event

Marshall, Michael. "A Million Years of Triassic Rain." *Nature*, vol. 576, no. 7787, pp. 26–28. 2019.

Simms, Michael J., and Alastair H. Ruffell. "The Carnian Pluvial Episode: From Discovery, through Obscurity, to Acceptance." *Journal of the Geological Society*, vol. 175, no. 6, pp. 989–992. 2018.

Petrified Forest National Park, Arizona

Lubick, George M. *Petrified Forest National Park*. Tucson: University of Arizona Press, 1996.

An Early Dinosaur: Coelophysis

Benton, Michael J., et al. "The Carnian Pluvial Episode and the Origin of Dinosaurs." *Journal of the Geological Society*, vol. 175, pp. 1019–1026. 2018.

Bernardi, Massimo, et al. "Dinosaur Diversification Linked with the Carnian Pluvial Episode." *Nature Communications*, vol. 9, pp. 1–10. 2018.

Colbert, E. H. *The Little Dinosaurs of Ghost Ranch*. New York: Columbia University Press. 1995.

Dal Corso, Jacopo, et al. "Discovery of a Major Negative δ13C Spike in the Carnian (Late Triassic) Linked to the Eruption of Wrangellia Flood Basalts." *Geology*, vol. 40, no. 1, pp. 79–82. 2012.

The New Jersey Palisades and the Opening of the Atlantic Ocean

Blackburn, Terrence J., et al. "Zircon U-Pb Geochronology Links the End-Triassic Extinction with the Central Atlantic Magmatic Province." *Science*, vol. 340, no. 6135, pp. 941–945. 2013.

Cuffey, Roger J., et al. "Geology of the Gettysburg Battlefield: How Mesozoic Events and Processes Impacted American History." *Geological Society of America Field Guide 8*, pp. 1–16. 2015.

Marzoli, Andrea, et al. "Extensive 200-Million-Year-Old Continental Flood Basalts of the Central Atlantic Magmatic Province." *Science*, vol. 284, no. 5414, pp. 616–618. 1999.

The Discovery of Dinosaurs

Brusatte, Stephen. "The Unlikely Triumph of Dinosaurs." *Scientific American*, pp. 28–35. 2018.

Delair, Justin B., and William A. S. Sargeant. "The Earliest Discoveries of Dinosaurs." *ISIS*, vol. 66, no. 1, pp. 4–25. 1975.

Moore, Randy. *Dinosaurs by the Decades*. Santa Barbara, California: Greenwood, 2014.

Paul, Gregory S. *The Princeton Field Guide to Dinosaurs*. Princeton, New Jersey: Princeton University Press, 2010.

Thomson, Keith Stewart. "American Dinosaurs: Who and What Was First?" *American Scientist*, vol. 94, no. 3, pp. 209–211. 2006.

Dinosaur Tracks

Alexander, R. McNeill. "How Dinosaurs Ran." *Scientific American*, vol. 264, no. 4, pp. 130–136. 1991.

Hitchcock, E. H. "Ornithichnology—Description of the Footmarks of Birds, (Ornithichnites) on New Red Sandstone in Massachusetts." *American Journal of Science*, vol. 29, no. 1, pp. 307–340. 1836.

Lockley, Martin G., et al. "Limping Dinosaurs? Trackway Evidence for Abnormal Gaits." *Ichnos*, vol. 3, pp. 193–202. 1994.

Milnere, Andrew R. C., et al. "The Story of the St. George Dinosaur Discovery Site at Johnson Farm: An Important New Lower Jurassic Dinosaur Track Site from the Moenave Formation of Southwestern Utah." *New Mexico Museum of Natural History and Science Bulletin 37*, pp. 329–345. 2006.

Mossman, David J., and William A. S. Sarjeant. "The Footprints of Extinct Animals." *Scientific American*, vol. 248, no. 1, pp. 74–85. 1983.

Schumacher, Bruce, and Martin Lockley. "Newly Documented Trackways at 'Dinosaur Lake,' the Purgatoire Valley Dinosaur Tracksite." *New Mexico Museum of Natural History and Science Bulletin 62*, pp. 261–268. 2014.

Sellers, William Irvin, and Phillip Lars Manning. "Estimating Dinosaur Maximum Running Speeds Using Evolutionary Robotics." *Proceedings of the Royal Society B*, vol. 274, pp. 2711–2716. 2007.

The Bone Wars

Carpenter, Kenneth. "History, Sedimentology, and Taphonomy of the Carnegie Quarry, Dinosaur National Monument, Utah." *Annals of Carnegie Museum*, vol. 81, no. 3, pp. 153–232. 2013.

Jaffe, Mark. *The Gilded Dinosaur*. New York: Crown Publishing, 2000.

Jurassic-Cretaceous Transition

Wimbledon, William A. P. "The Jurassic-Cretaceous Boundary: An Age-Old Correlative Enigma." *Episodes*, vol. 31, no. 4, pp. 423–428. 2008.

10. Western Interior Seaway

Monument Rocks and the Western Interior Seaway

Everhart, Michael J. *Oceans of Kansas*. Second edition. Bloomington, Indiana: Indiana University Press, 2017.

Zakrzewski, Richard J. "Geologic Studies in Western Kansas in the 19th Century." *Transactions of the Kansas Academy of Science*, vol. 99, no. 3/4, pp. 124–133. 1996.

Farallon Plate: An Introduction

Flament, Nicolas, et al. "A Review of Observations and Models of Dynamic Topography." *Lithosphere*, vol. 5, no. 2, pp. 189–210. 2013.

Gurnis, Michael. "Rapid Continent Subsidence following the Initiation and Evolution of Subduction." *Science*, vol. 255, no. 5051, pp. 1556–1558. 1992.

Perkins, Sid. "Seismic Tomography Uses Earthquake Waves to Probe the Inner Earth." *Proceedings of the National Academy of Sciences*, vol. 116, no. 33, pp. 16, 159–16, 16161. 2019.

Siglock, Karin, and Mitchell G. Mihalynuk. "Intra-oceanic Subduction Shaped the

Assembly of Cordilleran North America." *Nature*, vol. 496, no. 7443, pp. 50–56. 2013.

Spasojevic, Sonja, et al. "The Case for Dynamic Subsidence of the United States East Coast since the Eocene." *Geophysical Research Letters*, vol. 35, L08305. 2008.

Spasojevic, Sonja, and Michael Gurnis. "Sea Level and Vertical Motion of Continents from Dynamic Earth Models since the Late Cretaceous." *American Association of Petroleum Geologists Bulletin*, vol. 96, no. 11, pp. 2037–2064. 2012.

Williams, M. L., et al. *Unlocking the Secrets of the North America Continent.* p. 78. 2010.

Suspect Terranes and the Growth of Continents

Coney, Peter J., et al. "Cordilleran Suspect Terranes." *Nature*, vol. 288, no. 5789, pp. 329–333. 1980.

Jones, David L., et al. "The growth of western North America." *Scientific American*, vol. 247, no. 5, pp. 70–85. 1982.

Tabor, Rowland W., and Ralph Albert Haugerud. *Geology of the North Cascades.* Seattle: Mountaineers, 1999.

Wells, Ray, et al. "Geologic History of Siletzia, a Large Igneous Province in the Oregon and Washington Coast Range: Correlation to the Geomagnetic Polarity Time Scale and Implications for a Long-Lived Yellowstone Hotspot." *Geosphere*, vol. 10, no. 4, pp. 692–719. 2014.

Splitting of Pangea and the Origin of Florida

Mueller, Paul A., et al. "The Suwannee Suture: Significance for Gondwana-Laurentia Terrane Transfer and Formation of Pangaea." *Gondwana Research*, vol. 26, pp. 365–373. 2014.

Texas Oil, Part II: The Cretaceous Story

Mann, P., L. Gahagan, and M. B. Gordon. "Tectonic Setting of the World's Giant Oil and Gas Fields." *Giant Oil and Gas Fields of the Decade 1990–1999, AAPG Memoir 78,* pp. 15–105. 2003.

Owens, William A. "Gusher at Spindletop." *American Heritage*, vol. 9, no. 4. June 1958.

Wright, Lawrence. "The Dark Bounty of Texas Oil." *The New Yorker*, January 1, 2018.

Marion King Hubbert, Peak Oil, and Oil Shale

Hubbert, M. King. *Nuclear Energy and Fossil Fuels.* Publication 96, Shell Oil Company. 1956.

Inman, Mason. *The Oracle of Oil.* New York: W. W. Norton, 2016.

Kilian, Lutz. "The Impact of the Shale Oil Revolution on US Oil and Gasoline Prices." *Review of Environmental Economics and Policy*, vol. 10, no. 2, pp. 185–205. 2016.

Moyer, Michael. "How Much Is Left?" *Scientific American*, vol. 303, no. 3, pp. 74–81. 2010.

Priest, Tyler. "Hubbert's Peak: The Great Debate over the End of Oil." *Historical Studies in the Natural Sciences*, vol. 44, no. 1, pp. 37–79. 2014.

Origin of Angiosperms: Fossilized Leaves in Kansas

Bolick, M. R., et al. "The Abominable Mystery of the First Flowers: Clues from Nebraska and Kansas." *Museum Notes, University of Nebraska State Museum*, no. 88. 1994.

Doyle, James A. "Molecular and Fossil Evidence on the Origin of Angiosperms." *Annual Review of Earth and Planetary Sciences*, vol. 40, pp. 301–326. 2012.

Friis, Else Marie. *Early Flower and Angiosperm Evolution.* Cambridge: Cambridge University Press, 2011.

Hall, J., and F. B. Meek. "Descriptions of New Species of Fossils, from the Cretaceous Formations of Nebraska." *Memoirs of the American Academy of Arts and Sciences*, vol. 5, no. 2, pp. 379–411. 1853.

Jansen, Robert K., et al. "Analysis of 81 Genes from 64 Plastid Genomes Resolves Relationships in Angiosperms and Identifies Genome-Scale Evolutionary Patterns." *Proceedings of the National Academy of Sciences*, vol. 104, no. 49, pp. 19369–19374. 2007.

Jud, Nathan A., and Leo J. Hickey. "*Potomacapnos apeleutheron* Gen. et Sp. Nov., A New Early Cretaceous Angiosperm from the Potomac Group and Its Implications for the Evolution of Eudicot Leaf Architecture." *American Journal of Botany*, vol. 100, no. 12, pp. 2437–2449. 2013.

Monger, J. W. H., and C. A. Ross. "Distribution of Fusulinaceans in the Western Canadian Cordillera." *Canadian Journal of Science*, vol. 8, pp. 259–278. 1971.

Sauquet, Hervé, et al. "The Ancestral Flower of Angiosperms and Its Early Diversification." *Nature Communications*, vol. 8, article number 16047. 2017.

Birds and Tyrannosaurus rex

Bakker, Robert T. "Dinosaur Renaissance." *Scientific American*, vol. 232, no. 4, pp. 58–78. 1975.

Benton, Michael J. *Dinosaurs Rediscovered.* London: Thames & Hudson, 2019.

Brown, Barnum. "Tyrannosaurus, a Cretaceous Carnivorous Dinosaur." *Scientific American*, vol. 113, no. 15, pp. 322–323. 1915.

Brusatte, Stephen. "Taking Wing." *Scientific American*, vol. 236, number 1, pp. 48–55. 2017.

Dingus, Lowell, and Mark A. Norell. *Barnum Brown: The Man Who Discovered Tyrannosaurus Rex.* Berkeley: University of California Press, 2010.

Erickson, Gregory M. "Breathing Life into *Tyrannosaurus rex.*" *Scientific American*, vol. 281, no. 3, pp. 42–49. 1999.

Osborn, Henry Fairfield. "Tyrannosaurus and the Other Cretaceous Carnivorous Dinosaurs." *American Museum of Natural History Bulletin*, vol. 21, article 14, pp. 259–265. 1905.

Ostrom, John H. "Osteology of *Deinonychus antirrhopus*, an Unusual Theropod from the Lower Cretaceous of Montana." *Bulletin of the Peabody Museum of Natural History*, vol. 30, pp. 1–65. 1969.

11. A Calamitous Event

Meteor Crater, Arizona

Barringer, Daniel Moreau. "Coon Mountain and Its Crater." *Proceedings of the Academy of Natural Science of Philadelphia*, vol. 57, pp. 861–886. 1906.

Gilbert, Grove K. "The Origin of Hypotheses, Illustrated by the Discussion of a Topographic Program." *Science, New Series*, vol. 3, no. 53, pp. 1–13. 1896.

Shoemaker, Eugene M. "Penetration Mechanics of High Velocity Meteorites, Illustrated by Meteor Crater, Arizona." *Proceedings of the International Geological Congress, 21st Session, Part 18*, Copenhagen, pp. 418–434. 1960.

Shoemaker, Eugene M. "Asteroid and Comet Bombardment of the Earth." *Annual Reviews of Earth and Planetary Sciences*, vol. 11, pp. 461–494. 1983.

History of Meteorite Research

Baldwin, Ralph B. *The Face of the Moon*. Chicago: University of Chicago Press. 1949.

Marvin, Ursula B. "Ernst Florens Friedirch Chladni (1756–1827) and the Origins of Modern Meteorite Research." *Meteoritics & Planetary Science*, vol. 42, no. 9, Supplement, pp. B3–B68. 2007.

Pillinger C. T., and J. M. Pillinger. "The Wold Cottage Meteorite: Not Just Any Ordinary Chondrite." *Meteoritics & Planetary Science*, vol. 31, pp. 589–605. 1996.

Terrestrial Impact Craters

Alvarez, Walter, et al. "Synsedimentary Deformation in the Jurassic of Southeastern Utah: A Case of Impact Shaking?" *Geology*, vol. 26, no. 7, pp. 579–582. 1998.

Buchner, Elmar, and Thomas Kenkmann. "Upheaval Dome, Utah, USA: Impact Origin Confirmed." *Geology*, vol. 36, no. 3, pp. 227–230. 2008.

Flamini, E., et al. *Encyclopedic Atlas of Terrestrial Impact Craters*. London: Springer, 2019.

Glass, Billy P., and Bruce M. Simonson. *Distal Impact Ejecta Layers*. London: Springer, 2013.

Kriens, Bryan J., et al. "Geology of the Upheaval Dome Impact Structure, Southeast Utah." *Journal of Geophysical Research*, vol. 104, no. E8, pp. 18867–18887. 1999.

Morrow, J. R., et al. "Late Devonian Alamo Impact, Southern Nevada, USA: Evidence of Size, Marine Site, and Widespread Effects." *Large Meteorite Impacts III, Geological Society of America Special Paper 384*, pp. 259–280. 2005.

Poag, C. Wylie, et al. "Meteoroid Mayhem in Ole Virginny: Source of the North American Tektite Strewn Field." *Geology*, vol. 22, pp. 691–694. 1994.

Rondot, J. "Charlevoix and Sudbury as Gravity-Adjusted Impact Structures." *Meteoritics & Planetary Science*, vol. 35, pp. 707–712. 2000.

Schmieder, Martin, and David A. Kring. "Earth's Impact Events through Geologic Time: A List of Recommended Ages for Terrestrial Impact Structures and Deposits." *Astrobiology*, vol. 20, no. 1, pp. 91–141. 2020.

Schmitz, Birger, et al. "An Extraterrestrial Trigger for the Mid-Ordovician Ice Age: Dust from the Breakup of the L-Chondrite Parent Body." *Science Advances*, vol. 5, eaax4184. 2019.

Varricchio, D. J., et al. "Tracing the Manson Impact Event across the Western Interior Cretaceous Seaway." *Large Meteorite Impacts and Planetary Evolution IV, Geological Society of America Special Paper 465*, pp. 269–299. 2010.

K-T Boundary: Iridium Anomaly and Glass Spherules

Alvarez, Luis W., et al. "Extraterrestrial Cause for the Cretaceous-Tertiary Extinction." *Science*, vol. 208, no. 4448, pp. 1095–1108. 1980.

Claeys, P., et al. "Distribution of Chicxulub Ejecta at the Cretaceous-Tertiary Boundary." *Catastrophic Events and Mass Extinctions: Impacts and Beyond. Geological Society of America Special Paper 356*, pp. 55–68. 2002.

Chicxulub Crater: Discovery and Cenotes

Gulick, S. P. S., et al. "Geophysical Characterization of the Chicxulub Impact Crater." *Reviews of Geophysics*, vol. 51, pp. 31–52. 2013.

Gulick, et al. "The First Day of the Cenozoic." *Proceedings of the National Academy of Sciences*, vol. 116, no. 39, pp. 19342–19351. 2019.

Hildebrand, Alan R., et al. "Chicxulub Crater: A Possible Cretaceous/Tertiary Boundary Impact Crater on the Yucatán Peninsula, Mexico." *Geology*, vol. 19, no. 9, pp. 867–871. 1991.

Perry, Eugene, et al. "Ring of Cenotes (Sinkholes), Northwest Yucatán, Mexico: Its Hydrogeologic Characteristics and Possible Association with the Chicxulub Impact Crater." *Geology*, vol. 23, no. 1, pp. 17–20. 1995.

Chicxulub Crater: Formation

Bourgeois, Joanne, et al. "A Tsunami Deposit at the Cretaceous-Tertiary Boundary in Texas." *Science*, vol. 241, no. 4865, pp. 567–570. 1988.

Day, Simon, and Mark Maslin. "Linking Large Impacts, Gas Hydrates, and Carbon Isotope Excursions through Widespread Sediment Liquefaction and Continental Slope Failure: The Example of the K-T Boundary Event." *Large Meteorite Impacts III, Geological Society of America Special Paper 384*, pp. 239–258. 2005.

Gulick, S. P. S., et al. "Geophysical Characterization of the Chicxulub Impact Crater." *Reviews of Geophysics*, vol. 51, pp. 31–52. 2013.

Hart, M. B., et al. "The Cretaceous/Paleogene Boundary Events in the Gulf Coast: Comparisons between Alabama and Texas." *Gulf Coast Association of Geological Societies Transactions*, vol. 63, pp. 235–255. 2013.

Matsui, T., et al. "Generation and Propagation of a Tsunami from the Cretaceous-Tertiary Impact Event." *Catastrophic Events and Mass Extinctions: Impacts and Beyond. Geological Society of America Special Paper 356*, pp. 69–77. 2002.

Poag, C. Wylie. "Shake *and* Stirred: Seismic Evidence of Chicxulub Impact Effects on the West Florida Carbonate Platform, Gulf of Mexico." *Geology*, vol. 45, no. 11, pp. 1011–1014. 2017.

Smit, J. "The Global Stratigraphy of the Cretaceous-Tertiary Boundary Impact Ejecta." *Annual Reviews of Earth and Planetary Sciences*, vol. 27, pp. 75–113. 1999.

Witts, James D., et al "A Fossiliferous Spherule-Rich Bed at the Cretaceous-Paleogene (K-Pg) Boundary in Mississippi, USA: Implications for the K-Pg Mass Extinction Event in the Mississippi Embayment and Eastern Gulf Coastal Plain." *Cretaceous Research*, vol. 91, pp. 147–167. 2018.

Tanis, North Dakota

DePalma, Robert A., et al. "A Seismically Induced Onshore Surge Deposit at the KPg Boundary, North Dakota." *Proceedings of the National Academy of Sciences*, vol. 116, no. 17, pp. 8190–8199. 2019.

12. Extinction

Discovery of Extinction: Georges Cuvier

Harward, James L. "Fossil Proboscideans and Myths of Giant Men." *Transactions of the Nebraska Academy of Sciences and Affiliated Societies*, vol. 12, pp. 95–102. 1984.

Levin, David. "Giants in the Earth: Science and the Occult in Cotton Mather's Letters to the Royal Society." *William and Mary Quarterly*, vol. 45, no. 4, pp. 751–770. 1988.

Rudwick, Martin J. S. *Georges Cuvier, Fossil Bones, and Geological Catastrophes*. Chicago: University of Chicago Press, 1997.

Simpson, George Gaylord. "The Discovery of Fossil Vertebrates in North America." *Journal of Paleontology*, vol. 17, no. 1, pp. 26–38. 1943.

Taylor, Edward, et al. "The Giant Bones of Claverack, New York, 1705." *New York History*, vol. 40, no. 1, pp. 47–61. 1959.

Thomas Jefferson and Extinction

Barrow, Mark V. *Nature's Ghosts*. Chicago: University of Chicago Press, 2014.

Rowland, Stephen M. "Thomas Jefferson, Extinction, and the Evolving View of Earth History in the Last Eighteenth Century and Early Nineteenth Century." *The Revolution in Geology from the Renaissance to the Enlightenment, Geological Society of America Memoir 203*, pp. 225–246. 2009.

Cuvier versus Lyell: Catastrophism versus Uniformitarianism

Ager, Derek V. *The New Catastrophism*. Cambridge: Cambridge University Press, 1995.

Bonney, T. G. *Charles Lyell and Modern Geology*. New York: Macmillan, 1895.

Eiseley, Loren C. "Charles Lyell." *Scientific American*, vol. 201, no. 2, pp. 98–106. August 1959.

Hallam, A. "Catastrophists and Uniformitarians." *Great Geological Controversies*. Second edition. Oxford: Oxford University Press, 1989.

Major Mass Extinctions

Brannen, Peter. *The Ends of the World*. New York: HarperCollins, 2017.

Newell, Norman D. "Catastrophism and the Fossil Record." *Evolution*, vol. 10, no. 1, pp. 97–101. 1956.

Newell, Norman D. "Crises in the History of Life." *Scientific American*, vol. 208, no. 2, pp. 76–95. February 1963.

Raup, David M., and J. John Sepkoski. "Mass Extinctions in the Marine Fossil Record." *Science*, vol. 215, no. 4539, pp. 1501–1503. 1982.

Sepkoski, J. John. "Patterns of Phanerozoic Extinction: A Perspective from Global Data Bases." *Global Events and Event Stratigraphy in the Phanerozoic*, pp. 35–51. 1996.

Stanley, Steven M. "Estimates of the Magnitudes of Major Marine Mass Extinctions in Earth History." *Proceedings of the National Academy of Sciences*, vol. 113, no. 42, pp. E6325–E6334. 2016.

Large Igneous Provinces: Columbia River Basalts

Bond, David P. G., and Paul B. Wignall. "Large Igneous Provinces and Mass Extinctions:

An Update." *Volcanism, Impacts, and Mass Extinctions: Geological Society of America Special Paper 505*, pp. 29–55. 2014.

Ho, Anita M., and Katharine V. Cashman. "Temperature Constraints on the Ginkgo Flow of the Columbia River Basalt Group." *Geology*, vol. 25, no. 5, pp. 40–406. 1997.

Tolan, Terry L., et al. "An Introduction to the Stratigraphy, Structural Geology, and Hydrogeology of the Columbia River Flood-Basalt Province." *Volcanoes to Vineyards: Geological Society of America Field Guide 15*, pp. 599–643. 2009.

Wells, Ray E., et al. "The Columbia River Basalt Group—From the Gorge to the Sea." *Volcanoes to Vineyards: Geological Society of America Field Guide 15*, pp. 737–774. 2009.

Wignall, P. B. "Large Igneous Provinces and Mass Extinctions." *Earth-Science Reviews*, vol. 53, pp. 1–33. 2001.

End of the Dinosaurs: Meteor Impact or Volcanic Eruption?

Archibald, J. David, et al. "Cretaceous Extinctions: Multiple Causes." *Science*, vol. 328, no. 5981, pp. 973–976. 2010.

Difley, Rose L., and A. A. Ekdale. "Footprints of Utah's Last Dinosaurs: Track Beds in the Upper Cretaceous (Maastrichtian) North Horn Formation of the Wasatch Plateau, Central Utah." *PALAIOS*, vol. 17, no. 4, pp. 327–346. 2002.

Keller, Gerta. "The Cretaceous-Tertiary Mass Extinction, Chicxulub Impact, and Deccan Volcanism." *Earth and Life*. London: Springer, 2012.

Lyson, Tyler R., et al. "Dinosaur Extinction: Closing the '3 M Gap,'" *Biology Letters*, vol. 7, pp. 925–928. 2011.

Renne, Paul R., et al. "Time scales of Critical Events around the Cretaceous-Paleogene Boundary." *Science*, vol. 339, no. 6120, pp. 684–687. 2013.

Renne, Paul R., et al. "State Shift in Deccan Volcanism at the Cretaceous-Paleogene Boundary, Possibly Induced by Impact." *Science*, vol. 350, no. 6256, pp. 76–78. 2015.

Richards, Mark A., et al. "Triggering of the Largest Deccan Eruptions by the Chicxulub Impact." *Geological Society of America Bulletin*, vol. 127, no. 11–12, pp. 1507–1520. 2015.

Schoene, Blair, et al. "U-Pb Constraints on Pulsed Eruption of the Deccan Traps across the End-Cretaceous Mass Extinction." *Science*, vol. 363, no. 6429, pp. 862–866. 2019.

Schulte, Peter, et al. "The Chicxulub Asteroid Impact and Mass Extinction at the Cretaceous-Paleogene Boundary." *Science*, vol. 327, issue 5970, pp. 1214–1218. 2010.

Vogt, P. R. "Evidence for Global Synchronism in Mantle Plume Convection, and Possible Significance for Geology." *Nature*, vol. 240, no. 5380, pp. 338–342. 1972.

13. How the Mountains Grew

Appalachians, Rocky Mountains, and Basin and Range

Bjerstedt, T. W. "Regional Stratigraphy and Sedimentology of the Lower Mississippian Rockwell Formation and Pulslane Sandstone Based on the New Sideling Hill Road Cut, Maryland." *Southeastern Geology*, vol. 27, pp. 69–94. 1986.

Eaton, Gordon P. "The Basin and Range Province: Origin and Tectonic Significance." *Annual Reviews of Earth and Planetary Sciences*, vol. 10, pp. 409–440. 1982.

Hughes T. H., et al. "Road Log for the Trip from Denver to the Grand Junction along Interstate 70 in Colorado." *AIPG 46th Annual Meeting*, p. 20. 2009.

Theories about the Origin of the Alps

Adams, Frank Dawson. *The Birth and Development of the Geological Sciences.* Toronto: General Publishing Company, 1938.

Alexander, David. "Dante and the Form of the Land." *Annals of the Association of American Geographers.* vol. 76, no. 1, pp. 38–49. 1986.

Bainbridge, Simon. *Mountaineering and British Romanticism.* Oxford: Oxford University Press, 2020.

Burnet, Thomas. *Sacred Theory of the Earth.* Glasgow: R. Urie, 1753.

Gould, Stephen Jay. *Time's Arrow, Time's Cycle.* Cambridge, Massachusetts: Harvard University Press, 1987.

Greene, Mott. *Geology in the Nineteenth Century.* Ithaca, New York: Cornell University Press, 1982.

Theories about the Origin of the Appalachians

Dana, James D. "On Some Results of the Earth's Contraction from Cooling, Including a Discussion of the Origin of Mountains, and the Nature of the Earth's interior." *American Journal of Science,* third series, vol. 5, pp. 423–443. 1873.

Gerstner, Patsy A. "A Dynamic Theory of Mountain Building: Henry Darwin Rogers." *Isis,* vol. 66, no. 1, pp. 26–37. 1975.

Hall, James. "On the Vertical Position of Certain Strata and Their Relations with Granite." *Transactions of the Total Society of Edinburgh,* vol. 7, no. 1, pp. 79–108. 1815.

Le Conte, Joseph. "Honors to James Hall." *Science,* new series, vol. 4, no. 98, pp. 699. 1896.

Murphy, J. Brendan, and R. Damian Nance. "Mountain Belts and the Supercontinent Cycle." *Scientific American,* vol. 266, no. 4, pp. 84–91. February 1992.

Ruffner, W. H. "The Brothers Rogers," *The Scottish in America.* Nashville: Barber & Smith, pp. 123–139. 1895.

Sevier Orogeny

Armstrong, Richard Lee. "Sevier Orogenic Belt in Nevada and Utah." *Geological Society of America Bulletin,* vol. 79, no. 4, pp. 429–458. 1968.

Maier, Analisa, et al. "Geology, Age and Field Relations of Hadean Zircon-Bearing Supracrustal Rocks from Quad Creek, Eastern Beartooth Mountains (Montana and Wyoming, USA)." *Chemical Geology,* vol. 312/313, pp. 47–57. 2012.

Laramide Orogeny: Origin of the Rocky Mountains

Copeland, Peter, et al. "Location, Location, Location: The Variable Lifespan of the Laramide Orogeny." *Geology,* vol. 45, no. 3, pp. 223–226. 2017.

Dickinson, W. R., and W. Snyder. "Plate Tectonics of the Laramide Orogeny." *Geological Society of America Memoir 151,* pp. 355–366. 1978.

Gutscher, Marc-Andrè. "Scraped by Flat-slab Subduction." *Nature Geoscience,* vol. 11, pp. 889–893. 2018.

Heller, Paul L., and Lijun Liu. "Dynamic Topography and Vertical Motion of the US Rocky Mountain Region Prior to and during the Laramide Orogeny." *Geological Society of America Bulletin,* vol. 128, no. 5/6, pp. 973–988. 2016.

Humphreys, Eugene. "Relation of Flat Subduction to Magmatism and Deformation in the

Western United States." *Backbone of the Americas, Geological Society of America Memoir 204*, pp. 1–14. 2009.

Ko, Justin Yen-Ting, et al. "Lower Mantle Substructure Embedded in the Farallon Plate: The Hess Conjugate." *Geophysical Research Letters*, vol. 44, pp. 10216–10225. 2017.

Liu, Lijun, et al. "The Role of Oceanic Plateau Subduction in the Laramide Orogeny." *Nature Geoscience*, vol. 3, pp. 353–357. 2010.

Redden, Jack A., and Ed DeWitt. "Maps Showing Geology, Structure, and Geophysics of the Central Black Hills, South Dakota." *Scientific Investigations Map 2777. US Department of the Interior. 1:100,000 scale.* 2008.

Sager, Willam W. "Massif Redo." *Scientific American*, vol. 322, no. 5, pp. 48–53. 2020.

Sun, Daoyuan, et al. "A Dipping, Thick Segment of the Farallon Slab beneath Central US." *Journal of Geophysical Research, Solid Earth*, vol. 122, pp. 2911–2928. 2017.

Tikoff, Basil, and Julie Maxson. "Lithospheric Buckling of the Laramide Foreland during Late Cretaceous and Paleogene, Western United States." *Rocky Mountain Geology*, vol. 36, no. 1, pp. 13–35. 2001.

Yonkee, W. Adolph, and Arlo Brandon Weil. "Tectonic Evolution of the Sevier and Laramide Belts within the North American Cordillera Orogenic System." *Earth-Science Reviews*, vol. 150, pp. 531–593. 2015.

Závada, P., et al. "Devils Tower (Wyoming, USA): A Lava Coulée Emplaced into a Maardiatreme Volcano?" *Geosphere*, vol. 11, no. 2, pp. 354–375. 2015.

Delamination: Basin and Range

Bird, Peter. "Continental Delamination and the Colorado Plateau." *Journal of Geophysical Research*, vol. 84, no. B13, pp. 7561–7571. 1979.

DeCelles, P. G. "Late Jurassic to Eocene Evolution of the Cordilleran Thrust Belt and Foreland Basin System, Western USA." *American Journal of Science*, vol. 304, pp. 105–168. 2004.

Porter, Ryan C., et al. "Dynamic Lithosphere within the Great Basin." *Geochemistry, Geophysics, Geosystems*. vol. 15, pp. 1128–1146. 2014.

Snow, J. Kent, and Brian P. Wernicke. "Cenozoic Tectonism in the Central Basin and Range: Magnitude, Rate, and Distribution of Upper Crustal Strain." *American Journal of Science*, vol. 3400, pp. 659–719. 2000.

Wernicke, Brian, and J. Kent Snow. "Cenozoic Tectonism in the Central Basin and Range: Motion of the Sierran-Great Valley Block." *International Geology Review*, vol. 40, no. 5, pp. 403–410. 1998.

West, John D., et al. "Vertical Mantle Flow Associated with a Lithospheric Drip beneath the Great Basin." *Nature Geoscience*, vol. 2, no. 6, pp. 439–444. 2009.

Delamination: Appalachians and Sierra Nevada

Busby, Cathy J., and Keith Putirka. "Miocene Evolution of the Western Edge of the Nevadaplano in the Central and Northern Sierra Nevada: Palaeocanyons, Magmatism, and Structure." *International Geology Review*, vol. 51, no. 7–8, pp. 670–701. 2009.

Byrnes, Joseph S., et al. "Thin Lithosphere beneath the Central Appalachian Mountains: Constraints from Seismic Attenuation beneath the MAGIC Array." *Earth and Planetary Science Letters*, vol. 519, pp. 297–307. 2019.

Ducea, Mihai N. "Fingerprinting Orogenic Delamination." *Geology*, vol. 39, no. 2, pp. 191–192. 2011.

Gallen, Sean F., et al. "Miocene Rejuvenation of Topographic Relief in the Southern Appalachians." *GSA Today*, vol. 23, no. 2, pp. 4–10. 2013.

Garzione, Carmala, et al. "Rise of the Andes." *Science*, vol. 320, no. 5881, pp. 1304–1307. 2008.

Gilbert, H., et al. "Imaging Lithospheric Foundering in the Structure of the Sierra Nevada." *Geosphere*, vol. 8, no. 6, pp. 1310–1323. 2012.

Hammond, William C., et al. "Contemporary Uplift of the Sierra Nevada, Western United States, from GPS and InSAR Measurements." *Geology*, vol. 40, no. 7, pp. 667–670. 2012.

Horton Jr., J. W., et al. "The 2011 Mineral, Virginia, Earthquake, and Its Significance for Seismic Hazards in Eastern North America—Overview and Synthesis." *Geological Society of America Special Paper 509*, pp. 1–25. 2015.

Li, Zhong-Hai, et al. "Lithosphere Delamination in Continental Collisional Orogens: A Systematic Numerical Study." *Journal of Geophysical Research*, vol. 121, no. B, pp. 5186–5211. 2016.

Mazza, S. E., et al. "Volcanoes of the Passive Margin: The Youngest Magmatic Event in Eastern North America." *Geology*, vol. 42, no. 6, pp. 483–486. 2014.

Wagner, Lara S., et al. "The Relative Roles of Inheritance and Long-term Passive Margin Lithospheric Evolution on the Modern Structure and Tectonic Activity in the Southeastern United States." *Geosphere*, vol. 14, no. 4, pp. 1385–1410. 2018.

Yang, Xiaotao, and Haiying Gao. "Full-Wave Seismic Tomography in the Northeastern United States: New Insights into the Uplift Mechanism of the Adirondack Mountains." *Geophysical Research Letters*, vol. 45, pp. 5992–6000. 2018.

Zandt, George, et al. "Active Foundering of a Continental Arc Root beneath the Southern Sierra Nevada in California." *Nature*, vol. 431, no. 7004, pp. 41–46. 2004.

Farallon Plate

Forte, A. M., et al. "Descent of the Ancient Farallon Slab Drives Localized Mantle Flow below the New Madrid Seismic Zone." *Geophysical Research Letters*, vol. 34, pp. L04308–L04308. 2007.

Ko, Justin Yen-Ting, et al. "Lower Mantle Substructure Embedded in the Farallon Plate: The Hess Conjugate." *Geophysical Research Letters*, vol. 44, pp. 10216–10225. 2017.

Sigloch, Karin. "Mantle Provinces under North America from Multifrequency P Wave Tomography." *Geochemistry, Geophysics, Geosystems*, vol. 12, Q02W08. 2011.

Sun, Daoyuan, et al. "A Dipping, Thick Segment of the Farallon Slab beneath Central US." *Journal of Geophysical Research, Solid Earth*, vol. 122, pp. 2911–2928. 2017.

Wells, Ray, et al. "Geologic History of Siletzia, a Large Igneous Province in the Oregon and Washington Coast Range: Correlation to the Geomagnetic Polarity Time Scale and Implications for a Long-Lived Yellowstone Hotspot." *Geosphere*, vol. 10, no. 4, pp. 692–719. 2014.

Slab Subduction, Mantle Plumes, and Mantle Avalanches

Burke, Kevin. "Plate Tectonics, the Wilson Cycle, and Mantle Plumes: Geodynamics from the Top." *Annual Review of Earth and Planetary Sciences*, vol. 39, pp. 1–29. 2011.

Condie, Kent C. "Juvenile Crust, Mantle Avalanches, and Supercontinents in the Last 1.6 Ga." *Gondwana Research*, vol. 2, no. 4, pp. 543. 1999.

French, Scott W., and Barbara Romanowicz. "Broad Plumes Rooted at the Base of the Earth's Mantle beneath Major Hotspots." *Nature*, vol. 525, no. 7567, pp. 95–99. 2015.

Fukao, Yoshio, et al. "Stagnant Slab: A Review." *Annual Review of Earth and Planetary Sciences*, vol. 37, pp. 19–466. 2009.

Garnero, Edward J., et al. "Continent-sized Anomalous Zones with Low Seismic Velocity at the Base of Earth's Mantle." *Nature Geoscience*, vol. 9, pp. 481–489. 2016.

Harlan, Stephen S., et al. "Gunbarrel Mafic Magmatic Event: A key 780 Ma Time Marker for Rodinia Plate Reconstructions." *Geology*, vol. 31, no. 12, pp. 1053–1056. 2003.

Heron, Phil, and Ed Garnero. "'What lies beneath? Thoughts on the Lower Mantle." *Geoscientist*, vol. 29, no. 3, pp. 10–15. 2019.

Jeanloz, Raymond, and Thorne Lay. "The Core-Mantle Boundary." *Scientific American*, vol. 268, no. 5, pp. 48–55. May 1993.

O'Neill, C., et al. "Earth's Punctuated Tectonic Evolution: Cause and Effect." *Continent Formation Through Time, Geological Society of London, Special Publications 389*, pp. 17–40. 2015.

14. The Great Lakes of Wyoming

Recovery from Extinction: End of the Mesozoic

Alfaro, Michael E., et al. "Explosive Diversification of Marine Fishes at the Cretaceous–Palaeogene Boundary." *Nature Ecology & Evolution*, vol. 2, pp. 688–696. 2018.

Donovan, Michael P., et al. "Rapid Recovery of Patagonian Plant–Insect Associations after the End-Cretaceous Extinction." *Nature Ecology & Evolution*, vol. 1, 0012. 2017.

Field, Daniel J., et al. "Late Cretaceous Neornithine from Europe Illuminates the Origins of Crown Birds." *Nature*, vol. 579, no. 7799, pp. 397–401. 2020.

Jarvis, Erich D., et al. "Whole-Genome Analyses Resolve Early Branches in the Tree of Life of Modern Birds." *Science*, vol. 346, issue 6215, pp. 1320–1331. 2014.

Longrich, Nicholas R., et al. "Mass Extinction of Lizards and Snakes at the Cretaceous-Paleogene Boundary." *Proceedings of the National Academy of Sciences*, vol. 109, no. 52, pp. 21396–21401. 2018.

Fossil Butte, Wyoming, and the PETM

Buchheim, H. Paul. "A Walk through Time at Fossil Butte: Historical Geology of the Green River Formation at Fossil Butte National Monument." *Technical Report NPS/NRGRD/GRDTY-98/01*, pp. 56–60. 1998.

Buchheim, H. Paul, et al. "Stratigraphic Revision of the Green River Formation in Fossil Basin, Wyoming: Overfilled to Underfilled Lake Evolution." *Rocky Mountain Geology*, vol. 46, no. 2, pp. 165–181. 2011.

Green River Formation, Wyoming

Dyni, John R. "Geology and Resources of Some World Oil-shale Deposits." *Oil Shale*, vol. 20, no. 3, pp. 193–252. 2003.

McInerney, Francesca A., and Scott L. Wing. "The Paleocene-Eocene Thermal Maximum:

A Perturbation of Carbon Cycle, Climate, and Biosphere with Implications for the Future." *Annual Review of Earth and Planetary Sciences*, vol. 39, pp. 489–516. 2011.

Smith, Michael Elliot, and Alan R. Carroll. "Introduction to the Green River Formation," *Stratigraphy and Paleolimnology of the Green River Formation, Western USA*. London: Springer, pp. 1–12, 2015.

San Andreas Fault, California

Baldridge, W. R., et al. "The Rio Grande Rift." *Developments in Geotectonics*, vol. 25, pp. 233–275. 2006.

Faulds, James E., et al. "Kinematics of the Northern Walker Lane: An Incipient Transform Fault along the Pacific–North American Plate Boundary." *Geology*, vol. 33, no. 6, pp. 505–508. 2005.

Lynch, David. *The Field Guide to the San Andreas Fault*. Topanga, California: Thule Scientific, 2015.

Sweetkind, Donald S., et al. "Geology and Geochemistry of Volcanic Centers within the Eastern Half of the Sonoma Volcanic Field, Northern San Francisco Bay Region, California." *Geosphere*, vol. 7, no. 3, pp. 629–657. 2011.

Wallace, Robert E. *The San Andreas Fault System, California. US Geological Survey Professional Paper 1515*. Washington, DC: US Government Printing Office, 1990.

Colorado Plateau

Blakey, Ron, and Wayne Ranney. *Ancient Landscapes of the Colorado Plateau*. Grand Canyon, Arizona: Grand Canyon Association, 2008.

Fillmore, Robert. *Geological Evolution of the Colorado Plateau of Eastern Utah and Western Colorado*. Salt Lake City: University of Utah Press, 2011.

Gao, Wei, et al. "Upper Mantle Convection beneath the Central Rio Grande Rift Imaged by P and S Wave Tomography." *Journal of Geophysical Research*, vol. 109, B03305. 2004.

Liu, Lijun, and Michael Gurnis. "Dynamic Subsidence and Uplift of the Colorado Plateau." *Geology*, vol. 38, no. 7, pp. 663–666. 2010.

Moucha, Robert, et al. "Deep Mantle Forces and the Uplift of the Colorado Plateau." *Geophysical Research Letters*, vol. 36, L19310. 2009.

Schmandt, Brandon, and Eugene Humphreys. "Complex Subduction and Small-Scale Convection Revealed by Body-Wave Tomography of the Western United States Upper Mantle." *Earth and Planetary Science Letters*, vol. 297, pp. 435–445. 2010.

Sine, C. R, et al. "Mantle Structure beneath the Western Edge of the Colorado Plateau." *Geophysical Research Letters*, vol. 35, L10303. 2008.

Thatcher, W. "GPS Constraints on the Kinematics of Continental Deformation." *International Geology Review*, vol. 45, no. 3, pp. 191–212. 2003.

Van Wijk, J.W., et al. "Small-Scale Convection at the Edge of the Colorado Plateau: Implications for Topography, Magmatism, and Evolution of Proterozoic Lithosphere." *Geology*, vol. 38, no. 7, pp. 611–614. 2010.

Wilson, David C., et. al. "High-resolution Receiver Function Imaging Reveals Colorado Plateau Lithospheric Architecture and Mantle-Supported Topography." *Geophysical Research Letters*, vol. 37, L20313. 2010.

Origin of the Grand Canyon

Blair, Will N., and Augustus K. Armstrong. "Hualapai Limestone Member of the Muddy Creek Formation: The Youngest Deposit Predating the Grand Canyon, Southeastern Nevada and Northwestern Arizona." *US Geological Survey Professional Paper 1111*, p. 32. 1979.

Cooke, Michele L., and Scott T. Marshall. "Fault Slip Rates from Three-Dimensional Models of the Los Angeles Metropolitan Area, California." *Geophysical Research Letters*, vol. 33, L21313. 2006.

Hamlin, W. K. "Late Cenozoic Lava Dams in the Western Grand Canyon," *Geological Society of America Memoir 183*, p. 139. 1994.

Karlstrom, Karl E., et al. "Formation of the Grand Canyon 5 to 6 Million Years Ago through Integration of Older Palaeocanyons." *Nature Geoscience*, vol. 7, pp. 239–244. 2014.

McKee, Edwin D., et al. *Evolution of the Colorado River in Arizona: An Hypothesis Developed at the Symposium on Cenozoic Geology of the Colorado Plateau in Arizona, August 1964. Museum of Northern Arizona, Bulletin 44.* Flagstaff, Arizona: Museums of Northern Arizona, 1967.

Powell, James Lawrence. *Grand Canyon*. New York: Pi Press, 2005.

Ranney, Wayne. *Carving the Grand Canyon*. Grand Canyon, Arizona: Grand Canyon Association, Second edition. 2012.

Ranney, Wayne. "Geologists through Time in the Grand Canyon: From Newberry to a New Century." *A Rendezvous of Grand Canyon Historians: Proceedings of the Third Grand Canyon History Symposium. Grand Canyon Association Monograph.* pp. 113–118. 2013.

Webb, Bob, and Peter Griffiths. "The Changing Rapids of Grand Canyon: Lava Falls Rapid." *Boatman's Quarterly Review*, vol. 14, no. 1, pp. 8–10. 2000.

Whole-Genome Duplication and the Evolution of Grasses

Chaney, Ralph W. "The Bearing of the Living *Metasequoia* on Problems of Tertiary Paleobotany." *Proceedings of the National Academy of Sciences*, vol. 34, pp. 503–515. 1948.

Edwards, Erika J., et al. "The Origins of C4 Grasslands: Integrating Evolutionary and Ecosystem Science." *Science*, vol. 328, no. 5978, pp. 587–591. 2010.

Fawcett, Jeffrey, et al. "Plants with Double Genomes Might have had a Better Chance to Survive the Cretaceous-Tertiary Extinction Event." *Proceedings of the National Academy of Sciences*, vol. 106, no. 14, pp. 5737–5742. 2009.

Jardine, Phil. "The Paleocene-Eocene Thermal Maximum," *Palaeontology Online*, vol. 1, article 5, pp. 1–7. 2006.

Jiao, Uyannian, et al. "Ancestral Polyploidy in Seed Plants and Angiosperms." *Nature*, vol. 473, issue 7345, pp. 97–100. 2011.

Kellogg, Elizabeth. "C4 Photosynthesis," *Current Biology*, vol. 23, no. 14, pp. R594-R599. 2013.

Lepage, Ben A., et al. *The Geobiology and Ecology of Metasequoia.* London: Springer, 2005.

Panopoulou, Georgia, et al. "New Evidence for Genome-Wide Duplications at the Origin of Vertebrates Using an Amphioxus Gene Set and Completed Animal Genomes." *Genome Research*, vol. 13, pp. 1056–1066. 2003.

Retallack, G. J., et al. "Reconstructions of Eocene and Oligocene Plants and Animals of Central Oregon." *Oregon Geology*, vol. 58, no. 3, pp. 51–69. 1996.

Retallack, Gregory J. "Late Oligocene Bunch Grassland and Early Miocene Sod Grassland Paleosols from Central Oregon, USA." *Palaeogeography, Palaeoclimatology, Palaeoecology*, vol. 207, pp. 203–237. 2004.

Van de Peer, Yves, et al. "The Evolutionary Significance of Ancient Genome Duplications." *Nature Reviews Genetics*, vol. 10, pp. 725–732. 2009.

Vanneste, Kevin, et al. "Tangled Up in Two: A Burst of Genome Duplications at the End of the Cretaceous and the Consequences for Plant Evolution." *Philosophical Transactions of the Royal Society B*, vol. 369, 20130353, p. 13. 2014.

15. A Drowned River at Poughkeepsie

Erratics, Striations, and Drifts

Hambrey, Michael J., and Jürg C. Alean. *Colour Atlas of Glacial Phenomena*. Boca Raton, Florida: CRC Press, 2017.

Hutton, Jane. "Distributed Evidence: Mapping Named Erratics." *Making the Geologic Now*. New York: Punctum Books, pp. 99–103. 2017.

Snow, R. Scott. "A Field Guide: The Kelleys Island Glacial Grooves, Subglacial Erosion Features on the Marblehead Peninsula, Carbonate Petrology, and Associated Paleontology." *Ohio Journal of Science*, vol. 91, no. 1, pp. 16–26. 1991.

Louis Agassiz

Evans, E. P. "The Authorship of the Glacial Theory." *North American Review*, vol. 145, no. 368, pp. 94–97. 1887.

Hallam, A. "The Ice Age." *Great Geological Controversies*. Second edition. Oxford: Oxford University Press, 1989.

Lurie, Edward. *Louis Agassiz: A Life in Science*. Baltimore, Maryland: Johns Hopkins University Press, 1988.

Great Glacial Lakes

Bretz, J. Harlen. "The Channeled Scabland of the Columbia Plateau." *Journal of Geology*, vol. 31, no. 8, pp. 617–649. 1923.

Bretz, J. Harlen. "The Lake Missoula Floods and the Channeled Scabland." *Journal of Geology*, vol. 77, no. 5, pp. 505–543. 1969.

Donnelly, Jeffrey P., et al. "Catastrophic Meltwater Discharge down the Hudson Valley: A Potential Trigger for the Intra-Allerød Cold Period." *Geology*, vol. 33, no. 2, pp. 89–92. 2005.

Fisher, Timothy G. "Megaflooding Associated with Glacial Lake Agassiz." *Earth-Science Reviews*, vol. 201, Article 102974. 2020.

Hanson, Michelle A., et al. "The Sequence and Timing of Large Late Pleistocene Floods from Glacial Lake Missoula." *Quaternary Science Reviews*, vol. 31, pp. 67–81. 2012.

Larsen, Curtis E. "Geological History of Glacial Lake Algonquin and the Upper Great Lakes." *US Geological Survey Bulletin 1802*. 1987.

Larson, Grahame, and Randall Schaetzl. "Origin and Evolution of the Great Lakes." *Journal of Great Lakes Research*, vol. 27, no. 4, pp. 518–546. 2001.

Waitt, Richard. "Case for Periodic, Colossal Jökulhlaups from Pleistoncene Glacial Lake Missoula." *Geological Society of America Bulletin*, vol. 96, no. 10, pp. 1271–1286. 1985.

Cenozoic Cooling

Anderson, Roger C. "Evolution and Origin of the Central Grassland of North America: Climate, Fire, and Mammalian Grazers." *Journal of the Torrey Botanical Society*, vol. 133, no. 4, pp. 626–647. 2006.

Edwards, Erika J., and Stephen A. Smith. "Phylogenetic Analyses Reveal the Shady History of C4 Grasses." *Proceedings of the National Academy of Sciences*, vol. 107, no. 6, pp. 2532–2537. 2010.

Gullap, M. Kerim, et al. "The Effect of Bovine Saliva on Growth Attributes and Forage Quality of Two Contrasting Cool Season Perennial Grasses Grown in Three Soils of Different Fertility." *The Rangeland Journal*, vol. 33, no. 3, p. 307–313. 2011.

Janis, Christine M., et al. "The Origins and Evolution of the North American Grassland Biome: The Story from the Hoofed Mammals" *Palaeogeography, Palaeoclimatology, Palaeoecology*, vol. 177, pp. 183–198. 2002.

Pagani, Mark. "Marked Decline in Atmospheric Carbon Dioxide Concentrations during the Paleogene." *Science*, vol. 309, no. 5734, pp. 600–603. 2005.

Raymo, M. E., and W. F. Ruddiman. "Tectonic Forcing of the Late Cenozoic Climate." *Nature*, vol. 359, no. 6391, pp. 117–122. 1992.

Retallack, Gregory J. "Cenozoic Cooling and Grassland Expansion in Oregon and Washington." *Paleobios*, vol. 28, no. 3, pp. 89–113. 2009.

Retallack, Gregory J. "Global Cooling by Grassland Soils of the Geological Past and Near Future." *Annual Reviews of Earth and Planetary Sciences*, vol. 42, pp. 69–86. 2013.

Ice Ages and Milankovitch Cycles

Bender, Michael, et al. "Gases in Ice Cores." *Proceedings of the National Academy of Sciences*, vol. 94, pp. 8343–8349. 1997.

Hays, J., J. Imbrie, and N. Shackleton. "Variations in the Earth's Orbit: Pacemaker of the Ice Ages," *Science*, vol. 194, pp. 1121–1132. 1976.

O'Dea, Aaron, et al. "Formation of the Isthmus of Panama." *Science Advances*, vol. 2, no. 8, e1600883. 2016.

Petit, Jean-Robert, and Dominique Raynaud. "Forty Years of Ice-Core Records of CO2." *Nature*, vol. 579, no. 7800, pp. 505–506. 2020.

Walker, Mike, et al. "Formal Definition and Dating of the GSSP (Global Stratotype Section and Point) for the Base of the Holocene Using the Greenland NGRIP Ice Core, and Selected Auxiliary Records." *Journal of Quaternary Science*, vol. 24, no. 1, pp. 3–17. 2009.

Cascadia, the Great Earthquake of 1700, and ETS

Atwater, Brian F., et al. "The Orphan Tsunami of 1700." *US Geological Survey Professional Paper 1707*. 2005.

Ludwin, Ruth S., et al. "Dating the 1700 Cascadia Earthquake: Great Coastal Earthquakes in Native Stories." *Seismological Research Letters*, vol. 76, no. 2, pp. 140–148. 2005.

McCaffrey, Robert, et al. "Fault Locking, Block Rotation and Crustal Deformation in the Pacific Northwest." *Geophysical Journal International*, vol. 169, pp. 1315–1340. 2007.

McCaffrey, Robert, et al. "Active Tectonics of Northwestern US Inferred from GPS-Derived Surface Velocities." *Journal of Geophysical Research, Solid Earth B*, vol. 118, pp. 709–723. 2013.

Nelson, Alan R., et al. "Radiocarbon Evidence for Extensive Plate-Boundary Rupture about 300 Years Ago at the Cascade Subduction Zone." *Nature*, vol. 378, no. 6555, pp. 371–374. 1995.

Spence, G. D., and D. T. Long. "Transition from Oceanic to Continental Crustal Structure: Seismic and Gravity Models at the Queen Charlotte Transform Margin." *Canadian Journal of Sciences*, vol. 32, no. 6, pp. 699–717. 1995.

Wells, Ray E., and Robert McCaffrey. "Steady Rotation of the Cascade Arc." *Geology*, vol. 41, no. 9, pp. 1027–1030. 2013.

Boring Volcanic Field and Cascade Volcanoes

Barber, Elizabeth W., and Paul T. Barber. *When They Severed Earth from Sky*. Princeton, New Jersey: Princeton University Press, 2004.

Evarts, Russell C., et al. "The Boring Volcanic Field of the Portland-Vancouver Area, Oregon and Washington: Tectonically Anomalous Forearm Volcanism in an Urban Setting." *Volcanoes to Vineyards: Geological Society of America Field Guide 15*, pp. 253–270. 2009.

Madin, Ian P. "Portland, Oregon, Geology by Tram, Train, and Foot." *Oregon Geology*, vol. 69, no. 1, pp. 73–92. 2009.

Wood, Charles A., and Jurgen Kienle. *Volcanoes of North America*. Cambridge: Cambridge University Press, 1992.

Yellowstone, Wyoming

Anderson, Alfred L. "Drainage Diversion in the Northern Rocky Mountains of East-Central Idaho." *Journal of Geology*, vol. 55, no. 2, pp. 61–75. 1947.

Dzurisin, Daniel, et al. "History of Surface Displacements at the Yellowstone Caldera, Wyoming, from Leveling Surveys and InSAR Observations, 1923–2008." *US Geological Survey Professional Paper 1788*. 2012.

Farrell, et al. "Tomography from 26 Years of Seismicity Revealing that the Spatial Extent of the Yellowstone Crustal Magma Reservoir Extends Well beyond the Yellowstone Caldera." *Geophysical Research Letters*, vol. 41, pp. 3068–3073, 2014.

Hurwitz, Shaul, and Jacob B. Lowenstern. "Dynamics of the Yellowstone Hydrothermal System." *Reviews of Geophysics*, vol. 51, pp. 375–411. 2017.

Izett, Glen A., and Ray E. Wilcox. "Map Showing Localities and Inferred Distributions of the Huckleberry Ridge, Mesa Falls, and Lava Creek Ash Beds (Pearlette Family Ash Beds) of Pliocene and Pleistocene Age in the Western United States and Southern Canada." *US Geological Survey Miscellaneous Investigations Series, Map I-1325*. 1982.

Link, Paul K., and E. Chilton Phoenix. *Rocks, Rails & Trails*. Second edition. Boise, Idaho: Museum of Natural History, 1996.

Nelson, Peter L., and Stephen P. Grand. "Lower-mantle Plume beneath the Yellowstone Hotspot Revealed by Core Waves." *Nature Geoscience*, vol. 11, pp. 280–284. 2018.

Ruppel, E. T. "Late Cenozoic Drainage Reversal, East-Central Idaho, and Its Relation to Possible Undiscovered Placer Deposits." *Economic Geology*, vol. 62, no. 5, pp. 648–663. 1967.

Smith, Robert B., and Lee J. Siegel. *Windows into the Earth: The Geologic Story of Yellowstone and Grand Teton National Park.* New York: Oxford University Press. 2000.

Smith, Robert B., et al. "Geodynamics of the Yellowstone Hotspot and Mantle Plume: Seismic and GPS Imaging, Kinematics, and Mantle Flow." *Journal of Volcanology and Geothermal Research*, vol. 188, pp. 26–56. 2009.

Vazquez, Jorge, et al. "A Field Trip Guide to the Petrology of Quaternary Volcanism on the Yellowstone Plateau." *US Geological Survey Scientific Investigations Report 2017-5022-Q.* 2017.

16. A World Bequeathed and the Great Acceleration

Gold

Chapin, Charles E. "Origin of the Colorado Mineral Belt." *Geosphere*, vol. 8, no. 1, pp. 28–43. 2012.

Cline, Jean S., et al. "Carlin-Type Gold Deposits in Nevada: Critical Geologic Characteristics and Viable Models." *Economic Geology 100th Anniversary Volume*, pp. 451–484. 2005.

Goldfarb, R. J., et al. "Orogenic Gold and Geologic Time: A Global Synthesis." *Ore Geology Reviews*, vol. 18, pp. 1–75. 2001.

Groves, David I., et al. "Controls on the Heterogeneous Distribution of Mineral Deposits through Time." *Mineral Deposits and Earth Evolution, Geological Society of London, Special Publication 248*, pp. 71–101. 2005.

Diamonds

Howard, J. M., and W. D. Hanson. *Geology of the Crater of Diamonds State Park and Vicinity, Pike County, Arkansas. Arkansas Geological Survey State Park Series 03.* 2008.

Kjarsgaard, B. A., et al. "The North America Mid-Cretaceous Kimberlite Corridor: Wet, Edge-Driven Decompression Melting of an OIB-Type Deep Mantle Source." *Geochemistry, Geophysics, Geosystems*, vol. 18, pp. 2,727–2,747. 2017.

Shirey, Steven B., and James E. Shigley. "Recent Advances in Understanding the Geology of Diamonds." *Gems & Gemology*, vol. 49, pp. 188–222. 2013.

Stern, Robert J. "Plate Tectonic Gemstones." *Geology*, vol. 41, no. 7, pp. 723–726. 2013.

Stern, Robert J., et al. "Kimberlites and the Start of Plate Tectonics." *Geology*, vol. 44, no. 10, pp. 799–802. 2016.

Megafauna, the Younger Dryas Event, and Agriculture

Brooke, John L. *Climate Change and the Course of Global History.* Cambridge: Cambridge University Press. 2014.

Carlson, Anders E., et al. "Geochemical Proxies of North American Freshwater Routing during the Younger Dryas Cold Event." *Proceedings of the National Academy of Sciences*, vol. 104, no. 16, pp. 6556–6561. 2007.

Grayson, Donald K., and David J. Meltzer. "Revisiting Paleoindian Exploitation of Extinct North American Mammals." *Journal of Archaeological Science*, vol. 56, pp. 177–193. 2015.

MacPhee, Ross D. E. *End of the Megafauna.* New York: W. W. Norton, 2019.

Martin, Paul S., and Richard G. Klein. *Quaternary Extinctions: A Prehistoric Revolution.* Tucson: University of Arizona Press, 1984.

Meltzer, David J. "Overkill, Glacial History, and the Extinction of North America's Ice Age Megafauna." *Proceedings of the National Academy of Sciences,* vol. 117, no. 46, pp. 28555–28563. 2020.

Ruddiman, William F. *Plows, Plagues, and Petroleum.* Second edition. Princeton, New Jersey: Princeton University Press, 2016.

Tankersley, Kenneth, et al. "Clovis and the American Mastodon at Big Bone Lick, Kentucky." *American Antiquity,* vol. 3, pp. 558–567. 2009.

The Great Acceleration

Doney, Scott C., et al. "Ocean Acidification: The Other CO2 Problem." *Annual Review of Marine Science,* vol. 1, pp. 212–251. 2020.

Elhacham, Emily, et al. "Global Human-made Mass Exceeds All Living Biomass." *Nature,* vol. 588, no. 7838, pp. 442–444. 2020.

Fischetti, Mark. "Climate Clincher." *Scientific American,* vol. 321, no. 5, pp. 86. November 2019.

IPCC. *Climate Change 2014: Synthesis Report.* Geneva, Switzerland: IPCC. 2014.

Mann, Michael E., et al. "Northern Hemisphere Temperatures during the Past Millennium: Inferences, Uncertainties, and Limitations." *Geophysical Research Letters,* vol. 26, no. 6, pp. 759–762. 1999.

Martin, Justin T., et al. "Increased Drought Severity Tracks Warming in the United States' Largest River Basin." *Proceedings of the National Academy of Sciences,* vol. 117, no. 21, pp. 11328–11336. 2020.

Neukom, Raphael, et al. "No Evidence for Globally Coherent Warm and Cold Periods over the Preindustrial Common Era." *Nature,* vol. 571, no. 7766, pp. 550–554. 2019.

Pierrehumbert, Ray. "Climate Change Conspiracy Theories Are Ludicrous." *Scientific American,* vol. 315, no. 5, p. 51. 2016.

Rudd, Michelle Andreadakis, and James A. Davis. "Industrial Heritage Tourism at the Bingham Canyon Copper Mine." *Journal of Travel Research,* vol. 36, pp. 85–89. 1998.

Steffen, Will, et al. "The Anthropocene: Are Humans Now Overwhelming the Great Forces of Nature?" *Ambio,* vol. 36, no. 8, pp. 614–621. 2007.

Swain, Daniel L., et al. "Increasing Precipitation Volatility in Twenty-First Century California." *Nature Climate Change,* vol. 8, pp. 427–433. 2018.

Wilkinson, Bruce H. "Humans as Geologic Agents: A Deep-Time Perspective." *Geology,* vol. 33, no. 3, pp. 161–164. 2005.

PETM Redux

Barnosky, Anthony D., et al. "Has the Earth's Sixth Mass Extinction Already Arrived?" *Nature,* vol. 471, no. 7336, pp. 51–57. 2011.

Bowen, Gabriel, et al. "Two Massive, Rapid Releases of Carbon during the Onset of the Palaeocene-Eocene Thermal Maximum." *Nature Geoscience,* vol. 8, pp. 44–47. 2015.

Brannen, Peter. "Earth Is Not in the Midst of a Sixth Mass Extinction." *The Atlantic,*

July 13, 2017.

Gerlach, Terry. "Volcanic versus Anthropogenic Carbon Dioxide." *EOS*, vol. 92, no. 24, pp. 201–203. 2011.

Keller, Gerta, et al. "Environmental Changes during the Cretaceous-Paleogene Mass Extinction and Paleocene-Eocene Thermal Maximum: Implications for the Anthropocene." *Gondwana Research*, vol. 56, pp. 69–89. 2018.

Zeebe, Richard E., and James C. Zachos. "Long-term Legacy of Massive Carbon Input to the Earth System: Anthropocene versus Eocene." *Philosophical Transactions of the Royal Society A*. vol. 371, no. 2001, 20120006. 2013.

A New Geologic Epoch

Bar-On, Yinon, et al. "The Biomass Distribution on Earth." *Proceedings of the National Academy of Sciences*, vol. 115, no. 25, pp. 6506–6511. 2018.

Crutzen, Paul J., and Eugene F. Storermer. "The 'Anthropocene.'" *Global Change Newsletter*, vol. 41, pp. 17–18. 2000.

Hancock, Gary J., et al. "The Release and Persistence of Radioactive Anthropogenic Nuclides." *Stratigraphical Basis for the Anthropocene, Geological Society of London, Special Publication 395*, pp. 265–281. 2014.

Malhi, Yadvinder. "The Concept of the Anthropocene." *Annual Reviews of Environment and Resources*, vol. 42, pp. 77–104. 2017.

Plotnick, Roy E., and Karen A. Koy. "The Anthropocene Fossil Record of Terrestrial Mammals." *Anthropocene*, vol. 29, pp. 1–11. 2020.

Waters, Colin N., et al. "Can Nuclear Weapons Fallout Mark the Beginning of the Anthropocene Epoch?" *Bulletin of the Atomic Scientists*, vol. 71, pp. 46–57. 2015.

Waters, Colin N., et al. "The Anthropocene Is Functionally and Stratigraphically Distinct from the Holocene." *Science*, vol. 351, no. 6269, pp. 137. 2016.

Waters, Colin N. "Global Boundary Stratotype Section and Point (GSSP) for the Anthropocene Series: Where and How to Look for Potential Candidates." *Earth-Science Reviews*, vol. 178, pp. 379–429. 2018.

Zalasiewicz, Jan, et al. "Human Bioturbation, and the Subterranean Landscape of the Anthropocene." *Anthropocene*, vol. 6, pp. 3–9. 2014.

Zalasiewicz, Jan, et al. "When Did the Anthropocene Begin? A Mid-Twentieth Century Boundary Level Is Stratigraphically Optimal." *Quaternary International*, vol. 383, pp. 196–203. 2015.

Zalasiewicz, Jan. "What Mark Will We Leave on the Planet." *Scientific American*, vol. 315, number 3, pp. 31–37. 2016.

Morality

Callicott, J. Baird. "Environmental Ethics in the Anthropocene." *Journal of Global Cultural Studies*, no. 13, 2018.

Ellis, Michael A., and Zev Trachtenberg. "Which Anthropocene Is It to Be? Beyond Geology to a Moral and Public Discourse." *Earth's Future*, vol. 2, pp. 122–125. 2013.

Lowe, Benjamin S. "Ethics in the Anthropocene: Moral Responses to the Climate Crisis." *Journal of Agricultural and Environmental Ethics*, vol. 32, no. 3, pp. 479–485. 2019.

Epilogue: Rugby, North Dakota

Carbon Sequestration
Bastin, Jean-Francois, et al. "The Global Tree Restoration Potential." *Science*, vol. 365, no. 6448, pp. 76–70. 2019.

Douglas, Edward M. *Boundaries, Areas, Geographic Centers and Altitudes of the United States and the Several States. US Geological Survey Bulletin 817.* 1930.

Peck, Wesley D., et al. "The North Dakota Integrated Carbon Storage Complex Feasibility Study." *International Journal of Greenhouse Gas Control*, vol. 84, pp. 47–53. 2019.

Rogerson, Peter A. "A New Method for Finding Geographic Centers, With Application to US States." *Professional Geographer*, vol. 67, no. 4, pp. 686–694. 2015.

The Future
Bounama, Christine, et al. "The Fate of Earth's Ocean." *Hydrology and Earth System Sciences*, vol. 5, no. 4, pp. 569–575. 2001.

Hoffman, Paul F. "Tectonic Genealogy of North America." *Earth Structure: An Introduction to Structural Geology and Tectonics.* New York: McGraw-Hill, pp. 459–464. 1997.

Laskar, J., and M. Gastineau. "Existence of Collisional Trajectories of Mercury, Mars, and Venus with the Earth." *Nature*, vol. 459, issue 7248, pp. 817–819. 2009.

Safonova, Inna, and Shigenori Maruyama. "Asia: A Frontier for a Future Supercontinent Amasia." *International Geology Review*, vol. 56, no. 9, pp. 1051–1071. 2014.

Stern, Robert, et al. "Introduction: Making the Southern Margin of Laurentia." *Geosphere*, vol. 6, no. 6, pp. 737–738. 2010.

Weisman, Alan. *The World Without Us.* New York: St. Martins, 2007.

ENDNOTES

1. THE RELICS OF HELL

1   Gneiss (pronounced "nice") is derived from an old German word used by miners that means "spark," a reference to the shiny mineral grains contained within it.

2   The Morton Gneiss is found in a surprising number of places. One of my favorites is in Baltimore near the police headquarters on Fayette Street, where the stone protects a pedestal for flagpoles. Another is an inverted L–shaped sculpture in front of the Harold Stassen Building in St. Paul, Minnesota. In Spokane this durable stone was once used to decorate a drinking fountain dedicated to the Women's Temperance Union in front of the Old National Bank. Unfortunately, the fountain and the decorative stone have recently been removed.

3   For uranium-235, the "235" signifies the total number of particles in the nucleus. Uranium always has 92 protons, and uranium-235 has 143 neutrons.

4   By comparison to the Morton Gneiss, it can be difficult to find examples of the Acasta Gneiss. The few samples of this stone that are on public view are usually a few inches in size and displayed only in museums. But a sizable boulder does rest at the Peter Russell Rock Garden at the University of Waterloo in Waterloo, Ontario. This rock garden has scores of other rocks collected in North America, including the Lorrain Quartzite and the Gowganda Conglomerate. It also has several examples of stromatolites, a form of early life.

5   The Geologic Time Scale is a hierarchical system officially decided by the International Commission on Stratigraphy, which is part of the International Council of Scientific Unions. At the top of the hierarchy are eons, such as the Hadean Eon and the Archean Eon. Eons are divided into eras. Eras are divided into periods. And periods are divided into epochs. The moment in time in which we are living is the Phanerozoic Eon, Cenozoic Era, Quaternary Period, and Holocene Epoch.

6   If you are wearing a gold band around a finger or a silver bracelet around a wrist, then you are wearing material that was made during the catastrophic collision of two neutron stars.

7   Many years ago it was once described to me as: "The grains kissed and remained together." This was before astronomers uncovered evidence for many catastrophic collisions during the earliest part of the formation of the solar system.

8   There is observational evidence that supports the idea of the Grand Tack. As Jupiter and Saturn interact gravitationally and a resonance develops between their orbital motions, the dust that remains around the Sun should form into a ring. Such rings have been seen around young planetary systems. For example, in 2012, the first detailed images of a planetary system around the young star LkCa 15—shortened from the full name Lick Calcium 15—revealed a ring around the star with two Jupiter-sized planets orbiting near the inside edge of the ring. Another example is the young star Beta Pictoris that has at least one Jupiter-sized planet orbiting around it, following a path along the inside edge of a dust ring.

9   There is also observational evidence of a planet-planet collision in a distant star system. The
      star is Fomalhaut, the brightest star in the constellation Piscis Austrinus. An image taken
      in 2020 by the Hubble Space Telescope showed a large planet near Fomalhaut that was
      traveling along an escape trajectory away from the star and a dust cloud around the same star
      that had a geometry consistent with debris formed by a planet-planet collision.

10   Additional evidence of a giant impact comes from the Moon's unusual orbit. The orbits
      of most moons lie close to the plane of the planet's equator. The Moon's orbital motion is
      inclined by 5° to the Earth's equator, a result of Theia hitting the early Earth at an angle.

11   The early Earth probably had a magma ocean but because of the Earth's higher gravity and
      higher water content, its magma ocean was less extensive than the one on the Moon. So far,
      no evidence of a terrestrial magma ocean has been found.

12   In 2019, a piece of the Earth's crust was found in one of the rocks collected by Apollo 14, but
      when it formed on the Earth is still debated.

13   It should be noted that great volumes of rock are also being cycled from the surface to the
      interior and back again through plate tectonics, though the cycling of rock is occurring over
      tens to hundreds of millions of years. By comparison, on average, a water molecule resides in
      the ocean for about a thousand years before it is evaporated, then cycles from the ocean to the
      atmosphere then precipitates out as rain or snow and returns to the ocean in about ten days.

2.   BOMBARDMENT AND BOTTLENECK

1   As of 2021, the International Commission on Stratigraphy has not formally defined the end
      of the Archean Eon as a geologic event. Instead, members have voted to define an age of
      2.5 billion years as the end. (This age is used because it was the approximate time when a
      certain type of early continental crust, known as granite-greenstone, which was important
      in the formation of cratons, stopped forming.) A more precise date and definition will be
      determined in the near future. It is possible that the event that will be chosen is the dramatic
      increase in the amount of oxygen in the atmosphere at the beginning of the Huronian
      Glaciation at 2.420 billion years. And so that age is used here as the end of the Archean and
      the beginning of the Proterozoic Eon.

2   The Weald will be familiar to those who have read the book *Watership Down*. The rabbits
      started their travels in the Weald, then headed south to a place called Watership Downs, a
      part of South Downs, the name of the cliff that forms the southern boundary of the Weald.
      (Downs is an old Celtic word that means "hill.")

3   There are a few notable exceptions. The crater Copernicus is 800 million years old and the
      crater Tycho is 110 million years old.

4   DNA is deoxyribonucleic acid and is the hereditary material in humans and almost all other
      organisms. RNA is ribonucleic acid and performs a host of functions including regulating
      the expression of genes, catalyzing chemical reactions, and assembling proteins.

5   The study of hyperthermophiles has made a major contribution to criminal justice. Before
      the 1980s, the examination of fingerprints was one of the best ways to identify who may
      have been at the scene of a crime. DNA sequencing of blood or other body material had been
      suggested, but usually only small amounts were recovered. And there was no way to copy the
      DNA quickly and produce a large quantity that could be analyzed. Then, in 1983, biochemist
      Kary Mullis realized that an enzyme found in the bacteria *Thermophilus aquaticus*, which was
      recovered from the hot Mushroom Pool at Yellowstone National Park, replicated DNA at high
      temperatures, which meant that large amounts of DNA could be reproduced.

6   If the early history of Mars proves to have had too harsh an environment to produce life,
      then there is a place near Mars that was also a suitable and more benign alternative where
      carbon-rich compounds abound: the asteroid Ceres. Ceres, a dwarf planet, is about three

hundred miles in diameter, making it the largest object in a belt of asteroids that exists between Mars and Jupiter. In 2015, the NASA *Dawn* spacecraft began to orbit Ceres and discovered that the surface is covered with carbon-rich material. Two meteorites that have recently struck the Earth are also thought to be from Ceres. One landed near the town of Monahans in West Texas in 1998, interrupting a children's basketball game. The other fell near Zag, Morocco, in the Western Sahara, also in 1998. Both meteorites are rich in organics, or carbon. If a Ceres-like object is the source of life on the Earth, then such objects played two crucial roles: bringing life to the planet, or at least carbon-rich material, and bringing the water that now fills the ocean basins.

7　Steele made his discovery at a place known now as Petrified Sea Gardens, which unfortunately has been closed for several years. Stromatolites can be found about a mile north of the Petrified Sea Gardens at Lester Park.

3. THE CHILDREN OF UR

1　The age progression of volcanic islands is also used to estimate plate motion. The Hawaiian Islands are a long linear chain of volcanic islands, the youngest island of which is at the southeastern end of the chain. Knowing the ages of various islands and the distances between them, it is a simple matter to estimate that the Pacific Plate—the tectonic plate that the Hawaiian Islands sit on—is moving northwestward at about five inches each year.

2　To give another perspective on the rate of plate motion, the distance between New York and Paris increases by about an inch each year. That is, during an average human lifespan, the distance between the two cities increases by about the height of an average person. To give yet another perspective, since Columbus sailed across the Atlantic in 1492, the distance between Spain and the Bahamas has increased by about forty feet—about four car lengths.

3　On Staten Island, only the lowest part is exposed—the mantle material, which in this case was originally peridotite but after being infused with water and squeezed turned into a metamorphic rock, serpentinite.

4　The age of the oldest ophiolite is not the only indication that plate tectonics began during the Archean Eon about 2.5 billion years ago. The relative amounts of isotopes of neodymium and hafnium found in rocks indicate that large amounts of crustal material were being cycled through the mantle after that time. Also the positions of some continental masses, indicated by the direction of the magnetic fields on those masses, began shifting by 2.5 billion years ago. And finally, a recent program, known as Lithoprobe, which used earthquake waves to image the upper mantle beneath Canada, has revealed structures that look like subduction slabs that date from that period and later.

5　It should be noted that the transition from lid tectonics to plate tectonics did not occur as a sudden global event, and so it cannot be used as a marker to define a geologic period. The transition was gradual, occurring locally in a few regions, then over a few hundred million years it spread to more and more of the planet until plate tectonics was the dominant mode.

6　*Orogeny* is the geologic term for "mountain building," derived from the Greek words *óros* and *génesis*, which together mean "mountain origin."

7　The Morton Gneiss is one of the rock units of the Minnesota River Valley Terrane.

8　Some geologists refer to the first supercontinent as Nuna, an Eskimo name for lands that border the Arctic Ocean, because some of the early evidence for its existence was also found here.

4. GARDENS OF EDIACARAN

1　Such as sponges, which first appeared 2.5 billion years ago.

2　Such rapid microbial growth still occurs today though, of course, at a much smaller scale than during the breakup of a supercontinent. For example, a rapid microbial growth, or bloom, occurred during a recent eruption of Kīlauea volcano in Hawaii when, for three months in 2018, millions of tons of lava poured into the nutrient-poor waters around the islands. The result was a massive algal bloom that was more than a hundred miles long and twenty miles wide. Chemical analyses of the seawater where the lava was pouring in showed that it had high amounts of phosphates and nitrates.

3   The same process probably created the first Snowball Earth at the beginning of the Proterozoic Eon, but because there are fewer exposures of such rocks, much less is known about the first episode of Snowball Earth than the second.

4   Each of these sites was located along the edge of Laurentia and was a shallow sea after the breakup of Rodinia. Another accessible site is along Laurel Canyon in the Blue Ridge Mountains of Virginia, which was also along the edge of Laurentia. The glacial deposits there contain dropstones, but the carbonate rocks have been eroded away.

5   In 1999, excavation of a new convention center in St. John's uncovered many specimens of *Aspidella*, further indicating that this fossil lies under much of the city.

6   Fossil imprints of *Charnia masoni* and *Aspidella* might actually be recording the same type of individual. *Charnia* is the frond-like stalk and base that held the organism to the seafloor, and *Aspidella* might be the imprint that was made by the stalk.

7   Rocks of Newfoundland are from the early part of the Ediacaran Period, and so they are too old to contain these particular fossils.

8   Mitochondria are the locations within a eukaryotic cell where respiration takes place, where oxygen is combined with organic molecules to release energy that is used for the cell to grow, to multiply, and to move. Chloroplasts are where photosynthesis takes place, where carbon dioxide and water and the energy of sunlight produces oxygen and energy-rich organic molecules.

9   How Woese actually made a distinction among the three domains relied on determining the types of fat found in cell membranes.

10   The idea that eukaryotes are the result of the merging of bacterial and archaeal cells was championed and promoted for years by Lynn Margulis. The idea known as endosymbiosis was first proposed by the Russian botanist Konstantin Mereschkowski in the early 1900s. Margulis advanced it by pointing out physical similarities between mitochondria and chloroplasts and some bacteria. In 1966, she published a scientific paper explaining her views, a paper that was rejected more than a dozen times before it was finally accepted. The first clear experimental evidence that supported Margulis's view came in 1978 when it was shown that the mitochondria contains RNA that was similar to that in some bacteria.

11   The concept of "bottleneck and flush" came from the work done in the 1960s by Hampton Carson at the University of Hawaii. His work centered on studying the fruit fly, *Drosophila*, of which there are more than a hundred species in the Hawaiian Islands. His research indicated that the rate of evolution depended more on cycles of severe reduction and rapid increase in population size—bottleneck and flush—than on the slow accumulation of new mutations.

5. THE GREAT UNCONFORMITY

1   There are a few canyons that are larger than the Grand Canyon. The largest is Yarlung Tsangpo Canyon in Tibet, named for the river that flows through it. It has a length of about three hundred miles, same as the Grand Canyon in Arizona, but considerably deeper. The average depth of Yalung Tsangpo Canyon is 7,400 feet. The average depth of the Grand Canyon is 5,200 feet. However, there are no canyons larger than the one in Arizona that display the same characteristics that make the Grand Canyon so visually attractive: the cutting down from a broad horizontal plateau, the steep sides, the massive and multicolored horizontal layers. The closest comparison is Fish River Canyon in Namibia. It also cuts through a plateau, though is only a hundred miles long and less than two thousand feet deep.

2   Hutton's conclusion about the gray sandstone was based on the laws of succession outlined
    by Nicolaus Steno in the seventeenth century and briefly mentioned in the previous chapter.

3   The Potsdam Sandstone is sand eroded from mountains produced during the Grenville
    Orogeny, most likely from large islands in the general direction of the Adirondack
    Mountains of today.

4   The Great Unconformity made a cinematic appearance in the movie *Butch Cassidy and the
    Sundance Kid* (1969). The scene in which Butch and Sundance are arguing about jumping
    off a cliff was shot near Bakers Bridge along the Animas River north of Durango, Colorado.
    In the closeups of Butch and Sundance, they are hiding behind angular rocks with small
    mineral grains, the Bakers Bridge Granite, discussing whether to fight it out or to jump.
    When they look up in the direction of the men who are chasing them, they are looking at the
    layered rocks of a marine sandstone, the Ignacio Formation. They and the law are separated
    by the Great Unconformity.

5   Imagine a spinning basketball. It will can spin smoothly in any orientation. Now attach
    something to the outside, say a lump of clay. Spin it again. This time it will wobble and
    continue to change its spin direction—the "poles" of the basketball will shift—until the
    lump of clay lies along what would be the equator of the basketball. The same is true of a
    spinning planet or moon. Change the distribution of mass, and the object will wobble and its
    spin direction will change, causing the poles to wander.

6   There are three state parks near Cincinnati were fossils can be found in great abundance.
    Brachiopods and bryozoans are common at Caesar Creek State Park. Trilobites are found at
    Hueston Woods. The best of the three is Trammel Fossil Park, where thousands of fossils
    can be seen over an area of about ten acres.

7   Because rocks laid down during the Ordovician Period are often rich in marine fossils, they are
    often quarried and sold as decorative building stone. The Tyndall Stone, quarried near Tyndall,
    Manitoba, is especially rich in brachiopods, trilobites, and coral. Sinuous marks in the stone are
    trails left by critters that were burrowing through the sediment before it was compressed into
    limestone. Tyndall Stone can be found gracing the interior walls of the Canadian Parliament
    Buildings in Ottawa and the Empress Hotel in Victoria, British Columbia.

8   All except one, the bryozoans—small aquatic animals that, as mentioned, resembled small
    tree branches and lived in colonies.

9   The phrase "ontogeny recapitulates phylogeny" was first proposed by Ernst Haeckel in
    1866. It remained an attractive idea for almost a century, providing a simple explanation
    for a highly complex process. By the mid-twentieth century it was replaced by evolutionary
    developmental biology, or "evo-devo." Evo-devo is based on the idea that animal
    development is controlled by a special genetic system that involves Hox genes.

10 For example, the structures of the four fingers on a human hand are similar to each other
    and are determined by three genes from one of the sets of Hox genes. In the development
    of the thumb, however, these three genes are inactive, which is why a thumb has a different
    structure than the fingers.

11 Hox genes may be related to genetic disorders, such as leukemia, and to some cancers, such
    as prostate cancer. They have been linked to hand-foot-genital syndrome, a condition that
    affects the development of the hands and feet and that produces abnormally short thumbs
    and first (big) toes, and disorders of the reproductive system. Hox genes may also explain the
    close connection between humans and chimpanzees. Although humans and chimpanzees
    have almost identical genomes—there is a 99 percent similarity in their genes—they are
    dramatically different animals in brain size, bipedalism, speech capability, and cognitive
    understanding and susceptibility to diseases, such as the human immunodeficiency virus
    (HIV). This similarity of genes and the discrepancy in appearance and behavior may be due

to the Hox genes. That is, it is in the way the genes are expressed and not solely the content of the genes. To put it in another way: Humans are much more similar to chimpanzees than previously thought. The difference is in the Hox genes.

6. AN ANCIENT FOREST AT GILBOA

1   The Niagara Escarpment is such a prominent feature in the region of the Great Lakes that many people assume its origin must be related to the continental glaciers of the recent ice ages. But that is incorrect. The Niagara Escarpment is much older. It formed during the slow subsidence of the Michigan Basin that began in the Cambrian Period and continued through the Ordovician, Silurian, and Devonian periods; that is, it formed from about five hundred to four hundred million years ago. Why the Michigan Basin formed on the oldest and most stable part of the continent and why it subsidized over such a long period of time— and why the center dropped more than two miles—is still a mystery.

2   That this millipede must have been breathing air is indicated by tiny primitive breathing structures found on the outside of its fossilized body.

3   The Earth is usually portrayed as having no life-forms roaming across the land surface before the Silurian Period, but that may not be quite right. Slime molds, a strange gelatinous form of life that is neither plant nor animal, may have been creeping across the floors of shallow seas as early as two billion years ago. We know that they can also move across the land today. And so it is possible that, for hundreds of millions of years before the first animal stepped or scooted on land, patches of slime molds might already have been traversing across the landscape. Moreover, because slime molds can be quite colorful, it is intriguing to think that, if alien beings had visited the Earth during the Proterozoic Eon, they may have seen greasy-looking yellow or orange masses moving across the surface at rates of an inch or so a day.

4   Winifred Goldring was one of the first professional women paleontologists in the United States. As such, she often had to cope with other professionals who thought it was unseemly for a woman to be clambering around rocky outcrops picking through shattered rocks at a quarry site. Some told her that they were concerned about her personal safety. Others said she was at risk of being assaulted by men. She responded by learning to shoot and carrying a revolver. In 1918, she resigned for a short period from the museum over the issue of her low salary, which was less than half what a male stenographer at the museum was earning. In 1939, Goldring was appointed state paleontologist for the state of New York. And in 1949, she was the first woman to be elected president of the Paleontological Society, an international organization devoted to the science of paleontology.

7. FIRES, FORESTS, AND COAL

1   To clarify, the number of species of crinoids alive today is about six hundred, but these include those with long stalks and short stalks and those with essentially no stalks at all. By comparison, more than five thousand species of crinoids have been identified in the fossil record.

2   The stone blocks chosen for the Capitol Reflecting Pool are atypical of the Salem Limestone in that they were chosen for their high content of fossils, in particular fragmented corals, pieces of crinoids, bryozoans, and mollusk shells. The Lincoln Memorial was built of similar stones.

3   The Empire Hole was also the source of stone that was used in the 2001 renovations of the Empire State Building.

4   These are the same rock formations that were originally described and named as separate formations before geologists understood their connection.

5   Also no dinosaurs existed for almost a hundred million years.

6   *Cyclothem* is derived from two Greek words, *cyclos*, "cycle," and *thema*, a "deposit."

7   The seam at Devitt Field in Deer Park, Connecticut, is a segment of the Iapetus Suture, the point of contact between Avalonia and Laurentia. It is named for the Iapetus Ocean that separated Laurentia and Gondwana. The Iapetus Suture runs from Connecticut to Maine, and because Avalonia was later pulled away from Laurentia, the suture is also prominent in the British Isles. In particular, the course of the River Shannon, Ireland's largest river, follows the Iapetus Suture. It also crosses the Isle of Man and runs from Solway Firth to Lindisfarne on Great Britain, generally following the border between Scotland and England.

8   The bas-relief at Stone Mountain in Georgia depicts three Confederate leaders of the Civil War: President Jefferson Davis, General Robert E. Lee, and General Thomas J. "Stonewall" Jackson. The carving was done in the 1920s, designed and directed by Gutzon Borglum.

8.   THE GREAT DYING

1   There are oil seeps that reach the surface. Among the more accessible are those at the La Brea Tar Pits in Los Angeles, at Carpinteria State Beach, also in California, and in Titusville, Pennsylvania, where the first oil well in the Unites States was drilled.

2   For the geology tourist, rocks of the Wolfcamp Shale and the Bone Springs Formation are difficult to find because they are usually deeply buried. Surface exposures of the Bone Springs Formation can be seen from Highway 54 north of Van Horn. There is an exposure east of the highway near the 13 milepost and west of the highway near the 18 milepost.

3   The Guadalupe Peak Trail is more challenging and ends at Guadalupe Peak. An easier one, though still steep, is the Permian Reef Geology Trail, where such locations as "sponge paradise" are identified by trailside markers. Along both trails, more fossils are seen at higher elevations.

4   The full suite of rocks exposed at the Grand Canyon are arranged in the proper vertical sequence at the flagpole at Heritage Square, a brick plaza in downtown Flagstaff, Arizona. Larger rock samples are arranged along the Trail of Time that runs along the South Rim of Grand Canyon National Park just west of the Yavapai Geology Museum.

5   This is sometimes called the Hermit Shale, but most of this formation is not shale but siltstone and mudstone mixed with fine-grained sandstone, and so the official name is the Hermit Formation.

6   Rocks of the Fountain Formation are found at Red Rocks Park south of Golden, Colorado, and those of the Lyon Formation near Lyons, Colorado, where they are quarried for flagstone and building stone.

7   The Paradox Basin received its name from the Paradox Valley that is contained within it. And the Paradox Valley received its name from the apparently paradoxical course of the Dolores River that runs through it, crossing the valley across the middle instead of flowing along the length of the valley, hence the paradox.

8   Of particular note is the often-photographed, 600-foot-high sheer wall of De Chelly Sandstone that rises above the White House Ruins, one of the ancient cliff dwellings built by the Anasazi, an Ancestral Puebloan people.

9   It is also one of the most colorful landscapes on the continent, known for deep reds and purples. It was the colors and the erosional forms that inspired Georgia O'Keefe to make a set of paintings of Palo Duro Canyon in the 1910s. One of her many famous paintings is named "Red Landscape," an abstract expressionist oil painting of the bright red, Permian-age walls of Palo Duro Canyon. The painting is on display at the Panhandle-Plains Historical Museum in Canyon, Texas.

10   Named for the Guadalupian Epoch, which is one of the time divisions of the Permian Period. The GSSPs for the Guadalupian Epoch are found in the Guadalupe Mountains of Texas, hence the name.

11   One of the few sites in North America that is accessible and can be viewed from a distance
      is on the west side of State Highway 140 northwest of Winnemucca near the Quinn River
      Crossing in Nevada. Another is near Elpoca Mountain in Peter Lougheed Provincial Park
      near Banff National Park, about forty miles southwest of Calgary, Alberta.

12   Some scientists have proposed another way the atmosphere may have heated during the
      Permian extinction. They note that there are abundant ore deposits of nickel in Siberia, and
      that there is a spike in the concentration of nickel at the Permian-Triassic boundary in southern
      China. They also note that metals, such as nickel, are often the nutrient that limits the growth
      of microorganisms. It limits growth by being an essential chemical element in metabolic
      pathways. Moreover, a methane-producing organism, *Methanosarcina*, an archaea, made a
      sudden genetic change about 252 million years ago, at the time of the Permian extinction, a
      change that involved the use of nickel in a metabolic pathway. The genetic change worked in
      such a way that greatly improved the ability of this organism to produce methane, a greenhouse
      gas more efficient at heating the atmosphere than carbon dioxide. And so, just as cyanobacteria
      was responsible for the Great Oxygenation Event, *Methanosarcina* might have been responsible
      for the Permian extinction, another major turning point in the history of life.

13   Consider a discussion in the earlier part of this book in which both the end of the Cambrian
      and the end of the Ordovician Periods were defined by extinctions that were considerably
      less intense than the Permian extinction.

9.   A GRAND STAIRCASE

1   The Kaibab Limestone is exposed in the northwestern part of the park in Kolob Canyon. The
      mudstones and freshwater limestones of the Moenkopi and Chinle formations are seen near
      the southern entrance to the park near Rockville, Utah. And the Cretaceous-age Dakota
      Sandstone is found in northwestern part of the park near Horse Ranch Mountain.

2   The cross beds are especially intense in a region known as The Wave in northern Arizona.
      From the pattern of troughs and circular pits it is clear that the current erosion of The Wave
      is being caused by abrasion by the wind. The small parallel ribs are hard crusts produced by
      microbes lying just beneath exposed sandstone surfaces.

3   To put this in terms of physics, relative to rocks, water has a high heat capacity—that is, it
      can absorb a great deal of heat without a large increase in temperature. The heat capacity
      of rocky material is considerably less. Therefore, bodies of water remain at a more even
      temperature, while land temperature are more variable.

4   The lava is known as the Wrangellia Flood Basalts, which began to erupt 232 million years
      ago, the same time as the Carnian Pluvial Event, and which lasted about one million years.
      The basalts are as much as four miles thick and extend for more than fourteen hundred miles
      from Alaska, through the Yukon, and into British Columbia. The exposures in Alaska and the
      Yukon are mostly limited to slivers. One such sliver is found near Juneau. In British Columbia
      the basalt covers a substantial part of the center of Vancouver Island. In particular, many of the
      cliffs exposed at Schoen Lake and Strathcona Provincial Parks reveal these basalts.

5   That includes a pairing of the eruption of the Deccan Traps in India sixty-six million years
      ago and the demise of the dinosaurs at the end of the Cretaceous Period.

6   Excellent exposures of the lava can be seen at Passaic Falls in Paterson and in a large road cut
      a mile east of Roseland, both in New Jersey.

7   This includes the Deep River Basin of North Carolina, the Culpeper Basin of Virginia, the
      Gettysburg Basin of Pennsylvania, the Newark Basin of New Jersey, and the Hartford Basin
      of Connecticut. The Hartford Basin is noteworthy because, as the basin deepened, it was
      continuously filled with sediment that hardened into a brown sandstone known today as the
      Portland Formation. Beginning in the 1840s, this brown sandstone was cut and shipped to

New York as a fashionable (and cheap) building material, especially for houses in Brooklyn. The use was so widespread that the term brownstone has become synonymous with rows of adjacent houses with stone facades. Not everyone has seen beauty in this stone and its use. Novelist and New Yorker Edith Wharton wrote in her autobiography *A Backward Glance* that she detested the "chocolate-coloured coating" and considered the rock taken from the Portland Formation as "the most hideous stone ever quarried."

8   The first reported discovery of a fossilized bone that later work would show was almost certainly from a dinosaur was made in 1677 by Robert Plot, a professor of "chymstry" at Oxford University. Plot had been digging in a quarry in the Parish of Cornwell, presumably to add to his stock of chemicals, and had found what appeared to be part of a giant femur. From the size of the fossilized bone, he concluded that it "must have belonged to some greater animal than either an Ox or a Horse." He suggested that it was the bone of an elephant, perhaps one that had been brought to Great Britain when the island was governed by Rome.

9   As an example of paleontological irony, in 1992, Martin Lockley of the University of Colorado at Denver was working in the field near Page, Arizona, with Gerard Gierlinski, a Polish paleontologist, when Gierlinski had the misfortune of dropping a rock on his foot. He sustained a minor injury that left him with a limp for several days. Unbeknownst to Gierlinski, Lockley started measuring the trackway Gierlinski left behind on soft ground as he limped along, not telling Gierlinski what he was doing so that he would not bias the result. Lockley compared those results with trackways made by uninjured members of the group. He showed that Gierlinski, who had injured his right foot, was making longer steps from right to left than from left to right, confirming, in a real-time experiment, how the tracks of a biped dinosaur had changed after a foot injury.

10   These are the four best places to see dinosaur footprints in North America: At Dinosaur State Park near Rocky Hill, Connecticut, were several hundred footprints can be seen under the protection of a geodesic dome. In St. George, Utah, footprints and other impressions are protected by a large warehouse-style building. At Dinosaur Ridge west of Denver, a short trail takes one past several upended, rocky slabs that contain footprints. The three-toed tracks of *Acrocanthosaurus*, a small relative of *Tyrannosaurus rex*, are found along the Paluxy River near Glen Rose, Texas.

11   This was not the first discovery of ancient footprints. In 1828, the Reverend Henry Duncan, a minister of the Church of Scotland, found on a rock slab what looked like the impressions left by an ancient turtle ponderously moving across what was once wet beach sand. He decided to test the idea. He assembled several friends and, in their company, proceeded to march several turtles across a table covered with pie dough. The turtles became hopelessly stuck to the soft dough. He kneaded the dough afresh, added flour, and rolled it into a thin layer and resumed the experiment, satisfied that he could reproduce the impressions that he had seen on the rock slab.

12   Despite intensive searching, only seven skeletons and the impression of a single isolated feather of *Archaeopteryx* have been found, all recovered from the Solnhofen Limestone of Germany. The only display of an original fossil of *Archaeopteryx* in North America is at the Wyoming Dinosaur Centre in Thermopolis, Wyoming.

13   Any discussion of dinosaur tracks brings up the question: How fast could dinosaurs run? Could they move as fast as the dinosaurs in the movie *Jurassic Park*? Because of their sheer size and weight large dinosaurs, such as *Apatosaurus* or *Tyrannosaurus*, could not move quickly. *Apatosaurus* could probably manage a slow elephant-like run, but could not gallop or jump. *Tyrannosaurus* was more of a fast walker than a runner. Which means it was probably more of an ambush predator than a chaser, though its long stride meant it might move up to twelve miles per hour (about the top speed of a large elephant and one that a human could outrun) for

short distances. Small dinosaurs, such as *Coelophysis*, could probably maneuver quickly and run at speeds up to twenty-five miles per hour, about the speed of an Olympic sprinter.

14   The high concentration of dinosaur bones at the Carnegie Quarry is hard to envision without a visit to the quarry. Two specimens of the giant *Barosaurus* have been found. One was excavated by Douglass in 1923, which proved to be eighty-five feet long. A full-scale fiberglass model of this specimen is on display in the Theodore Roosevelt Rotunda at the American Museum of Natural History in New York. The actual fossilized bones are assembled into a complete skeleton at the Royal Ontario Museum in Toronto. The other specimen of *Barosaurus* found at the Carnegie Quarry is in the rock wall at Dinosaur National Monument. The *Stegosaurus* on display at the Carnegie Museum of Natural History is a composite of the fossilized bones of several individuals excavated from the quarry. A common type of dinosaur unearthed at the quarry is *Diplodocus*. Examples of *Diplodocus*, with fossilized bones, can be seen at the Denver Museum of Nature and Science in Denver, at the Carnegie Museum of Natural History in Pittsburgh and at the Smithsonian National Museum of Natural History in Washington, D.C.

15   The site where Lakes collected these fossils is along the western side of Dinosaur Ridge at Alameda Parkway in Morrison, Colorado.

10. WESTERN INTERIOR SEAWAY

1   Even though Kansas was then far from the shoreline, the remains of at least eight dinosaurs have been found in the Kansas chalk. These were animals that died on land, were swept into the seaway, then, because the carcasses were probably bloated with gas, floated a long distance before dropping onto the floor of the seaway.

2   This hard sandstone is the Dakota Sandstone. Look at it closely and compare it to the Navajo Sandstone and the Coconino Sandstone. The individual grains of those two sandstones were derived from large wind-driven dunes and are highly abraded and rounded and similar in size. The grains of the Dakota Sandstone, which were carried by streams and rivers, have a wide range of sizes and are angular. Footprints are numerous in the Dakota Sandstone because it lined the western shoreline of the Western Interior Seaway, a migratory route for dinosaurs.

3   This north-south trough moved across the continent as the Farallon Plate continued to slide eastward beneath North America. Today the trough lies along the East Coast of the United States, where, over the last thirty million years, the surface has dropped about three hundred feet. That is an average subsidence rate of about a tenth of an inch every thousand years, a minuscule amount compared to the current global rise of sea level of about one-tenth of an inch each year.

4   *Terrane* is a geologic term that refers to a region of rocks with similar histories that have moved great distances from where the rocks formed. *Terrain*, the homonym, refers to the topography or shape of the land.

5   These are more commonly known today as "suspect" or "accreted" terranes.

6   To be convinced that eastern Washington is part of Laurentia, examine Steptoe Butte about sixty miles south of Spokane. This steep hill with a road that spirals to the top is composed of a pink quartzite that formed 1.4 billion years ago. From the top, one can see other equally old remnants of Laurentia, including Kamiak Butte to the south and Tekoa Mountain to the north. Still not convinced? Go to Metaline Falls in the northeastern corner of Washington. Here water of the Pend Oreille River cascades down a cliff where the exposed rocks are of the same sequence and of the same age as those that lie atop the Great Unconformity in the Grand Canyon, making it part of Laurentia.

7   A significant part of Florida will again be below sea level in the near future. By the year 2100, the southern tip of Florida, from the city of Naples to Fort Lauderdale, will be flooded

by seawater. Twice daily, the city of Miami will lie a few feet below high tide. And these are conservative estimates based on the current rate of sea-level rise, which is accelerating, primarily due to the melting of polar ice caps.

8   This carbon-rich mud and silt is also the source of the belt of black shale that runs across the central parts of Mississippi, Alabama, and Georgia and into South Carolina. This shale, which formed close to the shoreline of a Cretaceous sea, weathered into a dark, rich black soil. It was the nature of this soil, high in carbon and in calcium, the calcium from the tiny shells that fell to the floor of the seas, that made this region highly productive in cotton. Without it, the large cotton plantations that existed before the Civil War might not have been possible. And without the plantations, the history of slavery in the United States might have been different.

9   Much smaller amounts of oil are being produced by fracking from the Barnett Shale west of Fort Worth, Texas, the Haynesville Formation in western Louisiana, and the Niobrara Formation in Wyoming. Though some oil is being produced by fracking in the Marcellus Shale and the Utica Shale beneath Pennsylvania and West Virginia, those areas are mainly producing natural gas.

10   The history of geology has produced an unfortunate confusion of names. *Shale oil* is liquid oil that is trapped in shale that requires a fracturing of the rock to allow the fluid to flow. *Oil shale* is a misnomer. It contains neither oil nor shale. The "oil" is a waxy material called kerogen that can be turned into oil by the correct application of heat. The "shale" is actually a limestone.

11   Meek had the additional challenge of coping with poor health. According to acquaintances, for most of his life he was an invalid, probably suffering from tuberculosis.

12   It should be noted that these ancient trees are not necessarily the ancestors of the modern versions that they seem to resemble. The pattern of veins in the fossilized leaves are not as complex as those in the modern ones. Also, the appearance of pollen grains are different.

13   Angiosperms are not only the dominant type of plant on the Earth today, they are also the most diverse group of plants to ever exist. According to a report released in 2016 by the Royal Botanic Gardens in Kew, England, there are about 391,000 species of vascular plants currently known. Of these, about 369,000 species, or 94 percent, are angiosperms. The number of species of conifers and other gymnosperms is about 1,100. In other words, almost all of the plants you see, except for conifers, ferns, mosses, and algae, are angiosperms.

14   The earliest fossilized flower discovered in North America was found in a collection stored at the Smithsonian Institution in Washington, DC. The fossil was unearthed during the Civil War when the Union Army was having a canal dug near the Dutch Gap along the James River in Virginia. This particular fossilized plant had originally been classified as a fern, but in 2013, Nathan Jud of the Smithsonian Institution recognized lobed leaves and a network of regular veins that indicated it was an angiosperm. He named the plant *Potomacapnos apeleutheron*. *Potomacapnos* is a reference to where the fossil was found, within the Potomac Group. *Apeleutheron* is a Greek word for freedmen, a word chosen by Jud to honor the former slaves who had recently been freed and who were forced to dig the canal. The fossilized plant has an age of about 120 million years. The earliest fossilized flower found anywhere in the world is *Archaefructus liaoningensis*, found in northeastern China in 1998. It has an age of about 125 million years.

15   The genome of the shrub *Amborella*, which today grows only in the cloud forests of New Caledonia, indicates it is the closest living relative, the so-called sister species, of all other living angiosperms. It produces cream-colored, dime-sized flowers without petals.

16   The great diversification of angiosperms during the middle of the Cretaceous Period had another consequence. Before this time, the number of animal species on land and in the sea were about equal. Afterward, during what is called the Cretaceous Terrestrial Revolution, the number of land species became five to ten times as diverse as in the sea, as it is today.

This diversity of land animals—mainly, beetles, spiders, insects, lizards, and mammals—is probably due to the explosion in the number of species of flowering plants.

17   This impression is illustrated by the first paragraph of Charles Dickens's *Bleak House*, first published in 1852. "Implacable November weather. As much mud in the streets as if the waters had but newly retired from the face of the earth, and it would not be wonderful to meet a megalosaurus, forty feet long or so, waddling like an elephantine lizard up Holborn Hill." This is the first mention of a dinosaur in a published work of fiction.

18   According to Scott Persons of the University of Alberta, who began studying this specimen intensively after it was recovered, the nickname was given by the crew that recovered this dinosaur. One night they decided to give a toast to the discovery, but the only liquor they had on hand was a bottle of scotch. Hence, they gave this *Tyrannosaurus rex* the name Scotty.

19   Other skeletal models of *Tyrannosaurus rex* composed of nearly complete skeletons can be seen at the Black Hills Institute of Natural History in Hill City, South Dakota; the Natural History Museum of Los Angeles County in Los Angeles, California; the Natural History Museum of Denmark in Copenhagen, Denmark; and the Naturalis Biodiversity Center in Leiden, the Netherlands. The one that was originally on display at the Museum of the Rockies in Bozeman, Montana, is now on loan and on display until 2063 at the National Museum of Natural History, Smithsonian Institution, in Washington, DC.

20   Why were dinosaurs so successful? There are a several reasons. Their legs were positioned directly beneath them, which allowed for more efficient and agile movement, unlike the side-to-side movement of lizards and crocodiles. They also had a faster rate of growth than lizards, indicated by the structure of their bones, which tended to be lightweight and porous, strengthened by layers of collagen fibers. They almost certainly had four-chambered hearts, like mammals and birds, which meant the transport of oxygenated blood was more efficient than in lizards. And they had a highly efficient method of breathing, more efficient than the method used by mammals. Mammals have lungs that must expand and contract as the animal inhales and exhales. In dinosaurs—and birds—air entered a large sac in the lower abdomen and then flowed through the lungs continuously in one direction. And so oxygen was being fed continuously to the blood system. This is the reason birds can maintain the high metabolic rate needed for flight; they are not constantly pausing to refill their lungs with oxygen-rich air. And so dinosaurs, especially the small ones, could move for long periods of time without resting.

11. A CALAMITOUS EVENT

1   Instead, one of Shoemaker's students, Harrison Schmidt, made the flight, the only geologist to stand on the surface of the Moon.

2   Eugene Shoemaker never achieved his dream of flying to the Moon. He died in an automobile accident in Australia on July 18, 1997. On July 31, 1999, a small amount of his ashes were carried to the Moon by the *Lunar Prospector* spacecraft.

3   The center of the Alamo impact was near the ghost town of Warm Springs, Nevada, at the junction of US Route 6 and Nevada State Highway 375. State Highway 375 is also known as the Extraterrestrial Highway, not because of the meteor impact, but because Area 51, the common name of a highly classified United States Air Force facility, where some claim the government scientists are working on alien spaceships, is nearby.

4   The Chesapeake impact is responsible for the field of glassy spherules, *tektites*, that covers the southeastern part of North America and most of the Caribbean. Because the field is strongly skewed to the south—the farthest north a tektite has been found is Martha's Vineyard in Massachusetts—the meteorite probably entered the atmosphere and struck the Earth's surface from the north. An impact crater known as Toms Canyon, located about 100 miles east of Atlantic City under the Atlantic Ocean, may have formed at the same time.

5   Another episode of increased impact cratering may have occurred about three hundred
     million years ago when impact craters formed near Middlesboro, Kentucky; Decaturville,
     Missouri; Kentland, Indiana; and Serpent Mound, Ohio. Yet another episode occurred
     about 220 million years ago when impact craters formed near Saint Martin in Manitoba,
     Wells Creek in Tennessee, and Red Wing in North Dakota. The sixty-mile-wide crater of
     Manicouagan in Quebec, which is the fifth-largest impact crater known on the Earth, also
     formed at this time.

6   The debris from this collision is still falling upon the Earth, though at a much reduced rate
     than during the Ordovician Period. Almost a third of all of the meteorites that now fall
     upon the Earth are L-chondrites.

7   The Cretaceous Period is often abbreviated with the letter *K*, the first letter of the German
     word *Kreide* for chalk, for the extensive chalk deposits produced during this period. The
     geologic period that came after the Cretaceous Period—that is, the first geologic period of
     the Cenozoic—was originally the Tertiary, but, in 2004, the International Commission on
     Stratigraphy changed the name to Paleogene. And so, officially, the boundary between the
     Cretaceous and the Paleogene periods is the K-Pg boundary. Outside of scientific circles, it
     is still referred to as the K-T boundary, and that is why I have used it.

8   The two larger ones are the 160-mile-wide Sudbury Crater in Ontario, age 1.849 billion
     years, and the 185-mile-wide Vredefort Crater in South Africa, age 2.023 billion years.

9   One of the more popular is Cenote Yaxbacaltun, where stairs have been constructed to allow
     easy access to a pool of cool, crystal-blue water. At Cenote Cuzamà tropical vegetation hangs
     down to the water level. And at Cenote Chihuàn the pool of water extends a considerable
     distance underground and is lit from below by underwater floodlights.

10   The flight distance between Trinidad, Colorado, and the Chicxulub Crater is 1,400 miles.
     Rocks would have flown ballistically between the two points in about ten seconds.

11   Others on the continent include several exposures—not quite as spectacular as in Colorado—
     found south of Montgomery, Alabama, along a highway that runs between the small
     communities of Moscow Landing and Mussel Creek. With considerable effort, the same
     layer of glass beads can be found near Dogie Creek and Teapot Dome in Wyoming and
     among the gray and tan buttes of soft eroding rocks along the Red Deer River in Alberta and
     the Frenchman River in Saskatchewan. Off the continent of North America, the layer is, of
     course, present near Gubbio, Italy, where the Alvarezes collected their first samples. It is also
     exposed near the city of Murcia in southeastern Spain and in the sea cliff of Sevens Klint near
     Copenhagen in Denmark. The farthest afield that anyone has yet found this thin layer is near
     Woodside Creek in New Zealand, more than ten thousand miles from Chicxulub Crater.

12   Because the layer of iridium lies atop the deposit, the huge sea waves produced by the filling
     of the crater could *not* be responsible for the deposit because those waves would have taken
     several hours to reach the site. Instead, if the deposit did form immediately after the impact,
     it must have been by water waves produced by the strong ground shaking because those
     would have arrived several minutes after the impact.

12. EXTINCTION

1   Today these fossils are at the Jardin des Plantes in Paris, part of the Muséum National
     d'Histoire Naturelle, the French National Museum of Natural History.

2   Big Bone Lick is now a state park in Kentucky. It was formerly a swampy area surrounded by
     salt and sulfur springs. Such a concentration of salt was usually described as a "lick" because
     it attracted wild animals, hence the origin of part of the name. "Big Bone," of course, comes
     from the many bones of woolly mammoths that have been recovered at the site, these large
     animals becoming trapped and dying in the swamp.

3   Cuvier gave the name "mastodon" to the animal fossils collected by Longueuil at what is now
     Big Bone Lick. It is the same animal whose bones and tooth were collected in New York in
     1706. It was a forest-dwelling animal that gathered food by browsing and grazing. It should
     not be confused with an animal that looked similar, the "mammoth." A mammoth was
     covered by long hair and grazed on open plains.

4   Modern work shows that the mastodon disappeared from North America about eleven
     thousand years ago. The woolly mammoth also disappeared from the continent at about the
     same time, though a small population of woolly mammoths was alive on Wrangel Island off
     the northern coast of Russia in the Arctic Ocean as recently as four thousand years ago.

5   Even though the Columbia River Basalts are the smallest LIP, the size of individual lava
     flows is staggering. Individual flows are as large as five hundred cubic miles (two thousand
     cubic kilometers) and flowed for a week or so. By comparison, the most voluminous eruption
     in history is the 1783 eruption of Laki volcano in Iceland. That eruption produced four cubic
     miles (fifteen cubic miles) of lava and lasted one year. The 2010 eruption of Eyjafjallajökull,
     also in Iceland, that halted air travel across western and northern Europe for three months
     produced 0.05 cubic miles (0.2 cubic kilometers) of lava.

6   Of the five major mass extinctions originally defined by Raup and Sepkoski, only the first
     one, the Ordovician extinction, has not been associated with the eruption of an LIP. That
     extinction is also the only one of the five major mass extinctions that occurred when there
     was a dramatic cooling of the planet. And so it might be more appropriate to consider it a
     minor episode of a Snowball Earth–type event.

7   In 2011, a brow horn of a triceratops-type dinosaur was found in the Hell Creek Formation
     just five inches below the K-T boundary. What this discovery of a single bone means
     in terms of dinosaur extinction is still debated. But this fragment does have one clear
     distinction: It is the youngest piece of physical evidence for a dinosaur yet recovered.

8   The direction of impact is inferred from the asymmetry of the crater structure revealed in
     gravity, magnetic, and seismic surveys and from the distribution of ejecta.

13. HOW THE MOUNTAINS GREW

1   The volume of papers and books written about the origin of mountains could fill a small library.
     Every known human culture that developed within sight of a mountain range has stories,
     traditions, and legends that describe how mountains form. Philosophers and artists, naturalists
     and scientists have used them to reveal something fundamental about nature. One of the first
     written references to the origin of mountains comes from Pythagoras in the sixth century B.C.E.,
     who proposed that mountains grew by the release of winds that had been shut up in dark regions
     inside the Earth. The Roman writer Pliny the Elder suggested that everything in nature seemed
     to somehow have been made for human beings, except for mountains, which Nature had made
     for herself. Much more recently, in the eighteenth century, the Swiss geologist Horace Bénédict
     De Saussure suggested that mountain scenery was the most intimate association one could
     have with nature. He also was one of the first to suggest that mountains could be subjected to
     scientific study and, hence, they could be used to learn something about the Earth.

2   The exposure at Sideling Hill is so famous that a few details need to be given. First, two
     formations are exposed. The upper one is the Purslane Formation, which is mainly sand and
     gravel deposited by rivers. A thick coal bed can also be seen. The lower one is the Rockwell
     Formation, which was mud (now shale) deposited along a coastal plain. Both are Mississippian
     in age. The U-shaped fold is a *syncline* that was produced by compression during the collision
     of North America and Africa during the Alleghenian Orogeny.

3   In particular, Dana saw the Caribbean Sea as a place where a geosyncline was developing
     and, hence, where he thought future mountains would form.

4   The Sevier Orogeny is named for the area around the Sevier River in Utah.

5   As well as the style of deformation produced by the three orogenies—the Taconic, the Acadian, and the Alleghenian—that produced the Appalachian Mountains.

6   c.e. is Common Era, formerly a.d., *Anno Domini*.

7   Recall from a discussion about the imaging of the Farallon Plate in Chapter 10 that seismic tomography relies on differences in the speed of earthquake waves as the waves pass through different features inside the Earth. The speed of those waves is faster in rock that is colder, denser, or drier than the surrounding rock.

8   Geographers know this boundary as the Atlantic Seaboard Fall line, or Fall Zone, a nine hundred-mile-long escarpment that runs from Alabama to New Jersey.

9   Delamination is now the favored explanation for the persistence of the Blue Ridge Escarpment that runs from North Carolina and into southern Virginia and for the high elevation, at 6,684 feet (2,037 meters) of Mount Mitchell, the highest peak east of the Mississippi River. Moreover, an independent cold slab probably lies beneath the Adirondacks in New York. Above the slab is an especially hot region of the asthenosphere, which may explain why the Adirondacks, which have risen more than a mile in the last ten million years, continue to rise at a rate of about a tenth of an inch each year—comparable to the rate the Sierra Nevada are now rising.

10   Since the 1980s, Long Valley has had several intense earthquake swarms and the broad floor of the volcano—that is, the floor of the caldera—has risen more than two feet. Both are indications that magma has been added to a shallow reservoir, increasing the chance for an eruption.

11   Why large slabs break away from the lower lithosphere is still being debated. One of the possible processes being proposed is "eclogitization," which simply means that the dense rock *eclogite*, which forms from basalt when it is under high pressure, plays an important role. Eclogitization has been difficult to study because such rocks are rarely found on the surface. But some have been found as xenoliths—that is, rocks that have been torn off of solid crust as magma rushes to the surface. In particular, xenolithic eclogites have been found in volcanic rocks in the Sierra Nevada, indicating its presence under those mountains.

12   While the downward descent of slabs and the pulling of tectonic plates is most of the story, it is not the full story. There is a smaller component of plate motion that comes from another source. For example, the North American Plate is moving, and yet, there are no descending slabs—that is, there are no subduction zones—along its boundaries that are pulling on the plate. This plate is moving—and at a significantly slower rate than the Pacific Plate— because the base of the North American Plate is being *dragged* as the hot asthenosphere flows beneath it. The hot asthenosphere is flowing for a variety of reasons: It has been churned up by several large descending plates, including the Farallon Plate, and because hot plumes are rising from the top of the core through the mantle and up to the surface.

13   This also changes the nature of the Mid-Ocean Ridge that winds along the bottom of the ocean basins like the seams of a baseball, including the segment of the ridge that runs down the center of the Atlantic Ocean and marks the boundary between the North American and the African and Eurasian Plates. The Mid-Ocean Ridge was once thought to be an active feature where the edges of two huge convection cells were rising in the mantle. Today it is seen as a passive feature not related to activity deep inside the Earth.

14   To put the size of the Earth's core in perspective, the diameters of both the Moon and Mercury are substantially smaller than the Earth's core. That is, the Moon or Mercury could fit inside the Earth's core.

15   Pronounced "D-double prime."

16   In the parlance of crystallography, these increases in density are caused by "pressure-induced phase transitions."

17   The two blobs have been given informal names taken from two of the pioneers in the theory of plate tectonics. The one beneath Africa is "Tuzo" for J. Tuzo Wilson and the one beneath the Pacific is "Jason" for Jason Morgan.

18   The mechanism was based on the idea that the 410-mile boundary represented an *exothermic* phase change, meaning that material that passed through the boundary would release energy and heat up the surrounding mantle. The hotter mantle would be less viscous and the slabs that had accumulated would then be able to pass through the boundary.

14. THE GREAT LAKES OF WYOMING

1   To clear up possible confusion—*Tyrannosaurus rex* lived on the western part of North America, the part of the Earth most affected by the Chicxulub impact. It was a large animal and probably went extinct as a result of the impact. But the loss of this species may not have been part of the mass extinction that occurred near the K-T boundary that was global and that caused the extinction of all non-avian dinosaurs and many other animal groups and that was probably caused mainly by the eruption of the Deccan Traps.

2   Today the average global temperature is 59° Fahrenheit (15° Celsius).

3   There are many places where the San Andreas Fault can be found in the mountains north and east of the Los Angeles Basin. Along I-5 near the town of Gorman the fault is exposed on a hillside as an abrupt change from gray metamorphic rocks to brown sandstone and siltstone. A road cut along Highway 14 south of Palmdale reveals rocks highly contorted by movement along the San Andreas Fault. East of Appletree Campground along Big Pines Highway in the San Gabriel Mountains is a short section of badlands next to the highway. These badlands are composed of hard granite that was pulverized by movement along the San Andreas Fault. At Cajon Pass, the colorful rocks at Blue Cut along US Highway 66 lie close to the San Andreas Fault. East of San Gorgonio Pass in the Coachella Valley is a twenty-mile-long ridge, Indio Hills. Here the San Andreas Fault runs as two strands on either side of the ridge. The two strands merge at Biskra Palms Oasis, the fault forcing groundwater to the surface accounting for the oasis.

4   The San Gregorio Fault runs along the western side of Seal Cove Bluffs north of Pillar Point.

5   Additional evidence of this northward migration is seen in the migration of volcanic activity near the coast of Northern California. About thirteen million years ago a system of volcanoes known as the Queen Sabe volcanics in the Diablo Range east of Hollister, California, was active. About ten million years ago the volcanic activity was in the Berkeley Hills. (These lava flows can be seen in the large road cuts east of the Caldecott Tunnel on Highway 24 near Berkeley.) About five million years ago the activity produced the Sonoma volcanic field north of Napa Valley. Today the very limited volcanic activity, much of it in the form of hot geysers and vigorous steaming, is in the area south of Clear Lake, eighty miles north of San Francisco.

6   The level of seismic activity along the Intermountain Seismic Zone is lower than along the Walker Seismic Zone or the San Andreas Fault, but major earthquakes have occurred in 1959, near Hebgen Lake in southwest Montana near Yellowstone, and in 1983, at Borah Peak in central Idaho.

7   The exception, of course, is the Coconino Sandstone, the white band of rock near the rim of the Grand Canyon that formed from sand dunes on dry land.

8   This is the same system that is used to determine the location of cars and other vehicles, though much more sophisticated electronics are used to determine the shifting of the Earth's surface.

9   A nutcracker might be a better analogy: Baja California is the lever and the Los Angeles Basin lies at the business end of the nutcracker.

10   In writing his report about the expedition, Ives was the first to attach the adjective "grand" to the canyon, referring to it as the "Grand Valley." The first use of the term "Grand

Canyon" came a decade later when Solomon Carvalho, who had been on a separate expedition into the American West, exhibited a painting at his New York studio inspired by his travels entitled "The Grand Canyon of the Colorado River."

11 It should be noted that, in the decades after Newberry, both John Wesley Powell and Clarence Dutton, based on their fieldwork, proposed ideas about the formation of the Grand Canyon. Powell invented a term called "antecedence" to emphasize that the Colorado River was in place before various mountain ranges and the Colorado Plateau were uplifted—that is, the river existed before, or was "antecedent" to, the mountains and the plateau. Dutton recognized that strata of the Grand Staircase had once covered the region of the Grand Canyon, and that there had been a period of intense erosion, the "Great Denudation." He suggested that the exact course of the Colorado River through the Grand Canyon was controlled by the alignments of old mountain ranges that had been covered by the sedimentary rocks exposed in the canyon walls, an idea known as *superposition*. Neither antecedence nor superposition is supported by recent work. Also, Dutton, who was trained in the classics at Yale, is the person who began the tradition of naming prominent features within the Grand Canyon after figures in Eastern religions. To him is owed the naming of Shiva Temple, Buddha Temple, and Isis Temple, all visible from the South Rim. Naming of the rock formations of the Vishnu Schist and the Zoroaster Granite followed Dutton's lead.

12 Meadville is only twenty-five miles from Grand Canyon Skywalk, the horseshoe-shaped glass bridge that extends out over the rim of the canyon.

13 One indication of this regional tilt is the different appearance of side canyons on the northern and the southern sides. The North Rim of the Grand Canyon is about a thousand feet higher in elevation than the South Rim. That has caused more runoff from rain to enter the canyon from the northern side, allowing for more erosion by the side streams. For this reason, the northern side of the Grand Canyon has been eroded away from the river about twice as far as the southern side.

14 The idea that the modern Colorado River and the Grand Canyon formed by the integration of at least two separate rivers, one young and one old, was proposed in 1964 by Edwin McKee of the Museum of Northern Arizona in Flagstaff, Arizona.

15 These lava flows and ancient dams are still best seen from the river. On the North Rim is a feature called Vulcan's Throne, a tall cinder cone and source for some of the lava flows that drape the walls of the Grand Canyon. There are a few places along the South Rim where, after considerable effort either by hiking or by driving along rough roads where all-terrain vehicles are required, one can reach spots where Vulcan's Throne and other volcanic features can be seen. To approach it from the North Rim is also a challenge, requiring a drive along gravel roads, sixty miles from Kanab and ninety miles from St. George, both in Utah.

16 To give another perspective on how big a thousand cubic miles is, if all of the water that is flowing in all of the rivers today was poured into the Grand Canyon, the canyon would be less than half full.

17 The dawn redwood, *Metasequoia*, is one of the most abundant and easily recognized plant fossils found in the plant record of the Northern Hemisphere. It is common at many localities in the John Day region, including in the Painted Hills Unit of the national monument and behind the high school in the town of Fossil, Oregon. It has been found at more than a hundred sites in North America, indicating that it was once widespread. The youngest fossils have been found in rocks of the Pliocene Epoch, about five million years in age. Because no living examples were known, it was assumed that the dawn redwood was extinct. Then, in 1946, Tsang Wang of the Ministry of Agriculture announced that three trees had been found living in central China. In the next two years, several hundred additional trees were found. In 1948, paleobotanist Ralph Chaney of the University of California, Berkeley, led a ten-thousand-mile

expedition up the Yangtze River and across three mountains ranges and into a lush, fog-shrouded valley where a thousand dawn redwoods were growing. Chaney eventually obtained seeds and now individual trees and small groves of dawn redwood are growing in temperate climates around the world. They can be found at the New York Botanical Garden in the Bronx and in the Strawberry Fields section of Central Park, both in New York. The Missouri Botanical Garden in St. Louis, Missouri, and the Hoyt Arboretum in Portland, Oregon, also have several specimens. A small grove is, appropriately, growing in the Painted Hills Unit of John Day Fossil Beds National Monument. And so three types of redwoods—the coastal redwood (*Sequoia*), the giant sequoia (*Sequoiadendron*), and the dawn redwood (*Metasequoia*, a tree that originated and was dominant during the middle part of the Cretaceous Period when the climate was much warmer and more humid and when flowering plants were first coming into their own)—are still growing in small regions of the planet.

18   This is the reason all vertebrates have the same basic body plan. For example, all vertebrates have four limbs and they have five digits at the end of each limb.

19   Polyploidy—that is, having multiple copies of chromosomes—is common in plants, especially among ferns and flowering plants, both wild and cultivated species. Some varieties of wheat have six sets of chromosomes. Some varieties of sugarcane have ten sets.

20   There is an interesting addendum here: Recent research supports a long-held assumption by ranchers that the saliva of cattle can increase the fertility of soil and, hence, the growth rate of grasses. It is another measure of the complex co-evolution of plants and animals.

21   Grasses that use C3 photosynthesis include wheat, oats, and ryegrass, the so-called cool-season plants that are highly productive in the spring and fall but have reduced growth rates during the high temperatures of summer. The warm-season C4 plants include corn and sorghum and crabgrass and bermuda grass. Look at the weeds in your garden. The ones that grow quickly are C4 grasses.

15.   A DROWNED RIVER AT POUGHKEEPSIE

1   Other notable erratics are the Bleasdell Boulder, which is the size of a large house, near Trenton, Ontario; Whale Head Rock, which can be found along a short hiking trail near Kinnelon, New Jersey; and the Rocking Stone at the Bronx Zoo.

2   A field of erratics near Gloucester, Massachusetts, was put to an unusual use. In 1858, Henry David Thoreau commented on them as producing "the most peculiar scenery of the Cape." Here he "could see no house, but hills strewn with boulders, as though they had rained down, on every side." Much later, the area was purchased by Roger Babson, a Boston millionaire and one-time presidential candidate for the Prohibition Party. After his election defeat, Babson had numbers carved into the boulders. He later had workers paint short slogans of encouragement on the rocks to inspire people who walked through the woods. The slogans included "Courage," "Be Clean," "Help Mother," "Get a Job," Keep Out of Debt," and "Never Try, Never Win."

3   An interesting line of erratics, known as the Foothills Erratics Train, runs for hundreds of miles along the eastern flanks of the Rocky Mountains of Alberta. The line begins from just south of Highway 16 on the McLeod River in Canada, passes through Calgary, and ends near the area of Courts and Sweetgrass on the international border. The origin of these erratics is clear: They are gravels and boulders of quartzite that comes from the Athabasca River Valley of western Alberta. They were carried by an ice sheet that covered the mountainous western part of North America and were dropped by the sheet where it ran up against another ice sheet that was spreading outward from eastern Canada. The largest erratic along the line is known as "Okotoks" and located about twenty miles south of Calgary. It was once a single block of quartzite up 120 feet in length and it is estimated to weigh more than twenty thousand tons.

4   Other places to see megagrooves—though, admittedly, not as spectacular as on Kelleys
     Island—are at the mouth of the Lester River near Duluth, Minnesota, where grooves have
     been cut into basaltic lava flows erupted more than a billion years ago from the Midcontinent
     Rift, and in the limestone on Pelee Island, Ontario, the southernmost point in Canada.

5   Fields of drumlins are common in North America. Cumorah is one of thousands of similar
     hills in a field that extends across west-central New York from Syracuse to the Niagara
     River. Other fields include the Smith-Reiner Drumlin Prairie near Madison, Wisconsin,
     and the Peterborough Drumlin Field in Ontario, where there are more than four thousand
     drumlins in an area about twenty miles on a side. The city of Seattle is set in a drumlin field.
     Almost every hill in the city is a drumlin: Capitol Hill, Queen Anne Hill, First Hill, and
     Beacon Hill. Bailey Peninsula, which forms Seward Park, is a drumlin. A drumlin, Denny
     Hill, once sat in what is now downtown Seattle. It was removed in a series of regrades in the
     early twentieth century.

6   More specifically, the Harbor Hill Moraine ran across New York Harbor where the
     Verrazanno-Narrows Bridge stands today.

7   Other notable drowned rivers are the Thames River in England and the Ems River in
     Germany. The most dramatic change in direction of water flow is at Reversing Falls along
     the Saint John River in New Brunswick, Canada.

8   Though technically not a glacial lake, another large lake formed on North America at this
     time. The existence of Lake Bonneville is recorded as horizontal lines inscribed on nearly
     every mountainside in northern Utah. At its greatest extent, Lake Bonneville was a thousand
     feet deep over Salt Lake City. Only a puddle is left, the Great Salt Lake, which has an
     average depth of twelve feet.

9   Shadow Falls and Minnehaha Falls may have been formed by the erosion caused by the
     floodwaters produced by the draining of Lake Agassiz. The falls have since retreated along
     tributary streams.

10   The giant white M high on the hillside above the campus of the University of Montana is
     about two hundred feet below the surface of the lake.

11   Grand Coulee is one of the many valleys carved by the flood and is now the route followed
     by the Columbia River.

12   This meteorite is now the centerpiece of a display at the American Museum of Natural
     History in New York. It was found in the Willamette Valley (hence its name), but passed
     through Montana in an ice sheet before that.

13   This means that if you scoop up a million molecules from the air, 1,500 would be carbon
     dioxide. Today the concentration of carbon dioxide in the air is about 400 ppm.

14   The saber-toothed cat diverged from other cat lineages eighteen million years ago and has
     no living descendants. The so-called roaring cats, *Panthera*, such as lions, leopards, jaguars,
     and tigers, diverged from a separate lineage six million years ago. The domestic cat diverged
     from a diminutive wildcat in the Near East about ten thousand years ago. *Panthera* and
     the domestic cat do share a common ancestor that existed eleven million years ago. For
     comparison, the dog is descended from a wolf-like animal but is not a direct descendent
     of any living wolf population. The dog may have been domesticated as early as fourteen
     thousand years ago, thousands of years before domestication of the cat.

15   One can follow this transition from hothouse to coolhouse to icehouse by traveling around
     North America. The period of a hothouse climate (i.e., PETM) is best seen at Fossil
     Butte. A coolhouse climate is represented by the fossils at Florissant Fossils Beds National
     Monument in Colorado. The transition from coolhouse to icehouse conditions is recorded at
     Agate Fossil Beds National Monument in Nebraska, where the most common mammalian
     fossil is a small two-horned rhinoceros. The fossil plants and animals at Hagerman Fossil

Beds National Monument in Idaho are of a later time, about three million years ago, just before an ice cap began to develop over the North Pole.

16   This location of the beginning of the Holocene Epoch means that, to access it, one must obtain deep-ice-coring equipment and receive permission from the Greenland Home Rule through the Danish Polar Centre to drill. The actual core removed from NGRIP2 has been archived and is available for study at the University of Copenhagen. But both are difficult for most researchers to accomplish. And so several auxiliary sites have also been defined that are more accessible. The one defined for North America is at Splan Pond in southwestern New Brunswick. Within the sediments at the bottom of this pond, at a depth of about fifteen feet, is a thick, gray clay layer that marks the beginning of the Holocene Epoch.

17   What is the largest earthquake possible? The 2011 earthquake in Japan was a magnitude 9.0 event, the largest earthquake ever recorded in Japan. The largest earthquake recorded anywhere in the world—that is, the largest earthquake ever recorded since seismometers have been in use, which is the last 120 years or so—is the 1960 earthquake in Chile, which was a magnitude 9.5 event. If the entire length of the San Andreas Fault ruptured it would produce a magnitude 8.5 event. If *all* of the earthquake faults in the world ruptured *simultaneously*, it would be equivalent to a magnitude 11 event. (For comparison, the amount of seismic energy produced by the Chicxulub impact was equivalent to a magnitude 9.5 or 10 earthquake.)

18   For example, a writer in the port city of Kuwagasaki, Japan, recorded that a tsunami flooded the harbor at midnight on January 27, 1700. Using modern time zones and allowing for the ten hours it took for the wave to cross the Pacific Ocean, the earthquake must have occurred at about 9 p.m. local time on January 26, 1700, along the coasts of Oregon and Washington.

19   Among the more accessible ghost forests are the ones at Neskowin Beach State Recreation Site in Oregon and at Sandy Point along Willapa Bay in Washington. When at these beaches, be aware that the ocean tide can rise rapidly and flood the area.

20   Other notable "volcano cities" include Auckland in New Zealand, where there are more than fifty volcanic craters and cones within and near the city limits; Naples in Italy, which is between the volcano Vesuvius and the ever-steaming Phlegraean Fields; Kagoshima, which is at the southern end of the Aira Caldera in Japan; and Managua in Nicaragua.

21   An intriguing place to see several of these thick lava flows of Columbia River Basalt—and lava flows of the Boring Volcanic Series that erupted much more recently—is midway along the three-mile-long Robertson MAX Tunnel through the Tualatin Hills. During construction of the tunnel, dozens of drill holes were made and cores retrieved. One of these cores—four hundred feet long—is displayed along the walls of the Washington Park/Zoo station of the MAX (Metropolitan Area Express) Light Rail Service that connects to downtown Portland. A timeline along the wall marks the separation between the flood basalts and the more recent eruptions.

22   A drive east of downtown Portland along I-84 reveals the erosive power of the Missoula Floods. As I-84 leaves the city and the Willamette River, it runs along the bed of an old flood channel scoured by the Missoula Floods about fifteen thousand years ago. Near the intersection of I-84 and I-205 is Rocky Butte. The Missoula Floods swept volcanic material from the east side of Rocky Butte to the west side leaving a long gravel bar called the Alameda Ridge, which is now being quarried for gravel and sand. Rocky Butte, at ninety-seven thousand years, is Portland's youngest volcano.

23   Another short-lived eruption produced Battleground Lake north of Vancouver. The bowl that holds the lake is a volcanic feature known as a maar, which forms during a highly explosive, gas-charged eruption.

24   The sequence of eruptions of the same volcano can show a range of activity. For example, in 1980, the eruptions of Mount St. Helens were explosive and produced mainly volcanic ash. After 1980, the eruptions were less explosive and produced several volcanic domes.

25 A consequence of the migration of the crustal bulge and reversal of stream and river directions is that the Continental Divide has shifted eastward. Six million years ago the Continental Divide ran close to Pocatello, Idaho. Today it runs through Yellowstone National Park and is east of Grand Teton National Park.

26 Along the Hawaiian Islands the age progression is: Kauai, five million years; Oahu, three million years; Molokai and Lanai, two million years; Maui, one million years; and Hawaii, less than 0.5 million years.

27 No volcanic ash from Yellowstone has been found east of the Mississippi River, but calculations suggest that the amount of ash that fell over Chicago was about one-half inch, over New York about a tenth of an inch, and a noticeable dusting of ash may have occurred as far south and east as Miami and across the Caribbean.

28 An earlier dome, the Sour Creek Dome, formed in the northeastern section of the national park east of Canyon Village about six hundred thousand years ago—that is, soon after the latest major explosive eruption of Yellowstone. The profile of the Mallard Lake Dome forms the skyline behind Old Faithful.

16. A WORLD BEQUEATHED AND THE GREAT ACCELERATION

1 Remember: Gold atoms formed during the merging of neutron stars. Aluminum, nickel, copper, and iron atoms formed during the much more common explosion of massive stars and white dwarfs.

2 To be complete, gold deposits can also be found in placer deposits—that is, small nuggets and flakes that were weathered and washed down from hillsides, then swept into and settled onto the bottom of streams, lakes, and rivers. Most of the independent prospectors of gold in California and in the Klondike recovered gold from placer deposits.

3 Jade is found in the Granite Mountains west of Casper, Wyoming, and rubies and sapphires around Gem Mountain in western Montana.

4 It should be noted that the largest diamond ever found in the United States, the Uncle Sam Diamond weighing 40.23 karats, was found here in 1924.

5 Debate has raged for many years whether the Hiscock site near Byron, New York, represents a Pleistocene-age kill site. The remains of at least ten mastodons have been found along with the remains of a condor, a giant beaver, a now-extinct moose, and a caribou. But none of the stone tools found at the site can be directly associated with the death of these animals. It has been suggested that ancient people may have gathered at the site to scavenge dead animals, using the remains as a source of raw material for bone and ivory tools, but there is no indication that ancient people hunted or butchered these animals at the site.

6 This event is named for a species of wildflower, *Dryas octopetala*, that grew in an alpine-tundra environment and that is used to define the time period when the Younger Dryas Event occurred.

7 To put this change in starker contrast, the weight of all of the world's plastics is twice the weight of all of the marine and terrestrial animals. All of the buildings and infrastructure now in use outweigh all of the trees and shrubs.

8 This is the now-famous "hockey stick" graph that shows a rapid increase in global atmospheric temperature since 1950. After two decades of additional research, the details of this graph are now widely accepted.

9 The next assessment report by the IPCC is now being prepared. It should be published in 2022.

10 Among the famous paintings are *Adoration of the Kings in the Snow* by Pieter Bruegel (1567) and *The Castle of Muiden in Winter* by Jan Abrahamsz Beerstraten (1658).

11 About forty billion tons of carbon dioxide are being emitted in the atmosphere today. To put this in perspective, the amount being emitted annually by volcanoes is 0.4 billion tons,

or one percent of what is being contributed by human activities. A major explosive eruption, such as the 1980 eruption of Mount St. Helens, can, for a few hours, send as much carbon dioxide into the atmosphere as is done on a daily basis by the world's industries and its many cars and trucks. But a Mount St. Helens–type explosive eruption would have to be happening on the Earth several times a day every day to match the amount of carbon dioxide that comes from the current rate of the burning of fossil fuels.

12   Though much of the information about the PETM comes from cores recovered from holes drilled into the ocean floor, a detailed history of the warming has come recently from a core recovered from a drill hole in the Bighorn Basin near Powell, Wyoming. It showed an abrupt warming of about 12° Fahrenheit caused by not one by two distinct carbon release events, the first pulse lasting less than two thousand years and releasing carbon at the modern rate, equivalent to about forty billion tons of carbon dioxide per year.

13   By comparison, the worldwide population of dogs is about nine hundred million, of which about five hundred million are kept as pets. The worldwide population of cats is about six hundred million, of which about four hundred million are kept as pets.

14   Recall that a trace fossil—the burrows left by a wormlike creature, *Treptichnus pedum*—is used to defined the beginning of the Cambrian Period. And so it is not outrageous to think that the millions of holes drilled into the ground in the search for oil might be an indicator of the beginning of a geologic epoch.

15   To give a short inventory: About one cubic mile of lava is erupted from all of the volcanoes in the world. Those on the continents erupt one-quarter of that. Those in the ocean basins erupt the remainder.

16   Hundreds of sites have been considered. One of the more unusual ones is Fresh Kills Landfill on Staten Island in New York. This was once a repository for New York City refuse. It covers more than two hundred acres and the accumulation of waste is more than two hundred feet thick. It is one of the largest human-engineered formations in the world. It opened in 1948 and closed in 2002, the last debris coming from the rubble of the World Trade Center that collapsed on September 11, 2001. It has been rejected as a GSSP because of its complexity.

EPILOGUE: RUGBY, NORTH DAKOTA: THE FUTURE

1   This would be the final geologic eon in the Earth's history, a time when not only the oceans will have disappeared but the orbital paths that the planets follow around the Sun may again be chaotic. And so there is a measure of symmetry here. The Earth, as it is known today by its chemical composition, size, and mass, was created after an impact between a proto-Earth and a Mars-sized body, Theia, 4.51 billion years ago. In the future, a billion years or more from now, Jupiter will finally exert enough gravitational influence on the orbits of the four inner rocky planets, mostly by disturbing the orbit of Mercury. That could cause the orbits to scramble and lead to a collision of Mercury, Venus, or Mars with the Earth. Thus, as unsettling as it might be, both the beginning and the end of the Earth might be marked by two planet-planet collisions—by two cataclysmic events separated by several billion years.

INDEX